LONG-TERM DYNAMICS OF LAKES IN THE LANDSCAPE

LONG-TERM DYNAMICS OF LAKES IN THE LANDSCAPE

Long-Term Ecological Research on North Temperate Lakes

Edited by

John J. Magnuson

Timothy K. Kratz

Barbara J. Benson

LTER

OXFORD

UNIVERSITY PRESS

2006

UNIVERSITY PRESS

Oxford University Press, Inc., publishes works that further
Oxford University's objective of excellence
in research, scholarship, and education.

Oxford   New York
Auckland   Cape Town   Dar es Salaam   Hong Kong   Karachi
Kuala Lumpur   Madrid   Melbourne   Mexico City   Nairobi
New Delhi   Shanghai   Taipei   Toronto

With offices in
Argentina   Austria   Brazil   Chile   Czech Republic   France   Greece
Guatemala   Hungary   Italy   Japan   Poland   Portugal   Singapore
South Korea   Switzerland   Thailand   Turkey   Ukraine   Vietnam

Published by Oxford University Press, Inc.
198 Madison Avenue, New York, New York 10016

www.oup.com

Oxford is a registered trademark of Oxford University Press

Library of Congress Cataloging-in-Publication Data
Long-term dynamics of lakes in the landscape : long-term ecological
research on north temperate lakes / Edited by John J. Magnuson, Timothy K. Kratz,
Barbara J. Benson.
p.   cm.—(Long-Term Ecological Research Network series)
Includes bibliographical references and index.
ISBN-13 978-0-19-513690-6
ISBN 0-19-513690-X
1. Landscape ecology—Research.   2. Lake ecology—Research.   I.
Magnuson, John J.   II. Kratz, Timothy K.   III. Benson, Barbara J.   IV. Series.
QH541.15.L35L65   2005
577.63'0912'3—dc22   2004020527

9 8 7 6 5 4 3 2 1

Printed in the United States of America
on acid-free paper

Memorials

Long-term ecological science plays out not only over the dynamics of ecosystems at decade and century scales but also over the lifetimes of the participants. We dedicate our book to three special people who contributed to and led the establishment and development of the LTER network and of our North Temperate Lakes program. Their unique contributions were pivotal to this new kind of science. We are grateful.

John Langdon Brooks (1920–2000) was the director of the Biotic Systems and Resources Division at the National Science Foundation during the formative years of the Long-Term Ecological Research program. John Brooks viewed NSF investigators with a good spirit and such a supportive attitude. He demanded much of us. None of us wanted to let John down; he cared so much for the science and for us. His vision, foresight, and direction moved the LTER program from its conception as an idea in the 1970s to its birth in 1980 with the funding of the first six sites to its rebellious teenage years when LTER scientists first realized that significant network-level science was possible and debated what that science might be. John believed that network science would lead to much-needed new conceptual advances. The fruits of his beliefs were only beginning to be realized in his lifetime.

James Thomas (Tom) Callahan (1945–1999) was the first program officer in charge of the LTER program at NSF. He championed the LTER program both within and outside NSF. Tom shaped the program through his persistent challenging of the participating scientists in site reviews and proposal reviews and in his essays and justifications for the program. He was not at all passive in his approach to keeping the principal investigators on their toes and advocating for the program. Tom believed in this new kind of science and in our ability, in fact, to do it. The NTL program was a beneficiary of his leadership, enthusiasm, critiques, and occasional wisecracks.

John Langdon Brooks: John in his office in winter 1976 at the National Science Foundation, Washington D.C., when he was Deputy Division Director for Environmental Biology. (J. Magnuson)

James Thomas Callahan: Tom at an LTER meeting in winter 1985, in Albuquerque, New Mexico. (J. Magnuson)

Thomas Michael Frost: Tom in summer 2000 on Little Rock Lake, Wisconsin. (J. Miller, University of Wisconsin-Madison University Communications)

Thomas Michael Frost (1950–2000) directed the Trout Lake Station, the nexus of LTER research in northern Wisconsin, from 1981 to 2000, and the station flourished with Tom's leadership. Everyone who came to the Trout Lake Station remembers Tom and his personal legacy of friendship and helpfulness. He was genuinely interested in doing what he could to encourage and enhance the research efforts of each person. On his arrival, Tom immediately became a co-principal investigator on the North Temperate Lakes LTER program, contributing both personally to site and network science and through the research by his students. Additionally, Tom was co-leader of a truly long-term, whole-system experiment linked to our LTER— the Little Rock Lake Acidification Project (Chapter 9). He is sorely missed professionally and personally.

We are grateful to John Brooks, Tom Callahan, and Tom Frost. We would like to be able to put this book before them, and we believe that each of these past colleagues would be gratified and complimented by the venture and the science described here.

John J. Magnuson, Barbara J. Benson, and Timothy K. Kratz working on the book at Trout Lake Station in 2002. (Center for Limnology)

Preface

A ll scientists have beliefs about science. We as editors of *Long-Term Dynamics of Lakes in the Landscape* have three beliefs that are embodied in the book's title. Lakes and inland waters, in general, are worthy and essential components of the ecological sciences and the world on which humans and all life depend. Understanding long-term dynamics that reflect interannual, interdecadal, and intercentury variability and change, coupled with spatial extents greater than single lake ecosystems, reveals important new facets of ecological systems. These longer and broader scales are those that will reflect and determine the future of humankind.

As a site synthesis, we present research findings and perspectives from the North Temperate Lakes Long-Term Ecological Research program. However, we have referred to others and other studies to provide context and clarify findings.

This synthesis volume is one in an emerging series being published by Oxford University Press. A synthesis volume is planned for each LTER site. To date, syntheses are available for the Konza Prairie in Kansas and Niwot Ridge in Colorado. Others are in progress.

Madison, Wisconsin
July 2004

J. J. M.
T. K. K.
B. J. B.

Acknowledgments

W e first acknowledge the contributions of the many people who have been active at the North Temperate Lakes LTER site and have made it viable and productive: graduate and undergraduate students, staff, technicians, undergraduate hourly helpers, postdoctoral associates, and faculty. The Center for Limnology has been the home of our LTER program, and a very good home it has been. The people at the Center for Limnology's two lake-side field stations, the Laboratory of Limnology, on Lake Mendota, and the Trout Lake Station, in northern Wisconsin, provided a helpful, efficient, friendly, creative, and goal-oriented environment for the North Temperate Lakes LTER. The University of Wisconsin-Madison is characterized by few barriers to interdisciplinary research and teaching, and we have many colleagues in departments, centers, and institutes across the University who were absolutely essential for conducting an LTER program; the contributions of many of these colleagues can be found within the pages of our book.

One simply cannot do intersite or regional or global research without the enthusiastic participation of colleagues from far away. Many of these were contributors to the science discussed in this book. We greatly appreciate these science colleagues and friends who participated in the workshops that led to analyses and papers at extents beyond the edges of our North Temperate Lakes LTER site. In particular, we thank the participants in our workshops on Variability in North American Ecosystems, Global Lake and River Ice Phenology, Position in the Landscape, and Regionalization for Inland Lakes across the Great Lakes Region.

We acknowledge the National Science Foundation's Division of Environmental Biology for supporting the workshops in the 1970s that led to the creation of the LTER program at NSF. The program's creation resulted from an inspired effort of the leadership at NSF and from the thought and judgments of the ecological scientific community. Our site has been funded by a continuous sequence of grants over the past 20-plus years. We thank the peer reviewers, advisory committee members, and site visit teams for recognizing the strengths and the weaknesses of the set of proposals that structured our LTER program at the University of Wisconsin-Madison.

We acknowledge the shared support from other organizations such as the U.S. Environmental Protection Agency; the U.S. Geological Survey; the Wisconsin Department of Natural Resources; the University of Wisconsin-Madison's Graduate School and the College of Letters and Science; the Department of Zoology; the Department of Geology and Geophysics; the Department of Civil and Environmental Engineering; the Department of Rural Sociology; the Environmental Remote Sensing Center; the program on Science, Technology, Agriculture, Resources, and Environment; and the Applied Population Laboratory.

Many individuals made unusual and sustained efforts to help us create this book. William J. Feeny, the artist in the Department of Zoology, redrew and created new line art for the entire book; Jonathan W. Chipman, of the Environmental Remote Sensing Center, created many of the new maps. At the Center for Limnology, Pamela S. Fashingbauer did the references, Eric Frome gathered the permissions and helped compile the book, Kimberly Babcock ferreted out the sources of the quotations that begin each chapter, Linda M. Holthaus dug out the old administrative records needed, and David F. Balsiger checked and rechecked data and generated plots for new graphs. Patty Bonito at the network office helped with the index. Elizabeth A. Levitt, Paul C. Hanson, Denise K. Karns, Amanda Grell, Carol J. Schraufnagel, and many others jumped in and helped.

Photographs were taken or contributed by A. Barbian; P. Barbian; C. Bowser; Center for Limnology; T. Cummings; Dane County Regional Planning Commission; T. Frost; P. Jacobsen; A. Kent; J. Klug; T. Kratz; J. Magnuson; G. R. Marzolf and J. Miller, University of Wisconsin-Madison University Communications; Office of Information Services, University of Wisconsin-Madison; D. Robertson; M. Turner; M. Woodford; J. Riera; B. Roth; Sevilleta LTER program; C. Simenstad; D. Stamm; D. Schneider; University of Wisconsin Archives; G. Wagner; C. Watras; and K. Wilson.

We thank Carl Bowser for taking the cover photo. The photo was taken looking northwest over four of the northern LTER lakes on September 29, 1999. The view is across Crystal, Big Muskellunge, Allequash, and Trout Lakes.

We thank the following persons for critically reviewing individual chapters: Shelley E. Arnott, Karen S. Baker, Stephen R. Carpenter, Alan P. Covich, Clifford N. Dahm, John G. Eaton, Stuart H. Gage, James R. Gosz, David Greenland, Lance H. Gunderson, Thomas A. Heberlein, Thomas R. Hrabik, Donald A. Jackson, Marko Jarvenin, Kirk Jensen, Mathew B. Jones, Bradley S. Jorgensen, Kenneth E. Kunkel, David B. Lewis, David M. Lodge, G. Richard Marzolf, Lars G. Rudstam, Charles L. Redman, James A. Rusak, Patricia A. Sorrano, William M. Tonn, M. Jake VanderZanden, Robert B. Waide, Karen A. Wilson, Thomas C. Winter, and Norman D. Yan. Their reviews greatly improved the manuscripts. The editors also reviewed each chapter.

On individual chapters: for Chapter 4, we thank Martin Futter for statistical assistance and Norm Yan and D. Findlay for use of original data; for Chapter 8, we thank Ann S. McLain for early drafts and ideas; for Chapter 12, we thank Elizabeth A. Levitt for help with figures and the NERC Centre for Population Biology, Silwood Park, England, for its support and hospitality while the chapter was being written, and the Pew Foundation and the Wisconsin Department of Natural Resources for their support; for Chapter 13, we thank Dave Balsiger, Maryan (Robin) Stubbs, Joyce Tynan, Cheryan Jacob, Dan Smith, Chris Stolte, and Ramkumar Venkataraman for their contributions to the data management and information system over the years.

Finally, we thank Stephen R. Carpenter, lead investigator of the North Temperate Lakes LTER since July 2000, for his inspired leadership now and into the future. Our LTER program experienced a smooth transition of leadership without loss of continuity and with fresh vigor and ideas.

Contents

Contributors

Mary P. Anderson
Department of Geology
University of Wisconsin-Madison
Madison, WI 53706

David E. Armstrong
Environmental Chemistry and Technology
    Program
University of Wisconsin-Madison
Madison, WI 53706

Shelley E. Arnott
Department of Biology
Queen's University
Kingston, Ontario, K7L 3N6, Canada

Elena M. Bennett
Center for Limnology
University of Wisconsin-Madison
Madison, WI 53706

Barbara J. Benson
Center for Limnology
University of Wisconsin-Madison
Madison, WI 53706

Carl J. Bowser
Department of Geology and Geophysics
University of Wisconsin-Madison
Madison, WI 53706

Kathy Brasier
Department of Rural Sociology
University of Wisconsin-Madison
Madison, WI 53706

Patrick L. Brezonik
Water Resources Center
University of Minnesota-Twin Cities
St. Paul, MN 55108

Stephen R. Carpenter
Center for Limnology
University of Wisconsin-Madison
Madison, WI 53706

Jonathan W. Chipman
Environmental Remote Sensing Center
University of Wisconsin-Madison
Madison, WI 53706

Alison C. C. Colby
Department of Zoology
University of Wisconsin-Madison
Madison, WI 53706

Stanley I. Dodson
Department of Zoology
University of Wisconsin-Madison
Madison, WI 53706

Janet M. Fischer
Department of Biology
Franklin and Marshall College
Lancaster, PA 17604-3003

Thomas M. Frost
Deceased, see dedication

María J. González
Ecology Research Center
Miami University
Oxford, OH 45056

Paul C. Hanson
Center for Limnology
University of Wisconsin-Madison
Madison, WI 53706

Thomas R. Hrabik
Department of Biology
University of Minnesota-Duluth
Duluth, MN 55812

Bradley S. Jorgensen
School of Journalism and Communication
University of Queensland
Brisbane, Queensland, 4072
Australia

Timothy K. Kratz
Trout Lake Station
University of Wisconsin-Madison
Boulder Junction, WI 54512-9733

Richard C. Lathrop
Wisconsin Department of Natural
   Resources and Center for Limnology
University of Wisconsin-Madison
Madison, WI 53706

George H. Lauster
Center for Limnology
University of Wisconsin-Madison
Madison, WI 53706

John D. Lenters
Department of Geology and Physics
Lake Superior State University
Sault Ste. Marie, MI 49783

David B. Lewis
Center for Environmental Studies/Biology
Arizona State University
Tempe, AZ 85287

John J. Magnuson
Center for Limnology
University of Wisconsin-Madison
Madison, WI 53706

Timothy W. Meinke
Trout Lake Station
University of Wisconsin-Madison
Boulder Junction, WI 54512

Pamela K. Montz
Trout Lake Station
University of Wisconsin-Madison
Boulder Junction, WI 54512

David Nowacek
Department of Sociology
University of Wisconsin-Madison
Madison, WI 53706

Peter Nowak
Department of Rural Sociology
University of Wisconsin-Madison
Madison, WI 53706

Amina I. Pollard
National Center for Environmental
   Assessment
U.S. Environmental Protection Agency
Washington, DC 20460

Tara Reed
Department of Natural and Applied
   Sciences
University of Wisconsin-Green Bay
Green Bay, WI 54311

Joan L. Riera
Departament d'Ecologia
Universitat de Barcelona
Barcelona, Catalonia, 08028
Spain

Dale M. Robertson
Water Resources Division
U.S. Geological Survey
Middleton, WI 53562

Beth L. Sanderson
Office of the Science Director
Northwest Fisheries Science Center
Seattle, WA 98112–2097

Patricia A. Soranno
Department of Fisheries and Wildlife
Michigan State University
East Lansing, MI 48864

Richard C. Stedman
College of Agricultural Sciences
Pennsylvania State University
University Park, PA 16802

Carl J. Watras
Wisconsin Department of Natural
    Resources and Trout Lake Station
University of Wisconsin-Madison
Boulder Junction, WI 54512

Katherine E. Webster
Department of Biological Sciences
University of Maine
Orono, ME 04469

Karen A. Wilson
Department of Zoology
University of Toronto
Toronto, Ontario M5S 3G5
Canada

LONG-TERM DYNAMICS OF LAKES IN THE LANDSCAPE

1

The Challenge of Time and Space in Ecology

John J. Magnuson
Timothy K. Kratz
Barbara J. Benson

Time is but the stream I go a-fishing in.
—Henry David Thoreau (1854, p. 155)

The field cannot be well seen from within the field.
—Ralph Waldo Emerson (1909, p. 161)

All of us can sense change over time and differences across the landscape. We sense "the reddening sky with dawn's new light, the rising strength of lake waves during a thunderstorm, and the changing seasons of plant flowering as temperature and rain affect our landscapes" (Magnuson 1990, p. 495). We sense in a lake the patches of aquatic plants next to which the large fish lie, in a forest of northern Wisconsin we sense the brown-water lake with cranberries hanging along shore and the crystal-clear lake with beaches of sand, and across the Wisconsin region we sense the transitions from cropland to forest as we drive north. More difficult challenges for scientists who study long-term regional ecology are to sense quantitatively long-term change and dynamics and large-scale gradients and patchiness, and at a suitably fine resolution for analyses of ecological structure and function. The challenges are to achieve a predictive understanding of not just temporal and spatial pattern and variability but also the interactions between these two dimensions. The challenges are to develop and test models and to synthesize our knowledge about how the world works.

The invisible present and the invisible place (Fig. 1.1) are metaphors that represent the view that we are limited in our ability to understand what is occurring at a particular time in a place of interest without considering the long-term dynamics at multiple spatial scales (Magnuson 1990, Swanson and Sparks 1990). Slow change often evades our senses, thus blocking understanding. Increasing the number of years of observation in a time series is like opening or unveiling the invisible present (Fig. 1.2). Each extension of the record allows the analysis of

Figure 1.1. The metaphor of the invisible present and the invisible place are captured in the clock in the dark out of context with time and space. (J. Magnuson)

new features. In the Lake Mendota time series of ice-cover duration—for example, a single year, the winter of 1997–98—indicates only that the lake was ice covered for 47 days. Opening the series to 10 years reveals that the winter of 1997–98 was an extreme year and that ice-cover duration is highly variable, ranging from 47 to 119 days over only 10 years. Opening the series to 50 years reveals an association between winters of short ice cover and the onset of an El Niño year, which is driven by atmospheric pressure differences in the South Pacific. Opening the series to the 1850s, when the records began, reveals a slow reduction in the duration of ice cover associated with long-term warming. These long-term changes are portrayed as a pair of step changes in Figure 1.2; when fit with a linear regression, the slope of the line is 16 days per 100 years from the winters ending in 1856 through 1998. Given the magnitude of interannual variability, this slow change could not have been detected from short-term records. A long-term view reveals a rich array of slow change and interannual and interdecadal dynamics.

We often are lulled into believing that causes of change are local or can be explained from a square-meter patch or a single lake out of context from its surroundings. Long-term ecological research studies posit that such a patch, while often useful in analyses, essentially does not acknowledge the many processes that determine that patch's status (character) or its temporal dynamics (Swanson and Sparks 1990). In 1887, Stephen Forbes offered a lake as a near-perfect example of a microcosm

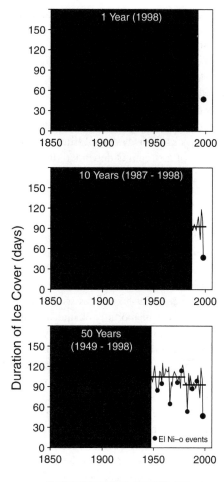

Figure 1.2. The significance of data from a single winter (1997–98) is revealed when a time series on Lake Mendota ice cover is opened from 1 year, to 10 years, to 50 years, and to the duration of record. Without such a time series, the data for that single winter are out of context and lost in the invisible present. Redrawn from Magnuson et al. 2000a with permission of E. Schweizerbart'sche Verlagsbuchhandlung Publishers, http://www.schweizerbart.de.

with processes that can be understood relatively independently of its surroundings (Forbes 1887, Real and Brown 1991). Later, Tansley (1935) named such holistic, interactive systems "ecosystems" (Golley 1993). Forbes (Real and Brown 1991) made his point about a lake as a microcosm clearly in his statement that lake organic complexes are isolated in "a little world within itself—a microcosm within which all the elemental forces are at work and the play of life goes on in full, but on so small a scale as to bring it easily within the mental grasp" (p. 14). Internal processes that occur in a lake are important to its character and dynamics (Chapter 10). However, Forbes also recognized external influences on the waters of Illinois nearly 100 years ago when he wrote "the productivity of a stream is dependent upon the extent and conditions of its backwaters and the period of its overflow, a fact which makes drainage district operations on river bottoms a menace to its productiveness" (Forbes 1912).

The idea that a longer and broader view is key to understanding complex natural systems is raised from quite different quarters. Aldo Leopold (1949), who developed the land ethic, made the point in *Thinking like a Mountain*, which was based on his last wolf kill in a predator control program in Arizona. He wrote: "Only the mountain has lived long enough to listen objectively to the howl of a wolf.—I thought that because fewer wolves meant more deer, that no wolves would mean 'hunters' paradise. But after seeing the green fire die [in the mortally wounded wolf's eyes], I sensed that neither the wolf nor the mountain agreed with such a view" (pp. 129–30). He went on to explain that the mountain had been around long enough to see that without predators, the herds of deer grew, and their overgrazing degraded the vegetation and also led to the collapse of the deer population. He did not make the point about a broad spatial view but could have from the vista a mountain provides. Peter Senge (1990) made the point in *The Fifth Discipline*, a book on the art and practice of operating a learning organization: "a fundamental characteristic of complex human systems: 'cause' and 'effect' are not close in time and space" (p. 63). The problem, he points out, is that "most of us assume, most of the time, that cause and effect are close in time and space" (p. 63). This broader view can lead to approaches for developing creative, learning organizations. Al Gore (1993) makes the point in *Earth in the Balance*, writing that using imagination to accelerate a slow change in the environment allows us "to see it in a more familiar frame and thus discern its meaning" (p. 43). He restated the Emerson quotation at the beginning of our chapter, as "it is helpful to stand at some distance from any large pattern we are trying to comprehend." Even the proverbial quote that "we cannot see the forest for the trees" can be considered in the sense of taking a broader spatial view. The realization reflected in these quotes is common to human experience and is philosophically fundamental to a long-term ecological research program like ours on north temperate lakes.

As an example of conducting ecological research within this broader context, we recognized that the status and dynamics of lakes are captured, in part, by the lakes' explicit positions in the landscape from high to low in hydrologic flow systems (Fig. 1.3) (Chapter 3) (Kratz et al. 1997b, Magnuson and Kratz 2000, Riera et al. 2000). In our northern Wisconsin lake district, upland lakes at the top of the landscape are small kettle-bog lakes largely isolated hydrologically from adjacent

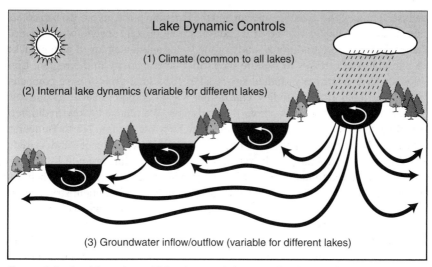

Lake Dynamic Controls

(1) Climate (common to all lakes)

(2) Internal lake dynamics (variable for different lakes)

(3) Groundwater inflow/outflow (variable for different lakes)

Figure 1.3. A wide variety of lake characteristics are related to a lake's position in the hydrologic flow system. Precipitation is relatively more important in lakes high in the flow system, whereas surface-water and groundwater inputs are more influential in lakes low in the flow system. (C. Bowser)

lakes. In other larger lakes high in the landscape, wave action prevents the formation of the sphagnum mats that typify bog lakes, and the water in these precipitation-dominated lakes is clear owing to low concentrations of nutrients. Farther down the flow field, lakes receive larger amounts of groundwater and associated nutrients from the landscape; connecting streams flow between these lower lakes. Some differences in physics, chemistry, and biology among lakes are related to these positional differences and associated processes in the landscape. Such a view moves one from believing that every lake is different to being able to understand the processes that underlie their differences, thus unveiling the invisible place (Swanson and Sparks 1990).

Ecologists who hoped that their science could move beyond the time-and-space constraints of the past helped the National Science Foundation (NSF) design the Long-Term Ecological Research Program (LTER) during the late 1970s (National Science Foundation 1977). Tom Callahan (1984) reaffirmed the point of view of the National Science Foundation that "Research in ecology has traditionally been funded for short periods and performed at single sites, conditions not conductive to projects addressing much greater time and geographic scales" (p. 363). Inherent in his statement was a goal to span multiple sites, not only among lakes as in our situation but also among disparate sites such as lakes and deserts to stimulate comparison and synthesis by a network of scientists who might otherwise pursue more narrow approaches.

The North Temperate Lakes LTER program, established in 1981, was one of the first six LTER sites. We set about designing a program of measurement and analysis to unveil the invisible present and the invisible place before we had coined the metaphors (Chapter 14). Several traditions were broken; several were retained. Rather than the more traditional study of the status and dynamics of a single lake, we chose a suite of lakes to capture the spatial heterogeneity in the region. Limnologists have intensively studied single lakes for a long time or have compared many lakes in short-term studies. Our design was to do both, that is, to study the long-term dynamics of lakes across the landscape. We retained a key tradition of limnology to be interdisciplinary; researchers who wrote the first North Temperate Lakes LTER proposal spanned the physical and biological sciences and shared a belief that interdisciplinary study, rather than single-discipline or multidisciplinary study, was required to understand lakes. Even with this belief, we at first behaved rather disciplinarily ourselves, as we tended to study physics, chemistry, or biology or, within biology, to study fish, plankton, or benthos. But we evolved toward increased interactions and linkages among these researchers and toward more interdisciplinary ideas and objectives (Chapter 14). To reflect the interdisciplinary nature of our program, we organized this book in a somewhat nontraditional way. The sections and chapters address a series of integrative concepts, rather than individual subdisciplines.

The North Temperate Lakes LTER project builds on a long history of limnological work in Europe and North America; useful reviews of this history are in Edgerton (1962), Frey (1966), Elster (1974), Golley (1993), Kalff (2002), and Magnuson (2002b). We discuss only some of the more relevant origins here. As mentioned earlier, Forbes developed the notion of a lake as a microcosm, that is, a well-bounded system. For the reasons realized by Forbes, we are able view a lake as an ecosystem more easily than we can a stream reach or a forested hillside. Raymond L. Lindeman (1942), in his classic paper on the trophic dynamics of Cedar Bog Lake, Minnesota, delineated the ecosystem concept in terms of the flow of energy among producers, consumers, and decomposers and the transformation of organic and inorganic matter. That this science was focused on a lake was no accident. The Canadian limnologist Donald S. Rawson (1939) conceptualized the external forces that act upon a lake, such as edaphic, climatic, topographic, and anthropogenic factors. His work and that of others placed a lake in the context of its drainage basin and the multitude of factors that affect the status of a lake that would change with geographic location. Borman and Likens (1967) and Likens et al. (1977) advanced the nutrient cycling approach in small watersheds at Hubbard Brook. More recently, longer-term dynamics were added to this conceptual foundation. In North America, the work of Edmondson (1991) on Lake Washington, Goldman (2000) on Lake Tahoe, and Likens (1985) on Mirror Lake and Hubbard Brook added the important temporal component. Comparative studies also led to advances in our understanding of many limnological topics, such as eutrophication from land-water interactions (Vollenweider 1968), local and regional processes that determine fish species structure in lakes (Tonn et al. 1990), and carbon dioxide exchange with the atmosphere (Cole et al. 1994), to mention but three. Experimental manipulations of entire lake ecosystems to discover how they work have become almost classic in limnology and ecology (Hasler 1966, Carpenter et al.1995). This approach has addressed issues of nutrient additions (Schindler 1974), acid rain

(Chapter 9) (Brezonik et al. 1993), and the role of consumers (Carpenter and Kitchell 1993), to mention only three.

The first book in which we noticed the phrase "lakes in the landscape" was by Burgis and Morris (1987). This readable general book about the different kinds of lakes around the world includes the following points:

- "... lakes enhance the scenic quality of landscapes" (p. 1).
- "... although a lake seems to be a clearly defined, self-contained entity, both on a map and on the ground, it is actually an integral part of the whole landscape and geographical area and should always be considered in that context" (p. 3).
- "The lake's regional geology is perhaps the most important influence on its life and style" (p. 5).
- "Underwater the 'landscape' of the lake is, to a considerable extent, a mirror of its surroundings" (p. 6).
- "There are regions where geological, or hydrological, events have left a legacy, not of huge lakes which dominate the landscape, but of many small lakes relatively close together. Such lake districts are particularly evident in areas affected by intense glacial activity . . ." (p. 19).

Our book contains many of these elements.

The major contribution of our North Temperate Lakes LTER program is a blending of long-term, comparative, experimental, and modeling approaches to attain new knowledge (Carpenter 1999). The treatment of temporal dynamics simultaneously with spatial heterogeneity of lakes across the landscape has allowed us to ask and answer new questions. We chose to study the long-term dynamics of a suite of lakes connected conceptually and mechanistically in a landscape. This choice has allowed us to study patterns of similarities and differences in the dynamics of adjacent lakes but also has enabled us to explain how the lakes' spatial settings influence their long-term dynamics. We have treated the landscape with the many embedded lakes as one of our units of analysis. We intuitively organized our thought about what is inside and what is outside our lakes' ecological systems and addressed analyses of interaction between internal (intrinsic) and external (outside) processes or drivers to help understand and predict the dynamics of lake ecosystems.

The choices we had to make in our first NSF proposal of which lakes to study for the long term made issues about the study of dynamics of lakes in the landscape very real and serious. The State of Wisconsin has at least 15,000 lakes embedded in a postglacial landscape. The number of lakes in the Laurentian Great Lakes region, in North America, and in temperate zones around the globe is staggeringly immense. The challenge of choosing a single lake to typify the essence of a north temperate lake was soon set aside, but not as soon as was the even more daunting challenge of studying them all. Even so, our goal was to understand the dynamics of lakes, rather than the dynamics of a particular lake, and we wanted to do that in a spatially explicit context.

Our program began with a heterogeneous set of seven lakes in the forested landscape of the Northern Highlands Lake District (Fig. 1.4, Table 1.1). Because E. A. Birge and C. Juday studied lakes in this lake district extensively in the 1930s and early 1940s (Frey 1966, Beckel 1987), we knew that these lakes differed from one another. They included lakes that were dimictic, monomictic, and polymictic; oli-

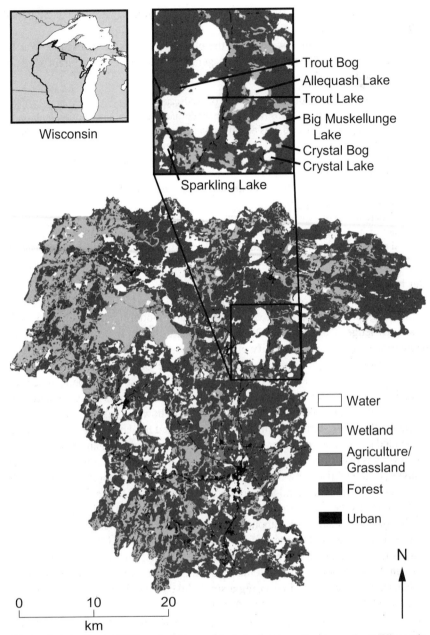

Wisconsin

Trout Bog
Allequash Lake
Trout Lake
Big Muskellunge
Lake
Crystal Bog
Crystal Lake

Sparkling Lake

Water

Wetland

Agriculture/
Grassland

Forest

Urban

N

0   10   20
km

Figure 1.4. Primary LTER study lakes and their landcover setting in northern Wisconsin.

Figure 1.5. Limnological field stations of the Center for Limnology, University of Wisconsin-Madison. Top: The Trout Lake Station in northern Wisconsin, May 1994. (White Swan Ltd., Winona, MN). Bottom: The Laboratory of Limnology on Lake Mendota in southern Wisconsin on the Madison campus, 1990. (Office of Information Services, University of Wisconsin-Madison)

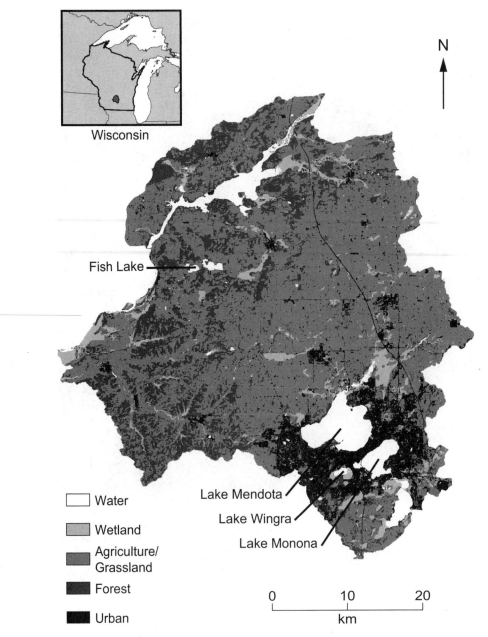

Wisconsin

N

Fish Lake

Water

Wetland

Agriculture/
Grassland

Forest

Urban

Lake Mendota

Lake Wingra

Lake Monona

0   10   20
              km

Figure 1.6. Primary LTER study lakes and their landcover setting in southern Wisconsin.

ecological research sites were embedded in the LTER Network's formation to gain new and more general theory, insights, and knowledge. In the 20 years after the network was founded, the diversity of sites increased as the number of sites increased from 6 to 24, additional new subdisciplines of ecology invigorated the research, and use of tools such as remote sensing, geographic information systems, and the Internet became standard. Information management (Chapter 13) became a strong component as LTER sites developed major long-term databases and an ability to network at the site, national, and international levels. Social sciences were included because the temporal dynamics across the landscape cannot be understood without knowledge of human behavior and influence. Although the LTER program began as a study of individual sites focused on ecosystem studies, it has developed into a network of long-term landscape and regional sites. Our colleagues expanded from a collection of site-specific, subdisciplinary experts to an interacting collection of disciplinary experts across a range of both similar and disparate ecosystems.

Long-Term Dynamics of Lakes in the Landscape chronicles an evolving inquiry, conducted at the North Temperate Lakes Long-Term Ecological Research sites in Wisconsin, into the long-term regional ecology of lakes. Our goals were to understand the dynamics of lakes and their interactions with the landscape at time scales of years to decades or longer, and spatial scales of individual lakes to lake districts or broader regions (Fig. 1.7). We believed that these were critical time and space scales to understanding how lakes respond to natural and human-induced processes, such as climatic change and variability, exotic species invasions, acid precipitation, eutrophication, and human settlement.

Our book has three themes. The first theme considers the conceptual approaches we have used or developed to study the dynamics of lakes in a landscape (Chapters 2–6). The second theme considers external and internal drivers of lake dynamics (Chapters 7–12). The third theme considers the experience and history we have had in developing and running an LTER program since 1980 and what we believe to be the legacies and the future (Chapters 13–15).

Summary

Dynamics of the ecological world play out and interact over time and space. The underlying dynamics and patterns are often invisible to the human observer because we tend to see the immediate present in a particular location. Long-term ecological research extends our abilities to see long-term and short-term dynamics across the landscape and the globe. The challenge is to observe and understand ecological dynamics across heterogeneous landscapes. Values of a long-term and broad view are embedded in the writings of philosophers such as R. W. Emerson, H. D. Thoreau, and A. Leopold. History of ecological science lays a foundation for such inquiry. Limnology has been a significant contributor to these foundations through the insights of S. A. Forbes, R. L. Lindeman, W. T. Edmondson, and many others. The U.S. National Science Foundation catalyzed long-term landscape ecology by es-

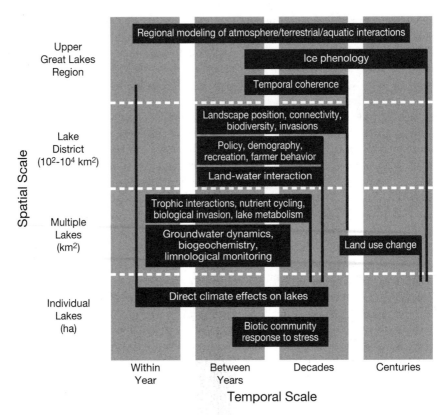

Figure 1.7. Multiple spatial and temporal scales considered within the North Temperate Lakes LTER program. The bar extending vertically from a box indicates the range of spatial scales for that box.

tablishing a formal program of long-term ecological research in the early 1980s. The North Temperate Lakes program, the focus of this book, is one of the sites established in that first round of support. Our book has three themes: emerging conceptual approaches to long-term landscape ecology, external and internal drivers of lake ecological dynamics, and the orchestration, history, and legacies of our program.

Part I

Conceptual Approaches

Understanding the long-term dynamics of lakes in the landscape requires that we develop an understanding of how processes acting over time and space interact to cause complex spatial and temporal patterns of lake dynamics. Each of the following chapters explores a theme or approach that we found especially useful. Chapter 2 discusses how water movement across the landscape links the terrestrial and aquatic processes. Chapter 3 discusses how the position of a lake in the landscape influences lake properties and dynamics. Chapter 4 discusses how the relative isolation of lakes as islands in a terrestrial sea influences biological structure and dynamics of lakes. Chapter 5 discusses how and why long-term interannual dynamics are synchronous or asynchronous between lakes, and Chapter 6 discusses how lakes can be compared generally with other ecosystem types.

The five approaches are clearly interrelated. For example, water movement across the landscape helps determine both how the landscape position of a lake influences its properties and why the interannual variability of certain lake features is synchronous for some lakes but not others. While lakes have been studied as individual ecosystems with great success, our consideration of a group of lakes as an object of study adds an important perspective to understanding dynamics of lakes.

2

Understanding the Lake-Groundwater System: Just Follow the Water

Katherine E. Webster
Carl J. Bowser
Mary P. Anderson
John D. Lenters

As many fresh streams meet in one salt sea . . .
—William Shakespeare, *Henry V*, 1.2.209 (Parrott 1953)

The availability, quality, and flux of water within and among compartments of the hydrologic system shape lake ecosystems. The hydrologic template for northern Wisconsin was set about 10,000 years ago when the glaciers retreated and left behind a thick layer of glacial drift dotted with kettle lakes, the remnants of ice blocks. The Long-Term Ecological Research (LTER) lakes in northern Wisconsin are among the thousands that make up one of the densest lake districts in the world; water covers 12% of the land surface in the Northern Highlands (Fig. 2.1). A striking feature of this population of lakes is the rarity of stream connections and the apparent hydrologic isolation of many lakes. In reality, this surface perspective hides a second legacy of the glaciers, that is, the conductive sands deposited so many years ago that form the matrix for saturated groundwater flow that links lakes across the landscape. Despite their surficial abundance, these kettle lakes in northern Wisconsin constitute less than 5% of the total water moving through the lake-groundwater system. These lakes are but a small, visible outcrop of the larger volume of water that moves through the landscape. Groundwater is the dominant flowpath that links lakes and that has provided a central organizing theme for understanding long-term change and temporal behavior of the multitude of lakes in the region.

The subsurface pathways of water through lakes and their surroundings are amazingly complex and are mostly invisible to even the well-trained eye. In the Eulerian sense you are an observer at a fixed spatial coordinate, watching barely

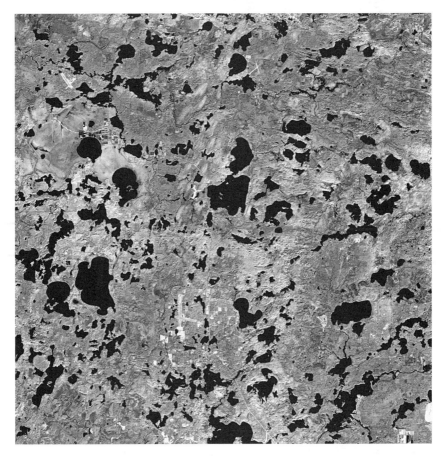

Figure 2.1. The high density of lakes and streams in the Northern Highlands of Wisconsin is illustrated in a Landsat image from October 6, 1999, of a 20 km by 20 km area.

perceptible changes brought about by the complex processes that control lake systems and their surroundings. The seasonal and interannual changes in lake water level might be your only visible access to the underlying complexity. If instead, in true Lagrangian fashion, you could follow the pathway of a single water droplet falling from the sky and moving through the landscape, a very different story could be told. For each droplet, an infinite number of pathways and subsequent transformations by chemical interactions with soils and the aquifer by biological uptake and by evapotranspiration are possible. The diverse pathways and processes that control the movement of water through the landscape are critical forces that shape the physical, chemical, and biological properties of lakes in northern Wisconsin.

To understand the relationship between lakes and their surroundings, one should follow the water using both Lagrangian and Eulerian perspectives. Each of the key reservoirs of water (the lake itself, the atmosphere, and the surrounding soil and groundwater) is connected intimately with the others, and together they constitute a restricted form of the hydrologic cycle. The relative size of each reservoir and the mass and rate of transfer among reservoirs all affect the movement of water to the lake. In turn, these water flowpaths provide a framework to evaluate the transport of solutes across the landscape. Short- and long-term climatic changes; human-induced changes such as acid precipitation, introduction of exotics, septic tank effluents, and road salt contamination; and terrestrial manipulations such as forest clear cutting, shoreline development, and road building all influence the temporal trajectories of lakes through their influence on water flowpaths and solute transport.

Over the past two decades, the importance of incorporating lake-groundwater interactions into a more comprehensive view of lake ecosystems has become increasingly apparent (Anderson and Bowser 1986, Winter et al. 1998, Devito et al. 2000). Lake-groundwater interactions have been an important element of our research, providing a key framework for interpreting the landscape-scale organization of lake features (Chapter 3) and for understanding dynamics (Chapters 5, 7). In this chapter, we describe the hydrologic template that links our study lakes with the groundwater system in the northern Wisconsin LTER lakes. Our three main goals are to examine this template from a variety of spatial scales from within-lake to the Upper Great Lakes region and temporal scales from season to century, to discuss how hydrology influences lake responses to regional stresses like acid rain and climate change, and, finally, to link the hydrology to the ecology.

Legacy of the Glaciers

Like many lake districts in north temperate regions, the lakes were formed by glacial forces during the Wisconsin glaciation in the Pleistocene Epoch that ended around 10,000 years ago. The Northern Highlands is a hummocky region of pitted outwash with low topographic relief that lies within the headwaters of two major river basins—the Wisconsin and the Chippewa. Within the glacial deposits are thousands of kettle lakes that, like our northern Wisconsin LTER lakes, were formed by the melting of ice blocks embedded in the outwash plains, moraines, drumlins, and eskers left by the glaciers (Fig. 2.2). Key landforms near the these LTER lakes include drumlins along the eastern shore of the south basin of Trout Lake and the extensive terminal Muskellunge moraine that borders the southeastern part of the watershed (Attig 1985).

As the glaciers receded across northern Wisconsin, they left behind 30- to 60-m-thick deposits of glacial drift. The area is unlike many glaciated landscapes where the scouring action of the glaciers exposed bedrock and removed overlying soils (Attig 1985). Most of these deposits are well- to poorly stratified sandy gravel and gravelly sand (Attig 1985) transported from Keweenawan or Cambrian sandstones to the north (Thwaites 1929). Underlying the glacial drift is a southern extension of

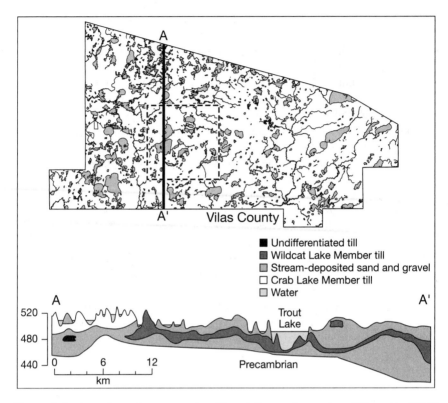

Figure 2.2. A north-south cross-section (A to A') near the northern Wisconsin LTER lakes showing glacial features. The dash-enclosed box on the plane view delimits the area in Fig. 2.4. Redrawn from Attig (1985) and Pint (2002) with permission of C. Pint.

the Canadian Shield formed of Precambrian bedrock (quartz-rich metasedimentary rock, granite, and amphibolite) (Attig 1985).

The lithological and mineralogical characteristics of the glacial drift in the Trout Lake region explain both the high hydraulic connectivity and the major ion chemistry of the lakes. The sandy glacial drift is highly permeable, although silt beds, usually less than 1 m thick, are occasionally present and can act as local confining beds for groundwater flow (Kenoyer and Bowser 1992a). Despite early reports of calcareous sediments, Kenoyer and Bowser (1992a) and Attig (1985) found no evidence of a surface or buried calcareous unit in Vilas County or of carbonate minerals within the dominant quartzose sands. This mineralogy helps explain the dilute nature and low carbonate content of many lakes within the Wisconsin glacial lobe (Eilers et al. 1986) and distinguishes them from lakes in other groundwater-dominated lake districts (Winter et al. 1998, Devito et al. 2000).

Surface flow is integrated poorly in the Northern Highlands because the lakes are embedded in this medium- to coarse-grained sandy matrix that is highly conductive to groundwater flow. As a result, a variety of lake hydrologic types is represented. Based on surface features, these lake types include seepage lakes that lack surface water inlets or outlets and drainage lakes that have outlets and inlets (Fig. 2.3). We can further classify lakes by the nature of their connections with the groundwater system.

- Recharge or hydraulically mounded lakes have a lake surface elevated above the water table and do not receive groundwater inputs. Water flows out of the lake and recharges the groundwater system.
- Flow-through lakes receive groundwater discharge in some areas and recharge the groundwater system in others.
- Discharge lakes have a lake surface below that of the water table, and groundwater flow is to the lake.

These hydrologic types highlight the diverse and complex nature of lake-groundwater interactions. Drainage lakes typically are connected to the groundwater system; in addition to direct groundwater discharge, their inflowing streams are fed by groundwater. Seepage lakes lie along a continuum of groundwater connection strength based on discharge patterns. Because precipitation exceeds evaporation in northern Wisconsin, even those lakes that receive no groundwater inflow maintain their water levels.

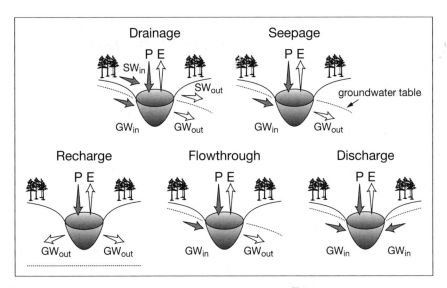

Figure 2.3. Hydrologic types of lakes: Top: Drainage lakes have a surface inlet and outlet, and seepage lakes do not. Bottom: Seepage lakes are classified by the patterns of groundwater flow. P = precipitation, E = evaporation, SW = surface water, GW = groundwater, in = into the lake, and out = out of the lake.

Hydrology guided the selection of the original seven primary lakes (see Chapters 1 and 14) as we aimed to represent the known diversity of lake types within the Trout Lake basin. Thus, the LTER lakes include drainage lakes (Trout and Allequash Lakes, with four and one inlets, respectively), seepage lakes (Big Muskellunge, Crystal, and Sparkling Lakes), and wetland-dominated seepage lakes (Crystal and Trout Bogs) (Table 2.1, Table 1.1). The selection, however, was predicated on surface features, as our knowledge of groundwater flowpaths at that time was limited. This set of lakes provided an informative range in complexity and a variety of lake-groundwater interactions not fully envisioned when our site was established.

Exposing Groundwater Flowpaths

Regional and Local Flowpaths and Chemical Evolution

To understand lake-groundwater interactions, substantial effort was devoted to modeling groundwater flowpaths and lake-groundwater dynamics. The regional hydrologic model of the Trout Lake basin is based on the groundwater flow code (MODFLOW) (McDonald and Harbaugh 1988), with the addition of a lake package to simulate fluctuations in lake level (Cheng and Anderson 1993, Merritt and Konikow 2000). Details on the testing and calibration of this model are described in Anderson and Cheng (1998), Hunt et al. (1998), Champion and Anderson (2000), and Pint et al. (2003).

Table 2.1. Hydrologic characterization of the North Temperate Lakes LTER primary lakes in northern Wisconsin.

Lake	Depth (m) Mean	Max	Surface Area (ha)	Water Residence Time (year)	GW_{in} (% of total)	Hydrologic Type[a] Groundwater	Surface	Surface Flow Inlets	Outlets
Trout Bog	5.6	7.9	1.1	na	na	GR	SE	0	0
Crystal Bog	1.7	2.5	0.5	*na*	*na*	GFT	SE	0	0
Crystal	10.4	20.4	36.7	12.7	0–10	GR & GFT	SE	0	0
Big Muskellunge	7.5	21.3	396.3	8.0	16	GFT	SE	0	0
Sparkling	10.9	20.0	64.0	10.4	25	GFT	SE	0	0
Allequash	2.9	8.0	168.4	0.5[b]	31[c]	GFT	DR	1	1
Trout	14.6	35.7	1607.9	4.6[b]	35[c]	GD	DR	4	1

[a]GR = groundwater recharge; GFT = groundwater flowthrough; GD = groundwater discharge; SE = seepage; and DR = drainage.

[b]Water residence time estimated from lake volume and surface outflow; groundwater outflow was assumed to be negligible.

[c]Percent groundwater input is a rough estimate based on extrapolation from the relationship between calcium plus magnesium concentrations and percent groundwater inflow for a set of seepage lakes (Webster et al. 1996).

Note: Data are from Webster et al. (1996) and the LTER Web site (lter.limnology.wisc.edu).

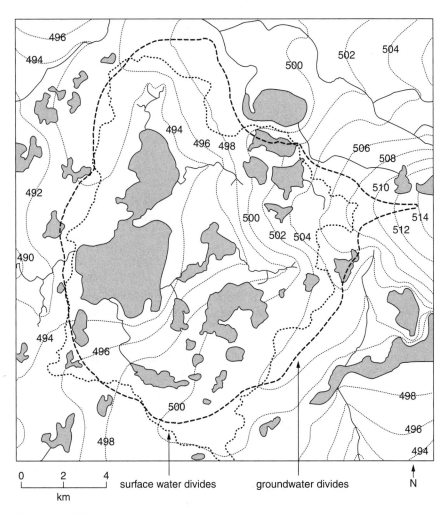

Figure 2.4. Water table map (2m contours above sea level) of the Trout Lake basin contrasting the groundwater divide and the surface water divide. Redrawn from Pint 2002 with permission of C. Pint.

In the Trout Lake basin, groundwater flows from southeast to northwest with Crystal Lake located near the groundwater divide and Trout Lake being the terminal groundwater discharge point for the basin (Fig. 2.4) (Hunt et al. 1998). The boundaries of the groundwater divide for the Trout Lake basin do not necessarily co-occur with surface divides (Hunt et al. 1998, Pint 2002, Pint et al. 2003). In this region of low topographic relief, definition of the topographic boundaries of the surface watershed is difficult. Moreover, groundwater divides and flowpaths cannot always be inferred precisely from surface features.

Local flowpaths are even more complex. Despite being separated by an isthmus only 120 m wide, Big Muskellunge Lake is 1 m lower than Crystal Lake in water elevation and has a higher concentration of total dissolved solids. Terrestrially derived recharge from precipitation mixes across this narrow isthmus, with the upper layers of a plume of lake water emanating from Crystal Lake (Fig. 2.5). The deeper part of the Crystal Lake plume remains relatively intact while traversing the isthmus, with isotope signatures showing a gradation of lake-derived water mixing with precipitation-derived water higher in the plume (Kim et al. 2000, Bowser and Jones 2002). In the upper 3 m of water below the water table, the plume structure is influenced by seasonal recharge events from terrestrially derived precipitation (Bullen et al. 1996, Kim et al. 2000).

As water flows through the sandy aquifer, chemical reactions with minerals transform solute concentrations, so that the older groundwater is more ion-rich because its exposure to minerals within the aquifer has been longer. Both the hydrology (Kim et al. 1999, 2000) and the chemical evolution of groundwater have been studied intensively across the isthmus between Crystal and Big Muskellunge Lakes (Kenoyer and Bowser 1992a, 1992b, Bullen et al. 1996, Bowser and Jones 2002). These studies describe the chemical evolution of water in contact with silicate-rich minerals over long time scales impractical to address in laboratory studies.

Crystal Lake water takes 24 to 98 years to cross the 120 m isthmus to Big Muskellunge Lake (Kim et al. 1999). During that time, a series of chemical reactions occur between water and silicate minerals (primarily plagioclase feldspar with some biotite and calcic amphiboles, and precipitation of kaolinite and beidellitic smectitic clays) such that product ion concentrations increase dramatically. These reactions control concentrations of many major ions. For example, calcium (1 and 5 mg/L), magnesium (0.3 and 2 mg/L), alkalinity (<100 and 400 μeq/L), and silicon (0.01 and 7.0 mg/L) are lower in Crystal Lake than in groundwater (Bowser and Jones 2002). Flowpath length or groundwater age thus determines the degree of chemical evolution of groundwater and the resulting solute concentrations of water that subsequently discharges into lakes. Groundwater provides the dominant reservoir from which these lakes acquire much of their characteristic major ion chemical fingerprint, which is based on the flow rate of groundwater, the rate of chemical weathering of silicate minerals, and the mineralogical controls on the water. The actual delivery of solutes to lakes by groundwater is influenced by interactions at the sediment water interface. For example, transformation and uptake by microbial processes along quite short distances within the hyporheic zone cause complex spatial and temporal variation in the flux rates of key carbon constituents such as dissolved organic carbon (DOC), carbon dioxide, and methane (Schindler and Krabbenhoft 1998).

Defining Groundwater Capture Zones for Lakes

Groundwater entering a lake may originate from a terrestrial source as precipitation that infiltrates the subsurface and recharges the groundwater system and then discharges directly to the lake. Groundwater inflow to a lake also may originate in

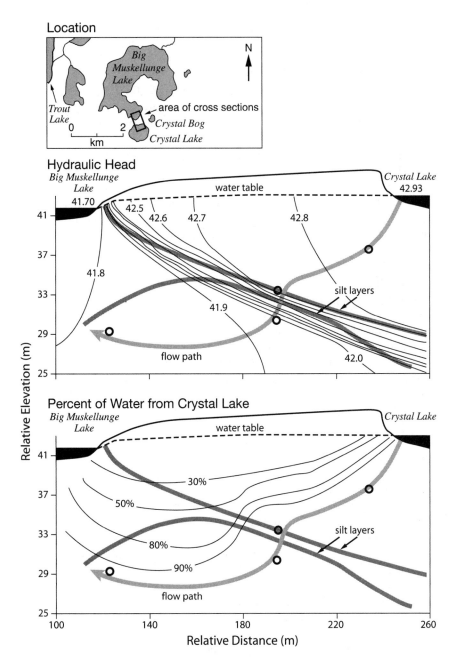

Figure 2.5. Hydrologic flow paths across the isthmus from Crystal Lake to Big Muskellunge Lake. Top: Location of the isthmus. Middle: Hydraulic head values across the isthmus. Bottom: Percent of water derived from Crystal Lake. o = well sample site. Redrawn from Bowser and Jones (2002) with permission of the American Journal of Science.

an up-gradient lake as water discharges from that lake and flows down-gradient. Differences in source area (terrestrial or lake) and the presence of intervening lakes that may or may not capture deep underflowing groundwater contribute to a complex mix of flow paths. Consequently, water of different ages and different water chemistries may discharge in close proximity (Pint et al. 2003). In effect, the lake-groundwater system in the Trout Lake basin acts as a conveyor moving water down the gradient toward Trout Lake so that water anywhere in the basin may have originated at the groundwater divide or anywhere in between.

The subsurface volume of porous media that directly contributes groundwater to a lake defines the capture zone. A capture zone is defined strictly by a three-dimensional surface (Townley and Trefry 2000), but we also can define a capture zone as the land surface area that contributes flow that discharges directly into the relevant lake or stream. Lake capture zones delineated using a groundwater flow model of the Trout Lake basin reveal that some lakes receive terrestrially derived groundwater that originates up-gradient of the next up-gradient lake in the system. For example, the capture zone for Allequash Lake includes an area up-gradient of Big Muskellunge Lake (Pint et al. 2003) that is 5 km from Allequash Lake. This up-gradient area appears to be the source of long, slow flowpaths that discharge water high in calcium to the southern shoreline area of Allequash Lake's lower basin.

Travel times within lake capture zones, calculated with the aid of the groundwater flow model for the basin, represent the time for water that has just entered the aquifer to reach the lake. Trout Lake, as the main discharge area in the basin, has the largest capture zone area with travel times greater than 200 years. Travel times are similar in the Big Muskellunge Lake capture zone but are less than 20 years in the smaller Crystal Lake capture zone (Fig. 2.6).

Tracking Water and Solutes from Land to Lake

Earlier, we focused on regional and local groundwater flowpaths, chemical reactions related to flowpath length, lakes as sources of recharge to the groundwater system, and capture zones of lakes. Here we consider the broader context of hydraulic connections between lakes and the groundwater system and how this influences lake hydrologic and chemical budgets. As mentioned previously, the permeable sands in the thick glacial drift preclude a strong network of surface water connections among lakes in the Trout Lake basin. These same hydrogeologic characteristics cause groundwater to be the dominant hydrologic connector across the landscape. From this perspective, lakes can be considered as windows into the groundwater system.

Much of the detailed hydrologic research on lakes at the LTER site in northern Wisconsin has focused on seepage lakes, defined by their lack of surface water inlets or outlets. This type of kettle lake is common in the hummocky glacial terrain of the Northern Highlands. Of the lakes larger than 10 ha in surface area, 47% are seepage lakes (Eilers et al. 1986). The nature of the connections between these seepage lakes and the groundwater has been of interest since E. A. Birge and C. Juday did their seminal limnological research in the Trout Lake area in the 1930s and 1940s.

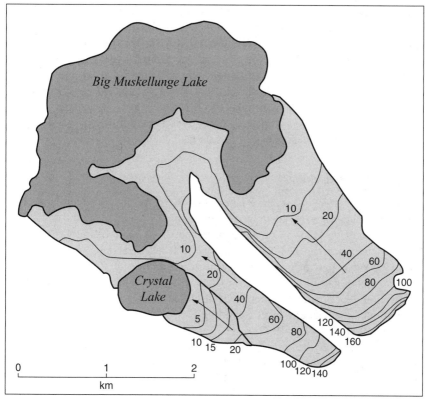

Figure 2.6. Groundwater capture zones for Big Muskellunge Lake and Crystal Lake with contours showing the years required for water to travel from the recharge point where it enters the groundwater to the discharge into the lake. Arrows = direction of flow. Redrawn from Anderson (2002) and Pint (2002) with permission of M. Anderson and C. Pint.

Juday and Birge (1933) used hydrology to explain heterogeneity in lake chemical composition; seepage lakes tended to have a more dilute chemistry, lower pH, and lower major ion content than drainage lakes, and the researchers identified groundwater connections as a cause of this divergence. Broughton (1941) and Juday and Meloche (1944) noted the large discrepancy in the water chemistry between seepage lakes and nearby groundwater and postulated that water-tight lake basin seals limited flow between lakes and ion-rich groundwater. They hypothesized that these bottom seals explained differences of 1 to 2 m in elevation between lakes located only 25 to 150 m apart.

Two decades of research on seepage lake hydrology paint a more complicated picture of lake-groundwater interactions. The impetus for understanding seepage lake hydrology in the Trout Lake basin came from concerns about the effects of acid deposition on lakes in the region. Similar to the results of Juday and Birge

(1933), chemical surveys in the late 1970s showed that seepage lakes were potentially sensitive to acidification owing to their low pH, dilute chemistry, and low acid neutralizing capacity (ANC) (Eilers et al. 1983). Because our understanding of the nature of lake-groundwater interactions was limited, hydrologic budgets were key to explaining lake sensitivity to acid rain and predicting lake responses to changes in deposition rates (Eilers et al. 1983, Anderson and Bowser 1986). Without an understanding of the flowpaths of water and solutes to seepage lakes, we could not predict responses of lakes to sulfur emission controls, nor could we fully understand temporal and spatial controls on lake chemical composition.

Seepage Lake Hydrology

Our hydrologic studies of seepage lakes in northern Wisconsin began in the mid-1980s, when budgets were determined for three slightly acidic (pH ≈ 6.0 and ANC ≈ 20 μeq/L) seepage lakes in or adjacent to the Trout Lake basin. Crystal Lake (Kenoyer and Anderson 1989) is a primary LTER lake; Vandercook Lake (Wentz et al. 1995) was part of a long-term acid rain monitoring study (Webster and Brezonik 1995), and Little Rock Lake (Rose 1993) was the site of a whole-basin acidification experiment (Chapter 9). Hydrologic studies also were conducted at the more alkaline seepage lake, Sparkling Lake, that is an LTER primary lake (Krabbenhoft et al. 1990a, 1990b).

In general, lake hydrologic budgets are quantified as the balance between inputs of precipitation, groundwater, and surface water and outputs of evaporation, groundwater, and surface water (Fig. 2.3):

$$\Delta S = (P + GI + SI) - (E + GO + SO)$$

where:

ΔS = the change in storage or volume of the lake

P = precipitation on the lake surface

GI = groundwater inflow to the lake

GO = groundwater outflow from the lake

E = evaporation from the lake surface

SI = surface water inflow to the lake

SO = surface water outflow from the lake

For lakes in the Northern Highlands, overland flow from diffuse water inputs from riparian areas was not observed and was considered inconsequential because of rapid infiltration into the sandy soils surrounding the lakes (Wentz et al. 1995). Because seepage lakes lack surface water connections, the preceding equation further simplifies to:

$$\Delta S = (P + GI) - (E + GO)$$

Studies of lake hydrology often include both groundwater inputs and outputs as part of the residual or error term, a practice that introduces significant uncertainty because errors for all other terms are included in the residual (Winter 1981). Because groundwater is so important, the Wisconsin studies quantified all components of the budget. Measurements of P and ΔS are relatively straightforward. Evaporation was estimated from pan evaporation, while the groundwater terms GO and GI were determined through a combination of flow-net analysis, direct measurements of seepage flux using seepage meters (Lee 1977), and application of Darcy's Law (Kenoyer and Anderson 1989, Krabbenhoft et al. 1990a, 1990b, Rose 1993, Wentz et al. 1995, Hunt et al. 2003, John et al. 2003). Through an integrative approach using stable isotopes of oxygen, Krabbenhoft et al. (1990a, 1990b) were able to solve independently for all terms in the hydrologic budget and estimate groundwater recharge and discharge for the intensively studied Sparkling Lake.

Hydrologic budgets for Crystal, Vandercook, and Little Rock Lakes indicated that most of their water was supplied by direct precipitation onto the lake surface, while direct evaporation was the major loss term. Groundwater inputs were only a fraction of the inputs in the water budget, ranging from ~3% for Little Rock Lake and Vandercook Lake (Wentz and Rose 1989, Rose 1993) to ~10% for Crystal Lake (Kenoyer and Anderson 1989). Krabbenhoft et al. (1990a, 1990b) estimated even higher groundwater inputs of 20% to 25% to Sparkling Lake, a groundwater flow-through lake. Groundwater inflow in these lakes occurs predominately in the littoral zone through conductive sandy sediments; deeper sediments contain thick organic layers that prevent flow.

Chemical budgets for these same lakes combined flow rates from hydrologic budgets with ion concentrations in precipitation, incoming groundwater, and groundwater out (set to lake concentrations). Evaporation is not part of the chemical budget calculation because only water is removed by the process. Despite being only a small source of water relative to precipitation, groundwater inputs have a disproportionately large influence on the water chemistry and ionic strength of these seepage lakes. Groundwater was the major source of silicon, calcium, magnesium, potassium, and acid neutralizing capacity, while precipitation was the primary source of H^+ and the mineral acid anions SO_4^{2-} and NO_3^-, not surprising owing to the acidic nature of precipitation in the region (Fig. 2.7). Groundwater also can be an important source of phosphorus to these lakes (Kenoyer and Anderson 1989), particularly from water flowing from deeper groundwater flowpaths.

Hydrologic studies on these seepage lakes provided evidence that even a small input of groundwater controls their major ion chemistry. As hypothesized by Eilers et al. (1983) and Anderson and Bowser (1986), the sensitivity of these seepage lakes to acidification by atmospheric deposition is a function of the amount of buffering substances they receive from groundwater. Further evidence from modeling studies suggested that both Crystal and Vandercook Lakes would acidify if even these small inputs of groundwater ceased (Garrison et al. 1987). In contrast to these acidic lakes, groundwater provided about 25% of the water input to nearby Sparkling Lake, supplying sufficient buffering capacity to make the lake insensitive to acidification by atmospheric deposition (Krabbenhoft et al. 1990a, 1990b, 1994). For these seepage lakes, it is impossible to understand their major ion chemistry without considering interactions with groundwater.

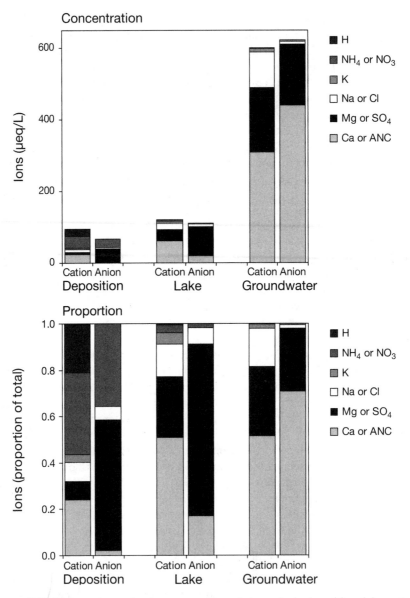

Figure 2.7. Major cation and anion composition of atmospheric deposition, lake water, and up-gradient groundwater for Vandercook Lake. Top: Concentrations. Bottom: Proportions of total. Redrawn from Wentz et al. (1995) with permission of the American Geophysical Union.

Differences among Seepage Lakes

Hydrologic studies of lakes in the Northern Highlands demonstrate four features of seepage lakes that distinguish them from drainage lakes and highlight the importance of lake-groundwater interactions. First, even small groundwater inputs (<5% of the total inputs) are the dominant sources of solutes generated by weathering reactions, for examples, Ca, Mg, Na, K, Si, and ANC, to seepage lakes. Second, water residence times (defined as the average time a water molecule resides in the lake) of seepage lakes are generally longer than those typically measured for drainage lakes, on the order of 4 to 6 years, but ranging up to 10 years for deep lakes like Sparkling and Crystal Lakes (Table 2.1). Third, these longer chemical residence times cause in-lake processes like microbe-mediated sulfate reduction to exert a larger influence on ion chemistry than is observed in lakes with shorter residence times (Wentz et al. 1995). Fourth, the strength of the lake-groundwater connection differs among the LTER seepage lakes, with groundwater inputs ranging from an average of about 5% for Crystal Lake to 25% for Sparkling Lake.

The hydrology of lakes is better described as a continuum from recharge lakes that receive no groundwater inflow, to flow-through lakes that receive groundwater inflow and discharge water to the groundwater system, to discharge lakes that have no groundwater outflow. This viewpoint provides the basis for the landscape position concept, discussed below and in detail in Chapter 3. Furthermore, as we discuss next, these hydrologic connections are dynamic across seasonal, annual, and decadal time scales leading to transient hydrologic states.

Development of a Dynamic Landscape Perspective

When the North Temperate Lakes LTER site was initiated, in 1980, we understood that hydrology led to differences among lakes in their major ion chemistry, but we had no framework for interpreting spatially explicit patterns at the regional scale. With regard to temporal dimensions, we assumed that groundwater systems were at steady state, with fairly stable flow patterns. The exciting insights from two-plus decades of research have come from considering lake-groundwater interactions from a dynamic landscape perspective.

Spatial Interactions between Lakes and Groundwater

Hydrologic studies within the Trout Lake basin contributed to the development of the landscape position concept, a powerful template for interpreting spatial patterns in lake features (Chapter 3). The hydrologic basis of this concept is that the strength of interaction between lakes and groundwater increases along a continuum from the basin divide to the regional discharge point. End members of the regional hydrologic flowpath within the Trout Lake basin are Crystal Lake, at a high landscape position near the groundwater divide, and Trout Lake, at the lowest position in the basin (Fig. 2.8, Table 2.1). The hydrology of lakes high in the landscape is dominated by precipitation, with minimal groundwater inputs from predominately local

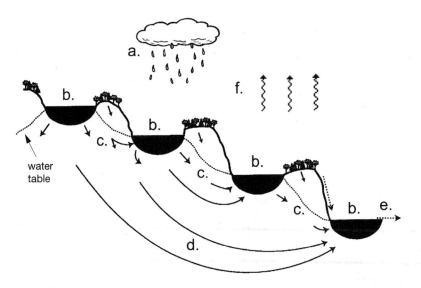

Figure 2.8. Diagram of the water budget for the landscape position continuum of the Trout Lake basin. Precipitation (a) falls onto the land and enters the groundwater system or directly falls on lakes (b) and streams. Within the groundwater system, water can follow short, local flowpaths (c) or long, regional flowpaths (d). Water leaves the basin through stream flow at the terminal discharge point (e) or through evaporation from either land or water surfaces (f). Solid arrows = groundwater flow. Dashed arrows = stream flow. Modified from Webster et al. (1996).

flowpaths. In lakes lower in the landscape, groundwater supplied from regional flowpaths becomes more important. This groundwater template is expressed most strongly in the lake concentrations of solutes that are the products of chemical reactions within the aquifer such as for calcium (Fig. 2.9) and for correlated chemical variables such as acid neutralizing capacity, pH, conductance, and silica. Similar heterogeneity in silica caused by groundwater was apparent for small kettle-hole peatland lakes in the Trout Lake basin (Kratz and Medland 1989).

Although these chemical patterns provide clues to the underlying hydrology of lakes, accurate quantification of groundwater inputs of both water and solutes is not trivial over a large scale. The traditional approach based on Darcy's law is time-consuming and costly, requiring temporally intensive measurements at many observation wells to describe the flow field in three dimensions. Spatial characterization of the hydraulic conductivity of the aquifer is difficult. Mass balance methods that use stable isotopes of oxygen have promise for providing estimates of groundwater inputs to lakes at a more regional scale (Krabbenhoft et al. 1990a, 1990b, 1994). The isotope mass balance method reflects a long-term water balance that precludes analysis of interannual shifts and assumes that the lake-groundwater system is in isotopic and hydrologic steady state. When this method was applied to LTER seepage lakes, groundwater inputs were estimated at about 5% for Crystal Lake, 17% for Big Muskellunge Lake, and 25% for Sparkling Lake (Krabbenhoft et al. 1994).

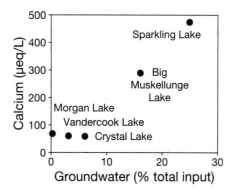

Figure 2.9. Lake calcium concentrations plotted against percent groundwater input for five seepage lakes in the Northern Highlands. Data from Webster et al. (1996).

Ackerman (1993) and Michaels (1995) extended the isotopic approach to all of northern and northwestern Wisconsin and provided groundwater input estimates for more than 40 lakes. Groundwater inputs ranged from only a few percent to nearly 50%.

At the within-lake scale, heterogeneous fluxes at the groundwater-lake interface can lead to complex distributions of aquatic organisms. For example, although water chemistry, wave exposure, and substrate type are heterogeneous within Sparkling Lake, aquatic plant distributions reflect groundwater flow patterns. Macrophyte biomass was highest where groundwater flow, either as recharge or discharge, was substantial (Lodge et al. 1989). Because macrophytes were dense in both inflow and outflow zones, nutrient supply appeared not to be important, but rather some feature of groundwater movement within the littoral sediments created conditions favorable to macrophyte growth. In contrast to macrophytes, benthic algal biomass was high in groundwater discharge zones but not in recharge areas (Hagerthey and Kerfoot 1998). Soluble reactive phosphorus and nitrate-nitrite concentrations and fluxes were higher in the discharge zones; thus, nutrient inputs appeared to directly stimulate benthic algal growth. Both studies suggest mechanisms by which heterogeneous groundwater flow patterns can directly influence spatial patterns of benthic plants and, by inference, other parts of aquatic food webs.

Temporal Interactions between Lakes and Groundwater

So far, we have provided a relatively static view of lake-groundwater interactions. However, lake-groundwater interactions are dynamic on seasonal, annual, and decadal time scales. Seasonal reversals in groundwater flow direction around lakes are driven by recharge during spring snowmelt and autumn rainfall (Anderson and Cheng 1993, Krabbenhoft and Webster 1995). Similar seasonal reversals have been documented near peatlands (Marin et al. 1990). At Crystal Lake, when water levels are relatively high, groundwater mounds form near the shore and cause groundwater to discharge to the lake from both shallow and deeper flowpaths (Fig. 2.10) (Anderson and Cheng 1993). While input from shallow flowpaths is a combination

Cross Section

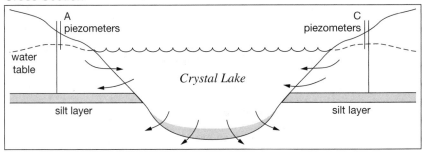

Planar View

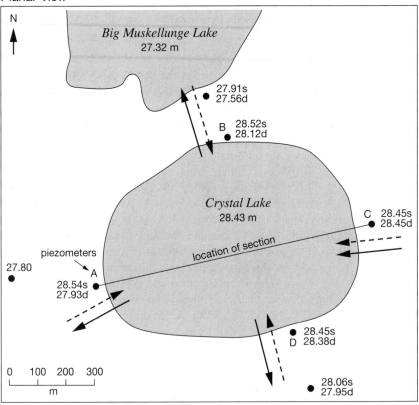

Figure 2.10. Formation of groundwater mounds leading to both shallow and deep inflow to Crystal Lake. Both diagrams are from the spring during high recharge. Top: Cross-section showing flow direction. Bottom: Planar view showing relative water elevations in meters in the lake and in shallow (s) and deep (d) observation wells located near the lakeshore. The dashed arrows (shallow) and solid arrows (deep) show flow direction. Redrawn from Anderson and Cheng (1993) with permission of the Journal of Hydrology.

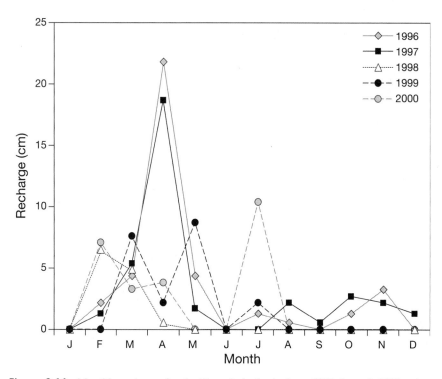

Figure 2.11. Monthly recharge for the Trout Lake basin from 1996 through 2000 calculated with a soil water balance model for the basin. Redrawn from Dripps et al. (2001) with permission of W. Dripps.

of rainfall and dilute groundwater and thus of little consequence for lake chemical budgets, deeper flowpaths supply ion-enriched groundwater. Such transient groundwater mounds can act to flush lake water back and forth across the lake-groundwater interface. For example, judging from stable isotopes (^{18}O and ^{2}H), water discharging from shallow areas into Nevins Lake in upper Michigan during spring was made up mainly of recycled lake water (Krabbenhoft and Webster 1995). In lakes like Crystal and Nevins Lakes that receive small groundwater inputs, these dynamics further complicate lake-groundwater studies because reversals are transient and the areal extent of the reversal is difficult to estimate (Anderson and Cheng 1993).

Temporal changes in groundwater flow paths in the isthmus between Crystal and Big Muskellunge Lakes were induced by fluctuations in groundwater recharge rates and to a lesser extent by fluctuations in lake levels (Kim et al. 2000). The transient fluctuations caused enhanced vertical mixing of an oxygen isotope plume emanating from Crystal Lake. The formation of seasonal groundwater mounds between the two lakes caused flow direction to change by as much as 160 degrees over the course of a year. These transient groundwater flow dynamics were driven by seasonal and annual variation in recharge rates (Fig. 2.11). Over the 5-year period

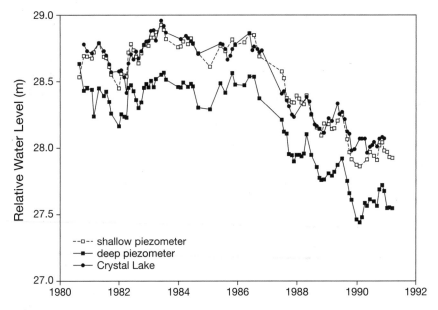

Figure 2.12. Water levels from 1980 to 1991 measured in Crystal Lake and in shallow and deep observation wells near the lakeshore at location B in Fig. 2.10. Redrawn from Anderson and Cheng (1993) with permission of the Journal of Hydrology.

from 1996 to 2000, annual recharge rates varied more than two-fold among years (Dripps et al. 2001).

Seasonal inputs of groundwater can be critical for maintaining cycles of aquatic organisms in lakes. For example, pulses of groundwater discharge following snowmelt stimulate large spring diatom blooms in Crystal Lake and account for about 50% of the annual silica input (Hurley et al. 1985). Diatoms require silica to construct frustules, part of their cellular structure. In oligotrophic, precipitation-dominated Crystal Lake, these seasonal pulses from groundwater accounted for nearly the entire annual external load of silica (Hurley et al. 1985).

Superimposed on these seasonal dynamics are hydrologic shifts that take place over annual or longer time scales. Shallow and deep groundwater flowpaths fluctuate significantly at interannual and decadal time scales (Anderson and Cheng 1993). Water levels in the deep and the shallow piezometers around Crystal Lake fluctuated by 1 to 1.5 m during the 1980s (Fig. 2.12) (Anderson and Cheng 1993). Most striking was an almost linear decline in water table elevation during the 1986–1990 drought, a clear long-term response of the groundwater system to a period of persistently low recharge. More visible were the parallel changes in lake water elevation. Like those of most seepage lakes in the region, lake water levels at Crystal Lake varied by as much as a meter in the 1980s and 1990s. High water levels flooded and eroded riparian zones at times, while large expanses of lake sediments became exposed during dry periods. In contrast to seepage lakes, lake level fluctuations in

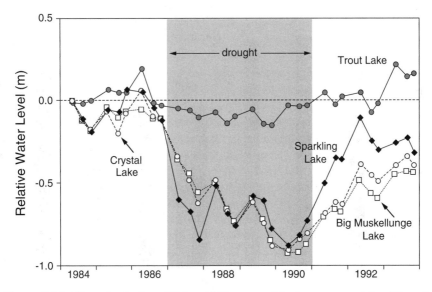

Figure 2.13. Water levels from 1984 to 1993 measured in three seepage lakes (Crystal, Big Muskellunge, and Sparkling Lakes) and a drainage lake (Trout Lake). Modified from Webster et al. (1996).

drainage lakes often are controlled by sills at the outlet and thus tend to be muted (Fig. 2.13).

Observed fluctuations in water table and lake water elevations reflect interannual dynamics in the hydrology of seepage lakes, driven in part by a prolonged drought in the mid- to late 1980s. During this drought, the balance between precipitation and evaporation was quite different from that of the pre- and postdrought years. For example, at Little Rock Lake, the ratio of precipitation to evaporation was about 1.1 during drought, much lower than the average of 1.7 measured between 1984 and 1986 or in 1990 (Fig. 2.14) (Rose 1993). More precise measurements calculated via energy budgets suggest that summertime evaporation from Sparkling Lake can exceed precipitation but that interannual variation leads to both positive and negative differences between precipitation and evaporation (Chapter 7) (Lenters et al. 2005). These estimates reflect only evaporation from the lake surface and do not include evapotranspiration over land, which can be equally important to the overall water budget.

A series of hydrologic shifts occurred between 1980 and 1990 for Crystal Lake (Anderson and Cheng 1993). When regional groundwater levels were high from 1983 to 1985, seasonally transient groundwater mounds consistently formed, typically for several weeks in late spring or early summer and again in late autumn. In the early years of the drought, in 1987 and 1988, when groundwater levels were still high relative to declining lake water levels, mounds still formed and persisted. However, the difference in elevation between the lake and the regional groundwater table became too small to generate discharge to the lake after 1988, mounds did not form, and Crystal Lake received essentially no groundwater.

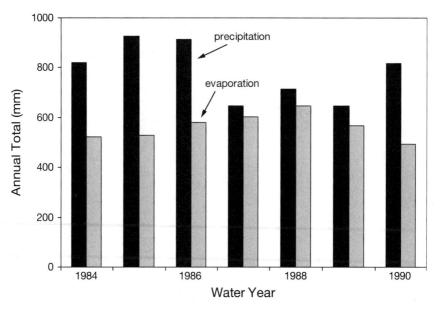

Figure 2.14. Annual evaporation and precipitation at the surface of Little Rock Lake calculated on a water-year basis. Precipitation was from a recording gauge near the lake. Evaporation was estimated from Class-A evaporation-pan data collected near Vandercook Lake. Data from Rose (1993).

Such hydrologic shifts and implications for the maintenance of lake buffering capacity also were demonstrated by the long-term hydrologic research at nearby Vandercook Lake, an LTER secondary lake. Three hydrologic states were identified for Vandercook Lake (Wentz et al. 1995). Between 1980 and 1983, when regional precipitation was 108% of normal, the lake was a flow-through system, receiving from 5% to 9% of its water from groundwater. This input provided acid neutralizing capacity in sufficient quantity to buffer incoming acids in atmospheric deposition. During the intermediate years 1984–1986, groundwater inputs averaged only 1%, despite the high precipitation of 114% of average. During this time, groundwater recharge was much lower than in 1980–1983, perhaps the result of different timing of precipitation events and thus recharge. During the final interval of low precipitation, 90% of average for 1986–1988, groundwater inputs to Vandercook Lake ceased, and the lake became mounded hydraulically (Fig. 2.15). As a result, the buffering system of the lake switched from control by inputs of ANC derived from chemical reactions in the groundwater system to control by in-lake alkalinity-generating processes like sulfate reduction. Similar patterns occurred in Little Rock Lake (Rose 1993), which received intermittent groundwater inputs of around 1% between 1983 and 1986 but no groundwater inputs from 1986 to 1990. These transitions highlight the dynamic nature of the hydrology of seepage lakes like Crystal, Vandercook, and Little Rock Lakes that have relatively small and transient contributions of groundwater and can switch from recharge to flow-through conditions.

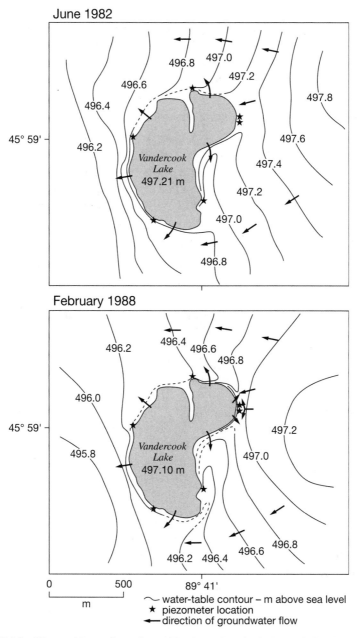

Figure 2.15. Water-table configuration at Vandercook Lake in June 1982, when the lake was receiving groundwater inflow in the northeastern bay, and in February 1988, when flow was reversed and the lake received no groundwater. Data from Wentz et al. (1995).

Lake-groundwater interactions are quite dynamic, with variation on seasonal, annual, and decadal time scales. Hydrologic studies conducted over short time intervals of even two to three years would not represent adequately the long-term water balance and thus could yield misleading results if applied to long-term management (Anderson and Cheng 1993, Wentz et al. 1995).

Responses to Drought and Sulfate Deposition

Since 1980, northcentral Wisconsin has experienced two long-term regional changes that have influenced hydrology and the transfer of solutes across the landscape. First, deposition rates of SO_4^{2-} (Fig. 2.16, top) and H^+, the key ingredients of acid rain, decreased steadily from the early 1980s through the 1990s. This decline was stimulated by Wisconsin legislation passed in 1984 that mandated a 50% reduction in sulfate emissions. Second, the prolonged drought, likely initiated by a low snow year in the winter of 1986–87, set off a chain of hydrologic shifts that directly influenced lake hydrology. This drought persisted from 1987 until 1990 (Fig. 2.16, bottom). The interaction between sulfate deposition and drought has provided interesting insights into the nature of lake-groundwater interactions and implications for ecological properties of lakes.

The extent of, and potential for, further acidification of sensitive lakes with low acid neutralizing capacity and pH was a central interest of our LTER program at the start and an impetus for the Little Rock Lake acidification experiment (Chapter 9). Hydrologic studies (Kenoyer and Anderson 1989), combined with a geochemical speciation and reaction path model (Anderson and Bowser 1986), suggested that even small amounts of groundwater were sufficient to neutralize the inputs of mineral acids received by lakes in northern Wisconsin. This conclusion was based on our initial calculations for Crystal Lake of 10% groundwater input, an estimate on the high end because this lake switches between flow-through and recharge hydrology both seasonally and annually (Anderson and Cheng 1993). Seepage lakes are most sensitive to acid inputs because they are located high in the flow system where groundwater inputs are limited or transient and are dominated by local flowpaths carrying young groundwater. Even the small and transient inputs of ANC-enriched groundwater to Crystal Lake likely were sufficient to limit major changes in buffering capacity in response to the level of acid loading in the region (Anderson and Bowser 1986, Garrison et al. 1987).

An associated set of long-term observations reinforces the importance of groundwater inputs in maintaining sufficient buffering capacity to counteract incoming mineral acids. Between 1983 and 1988, Nevins Lake, a transitional flow-through lake in upper Michigan, acidified as acid neutralizing capacity declined from 175 to 50 μeq/L and pH declined from 7 to 6 (Webster et al. 1990). This response was unexpected as rates of mineral acid deposition were declining in the region. The acidification was not from increases in either sulfate or nitrate, which remained stable, but from declines in the base cations, calcium and magnesium, both markers of groundwater inputs. Further hydrologic studies of Nevins Lake supported the hypothesis that a decline and eventual cessation of groundwater inputs related to drought decreased inputs of buffering materials, for examples, cations and ANC,

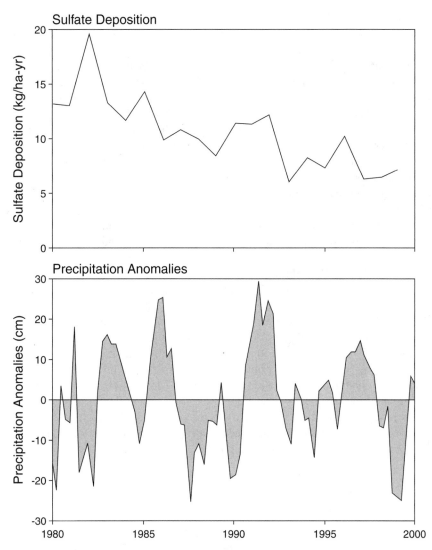

Figure 2.16. Changes in sulfate deposition (top) and precipitation (bottom) from 1980 to 2000. Sulfate deposition was measured at the National Atmospheric Deposition Program monitoring site at Trout Lake. Precipitation anomalies are deviations from the 20-year mean annual precipitation of 76 cm at the weather station at Minocqua Dam. They are plotted quarterly as running means of total precipitation over the previous 12 months.

to the point that the lake acidified (Krabbenhoft and Webster 1995). The response of Nevins Lake was extreme. Even though groundwater inputs to Crystal and Vandercook Lakes ceased during the drought, a strong chemical response was not apparent, although the mass of cations in Vandercook Lake did decrease (Webster et al. 1996). Similarly, small declines in acid neutralizing capacity and pH were observed in the reference basin of Little Rock Lake during the drought (Brezonik et al. 1993). With its short water residence time (<2 years) Nevins Lake responded more quickly to the 4-year drought than did the other two lakes, which had residence times on the order of 4 to 10 years. Extended periods of dry and warm conditions that override the short-term increase in solutes caused by evapo-concentration likely would cause similar acidification responses in lakes like these that have longer hydraulic residence times (Garrison et al. 1987, Kenoyer and Anderson 1989, Wentz et al. 1995).

These lakes are all located high in their respective flow systems. To broaden our understanding of lake-groundwater interactions, Cheng and Anderson (1994) used the numerical models described earlier to simulate dynamic groundwater flow around hypothetical flow-through lakes located at high, intermediate, and low positions along a flow system with hydrologic features similar to the Trout Lake basin. The model simulated steady state, transient, and high-intensity recharge conditions. Seepage lakes higher in the flow system, analogous to Crystal Lake, receive lower total groundwater fluxes during steady state simulations (Fig. 2.17) and more variable groundwater inputs during transient simulations. Further, high-intensity recharge events like spring snowmelt cause a switch from flow-through to discharge conditions as stagnation points form and groundwater mounds develop. The groundwater that enters lakes high in the flow system during these transient events is supplied by local flowpaths, where contact time between precipitation recharging the system and the surrounding aquifer is limited. In contrast, seepage lakes lower in the basin, like Sparkling Lake, have much higher groundwater in-

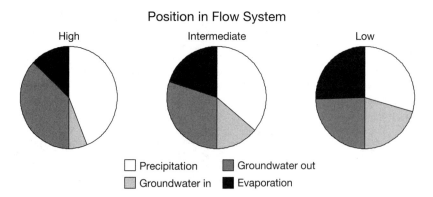

Figure 2.17. Modeled hydrologic budgets of seepage lakes that are high, intermediate, and low in the flow system. Redrawn from Cheng and Anderson (1993) with permission of the National Groundwater Association, copyright 1993, *Ground Water*.

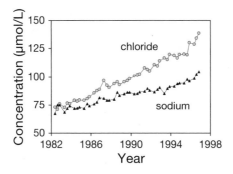

Figure 2.18. Volume-weighted concentrations of chloride and sodium measured from 1982 to 1998 in Sparkling Lake. Values for spring, summer, and autumn.

puts (Fig. 2.17) and more stable groundwater flowpaths. Because these lakes intercept deeper regional flowpaths that have more extended contact time with the aquifer, concentrations of solutes supplied by weathering are higher, further increasing chemical loading rates.

Because groundwater is a conduit for more than the products of mineral reactions, lake location relative to other sources of solutes can lead to spatial patterning among lakes. For example, groundwater transports salt to lakes located near roads that receive deicer applications during the winter months. Sparkling and Trout Lakes, both located near major highways, have elevated levels of chloride and sodium compared to the other LTER lakes, presumably owing to inflows of up-gradient groundwater contaminated by road salt runoff (Fig. 2.18) (Bowser 1992). In 1995, average chloride concentrations in the LTER lakes ranged from 4.3 mg/L in Sparkling Lake to 0.46 mg/L in the more remote Crystal Lake. Chloride is accumulating over time in Sparkling Lake, at a rate of 0.15 mg/L·yr, and translates to a loading rate of approximately 1,200 kg/yr. Interestingly, sodium has not accumulated at the same rate as chloride, suggesting active removal processes in the groundwater system through ion exchange for calcium, magnesium, potassium, and ammonium in the surrounding glacial drift (Bowser 1992).

Linking Hydrology and Management

Seasonal, interannual, and multiyear hydrologic shifts influence lake ecosystems and organisms in many ways. Spring pulses of silica stimulate diatom blooms. Groundwater flow through near-shore sediments influences the distribution and abundance of a mosaic of macrophyte and benthic algal communities. Even small groundwater inputs provide adequate buffering to maintain the pH and acid neutralizing capacity of lakes, including those located high in the flow system. Groundwater supplies calcium and other cations necessary for the construction of snail shells and crayfish exoskeletons. For example, snail species richness and composition are related to lake position in the flow system (Lewis and Magnuson 2000). Lake-level changes influence littoral habitats, exposing macrophyte beds or flooding neighboring riparian zones. The long water residence times typical of seepage lakes increase the importance of in-lake processes and temper interannual and intraannual variation in chemistry. Position along the hydrologic flow system influences the type of interaction with groundwater and determines the major ion composition of lakes and the dynamics of water flow. As discussed in Chapter 3, this template of landscape position has enriched our understanding of the spatial variation in the composition and diversity of biotic communities in lakes (Kratz et al. 1997b, Riera et al. 2000, Lewis and Magnuson 2000, Hrabik et al. 2005).

The seasonal and spatial dynamics in the Trout Lake flow system (Fig. 2.19) provide four useful insights for research and management in the future. First, and most obvious, is that 1- to 2-year studies do not necessarily represent the average or long-term hydrologic budgets of groundwater-influenced lakes. Interannual dynamics can be large, and flowpaths are much more transient than we originally thought. Because of this variability, assumptions of steady-state conditions are not always appropriate for these systems. Second, areas of groundwater inflow can be localized and transient, making their hydrologic contribution difficult to quantify. From a chemical budget perspective, quantifying deeper flowpaths may be more

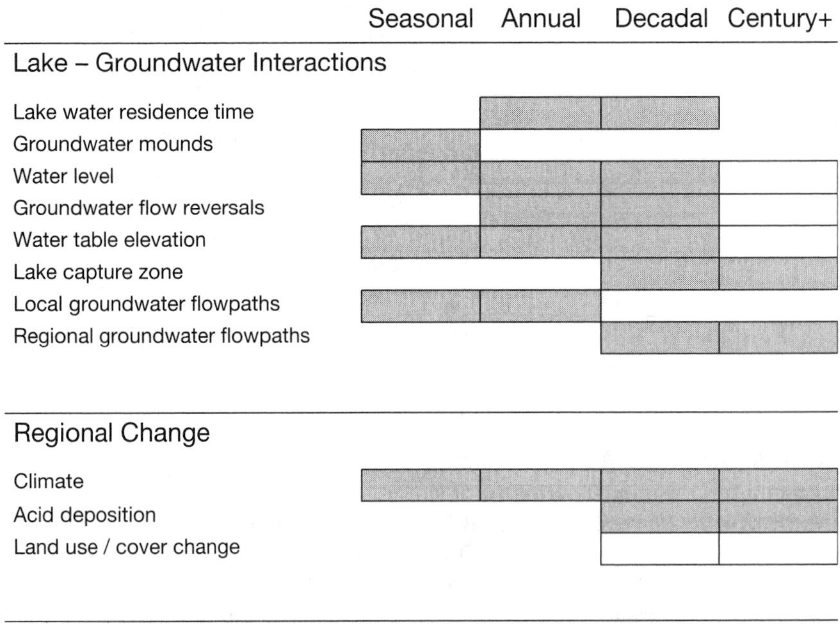

	Seasonal	Annual	Decadal	Century+
Lake – Groundwater Interactions				
Lake water residence time		■	■	
Groundwater mounds	■			
Water level	■	■	■	
Groundwater flow reversals		■	■	
Water table elevation	■	■		
Lake capture zone			■	■
Local groundwater flowpaths	■	■		
Regional groundwater flowpaths			■	■
Regional Change				
Climate	■	■	■	■
Acid deposition			■	■
Land use / cover change			□	□

Figure 2.19. Relevant time scales for hydrologic properties of lake-groundwater systems in the Northern Highlands Lake District (top) and of three drivers of regional change (bottom). Gray boxes indicate time scales considered from 1980 to 2001 in hydrologic studies by the North Temperate Lakes LTER site. White boxes indicate time scales of interest for future research.

important than quantifying shallow flow that can be dominated more by younger groundwater carrying recycled lake water. Third, the traditional catchment view of land-lake interactions may be inappropriate for groundwater-dominated systems where contributing areas do not necessarily match topographic boundaries. From a management perspective, focusing on land use alterations within groundwater contributing areas, even those located far from the lakeshore, may be more important than focusing on down-gradient areas located closer to the lake. Fourth, the effects of drought and interaction with other regional disturbances like acid rain highlight the value of regional and long-term approaches toward the study of multiple stresses (Frost et al. 1999b).

Future studies linking the regional hydrologic model with biogeochemical and ecological processes can enhance greatly our ability to predict the effects of landscape-scale change in climate, land use, and atmospheric inputs on the lakes. The lake-groundwater system of the Trout Lake basin acts as a conveyor moving water down-gradient toward Trout Lake, the basin discharge point. As a result, water at any given location in the basin originally could have entered the system at any of an infinite number of points from immediately up-gradient of that point to the basin groundwater divide. Groundwater flowpaths essentially slow the flow of water through the Trout Lake basin, lengthening the time for interactions with the catchment and creating a system with a long memory. Because water that follows pathways within groundwater contributing zones may take upwards of 100 years to reach an individual lake, we need to consider the effects of historical legacies of past land use and climate on contemporary lake ecosystems.

Summary

A hydrologic template that considers lakes as integral and interacting components of groundwater flow systems provides a powerful framework for interpreting the spatial heterogeneity and dynamics of lake ecosystems. Within the thick glacial drift of northern Wisconsin, lakes are outcrops of a complex groundwater flow system that acts as a conveyor belt that slowly moves water through the basin. Water supplied by precipitation travels through a complex mix of flowpaths with an infinite array of source points from the groundwater divide to immediately up-gradient of the discharge point. Water flowing through the groundwater system is chemically transformed by chemical reactions with minerals. Older groundwater follows longer flowpaths and contains higher dissolved solutes, such as Ca, Mg, Si, and acid neutralizing capacity, compared to younger water that follows shorter local flowpaths. Lake-groundwater connections vary across the landscape in the percentage of a lake's water budget supplied by groundwater and the age of the contributed water. Lakes near the groundwater basin divide tend to be precipitation-dominated, with weaker connections to groundwater and no surface water connections. As a result, they have a more dilute chemistry than lakes lower in the flow system. Lakes closer to the regional discharge point of the basin have extensive groundwater capture zones, including up-gradient lakes (groundwater flow-through seepage and drainage lakes), and stronger connections to the groundwater system. They also contain waters with higher ionic strength. This landscape-scale heterogeneity leads to spatial patterns in lake chemistry, biotic assemblages, and sensitivity to regional disturbances such as acid rain and climate variability. The spatial template of the lake-groundwater system is not static but exhibits dynamics at a variety of temporal scales. Seasonally transient groundwater mounds form adjacent to the lakeshore following snowmelt, temporarily reversing local and regional groundwater flowpaths toward lakes. At the interannual scale, prolonged drought can shift lakes high in the flow system from groundwater flow-through lakes to groundwater recharge lakes, decreasing the supply of ion-rich groundwater. Over centuries, historical legacies of events that influence the terrestrial source area are preserved in the lake-groundwater system. These dynamics and spatial heterogeneity influence lake ecology in ways we are just beginning to understand.

3

Making Sense of the Landscape: Geomorphic Legacies and the Landscape Position of Lakes

Timothy K. Kratz
Katherine E. Webster
Joan L. Riera
David B. Lewis
Amina I. Pollard

Mirrors in a room, water in a landscape, eyes in a face—those are what give character.
—Brooke Astor (1982, p.36)

At first thought, one might expect that lakes within a lake district, such as the Northern Highlands Lake District, would be relatively similar to one another. These lakes were formed by the same processes, share a similar geologic setting, have access to the same regional species pool, and experience nearly identical climate and weather. Yet, our experience suggests that lakes that are within a few hundred meters of each other can differ dramatically, not only in average values of individual variables but also in the dynamics of these variables. Understanding the mechanisms that lead to these differences provides insight into the relative importance of the various processes influencing lake characteristics and dynamics. In this chapter, we explore how landscape-level processes lead to predictable spatial patterns in some lake characteristics. In short, we try to understand how location in the landscape provides a basis for predicting general relationships among lakes. We attempt to make sense of the landscape.

Historical Perspective

Traditionally, limnologists have studied patterns and processes that occur within single lakes (Chapter 1) (Likens 1985, Edmondson 1991, Goldman 2000) or between lakes and their catchments (Hasler 1975). When multiple lakes within the same lake district or multiple lake districts have been used in comparative research as the unit of study, the focus has usually been to discern patterns at essentially the spatial scale of an individual lake. For example, the development of phosphorus-chlorophyll relationships by Dillon and Rigler (1974) and the establishment of the relative roles of grazing versus nutrients in controlling primary production (Carpenter et al. 1991) were both ultimately designed to develop an understanding at the scale of individual lakes and their reflection of surrounding landscapes. Despite the predictive relationships between lake characteristics and, for example, lake depth or the ratio of catchment area to lake area, lakes still are considered most often as individual units within the larger lake district or region.

Stream limnologists have long considered individual stream units as connected in a spatially explicit way through the River Continuum Concept and similar constructs (Vannote et al. 1980, Newbold et al. 1981, Ward and Stanford 1983, Wiley et al. 1990, Thorp and Delong 1994). These concepts describe predictable patterns from first-order headwater streams to large floodplain rivers in ecological properties ranging from species composition to carbon cycling. In a sense, this spatial context is apparent from the longitudinal nature of, and explicit hydrologic connections between, streams. In contrast, the shorelines of lakes make them more readily identifiable as individual units, obscuring the perspective of hydraulically connected units like rivers and streams. To put lakes in this same fluvial framework, one can think of arraying aquatic systems as landscape units on the basis of their degree of hydrologic connectivity (Fig. 3.1) (Magnuson and Kratz 2000). Seepage lakes represent the isolated end of the gradient. Some seepage lakes, like Crystal Lake, are precipitation dominated, and others, such as Sparkling Lake, receive significant amounts of groundwater (Chapter 2). Lakes connected to each other via wetlands or ephemeral streams are next along this gradient. Farther down a flow system, streams connect lakes, or rivers connect a sequence of reservoirs. Finally, streams and rivers represent the highly connected end of the gradient. This isolation-connection gradient maps generally onto a spatial gradient; isolated systems tend to be higher in the flow system, whereas connected systems such as streams and rivers tend to be lower in the flow system.

The importance of hydrology in the landscape ecology of both lakes and wetlands has been identified in a number of landscape settings. For example, Winter (2000) discussed the critical importance of hydrologic linkages between wetlands and surrounding uplands, as well as the importance of considering hydrologic units when trying to understand lake ecology (Winter 2001). Hershey et al. (1999) developed a geomorphic-trophic model to describe among-lake differences in arctic food webs. Hillbricht-Ilkowski (2002a, 2002b) described the importance of landscape-level hydrology in influencing nutrient dynamics and loading to a chain of lakes in Poland. Collectively, these and the other recent studies described later indicate a growing awareness of and interest in the role that spatially explicit land-

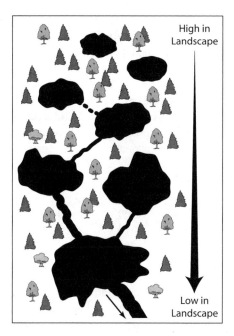

Figure 3.1. From high to low in a landscape, hydrologic connectedness ranges from isolated seepage lakes, to lakes connected by streams, to large river systems.

scape processes play in shaping differences in the characteristics and dynamics of neighboring lakes.

One of the hallmarks of our North Temperate Lakes Long-Term Ecological Research (LTER) program is that we have designed our research to consider multiple spatial scales, ranging from within lake, to whole lake, to multiple lakes, to lake districts, and to regions. Although we use multiple lakes, that is, subsets of lakes within a lake district, as the unit of study, the key distinguishing feature of our research is that we explicitly consider spatial positioning of individual lakes relative to each other. Here we further develop the concept of landscape position to describe patterns of spatially explicit variation of lake properties and dynamics as a function of the lakes' positions in the landscape. Thus, we provide a framework intermediate in scale between the individual lake and the lake district that recognizes the importance of the specific locations of lakes within the landscape. This emphasis on spatial gradients is similar conceptually to but quite different in detail from the River Continuum Concept discussed earlier.

Geomorphic Legacies and the Concept of Position in the Landscape

The abundance, distribution, origin, and hydrologic connection among lakes in the Northern Highlands is a legacy of the glacial activity that occurred in the region more than 10,000 years ago (Attig 1985). The hydrologic and geomorphic template left by the glaciers acts as a constraint on the expression of many limnological features and imposes a characteristic spatial pattern to this lake district.

This legacy has many aspects, but one of the more interesting ones is the hydrology at the scale of the lake district. The low level of elevational relief in conjunction with the thick sandy tills and outwash has led to a hydrologic setting where precipitation, surface water, and groundwater play important roles in the regional hydrology (Chapter 2). Because the position of a lake in the flow system influences the relative hydrologic balance among precipitation, groundwater, and surface water inputs and because this hydrologic balance may influence lake chemistry and a suite of chemically dependent biological variables, we expect that spatial positioning of a lake within the flow system will determine, in part, various lake characteristics.

We define a lake's landscape position as its explicit location relative to other lakes within the hydrologic flow system. A lake high in the landscape receives most of its water from the atmosphere, whereas a lake low in the landscape receives a substantial percentage of its water from surface inlets or groundwater. What makes this concept useful in a landscape context is that lake characteristics often are patterned spatially along a landscape position gradient. For example, as discussed later, the surface area of lakes tends to be larger in lakes lower in the landscape. This spatial patterning imparts an organizing framework for understanding similarities and differences among lakes arrayed along a landscape position gradient and linked to each other hydrologically through the flow system.

Spatial Organization of Lake Properties

Patterns among LTER Lakes

A strong linkage between hydrology and lake chemistry was recognized early on by Juday et al. (1935), who showed that seepage lakes, lacking stream inflows or outflows, had more dilute water than drainage lakes with stream inflows and/or outflows. Eilers et al. (1983) took this one step further and concluded that the sensitivity of lakes to acidification was related to a lake's hydrologic setting. Lakes that received most of their input waters from precipitation were more sensitive to acidification than lakes that received significant amounts of water from streams or groundwater. Although they recognized the importance of hydrologic setting to water chemistry, both of these studies considered lakes as independent entities in the landscape; the spatial positioning of the lakes relative to each other was not considered or analyzed. Yet, the importance of hydrologic setting and the high densities of lakes in many lake districts suggest that spatial patterning of lake characteristics may be a defining feature of ecological relationships. In short, the landscape can be viewed as organized into collections of lakes that occupy different positions in a common hydrologic flow system.

Our seven primary study lakes in northern Wisconsin are arrayed along a hydrologic flow system in the Trout Lake basin (Fig. 3.2) (Chapter 2). This choice of study lakes has allowed us to understand lakes in the context of their landscape

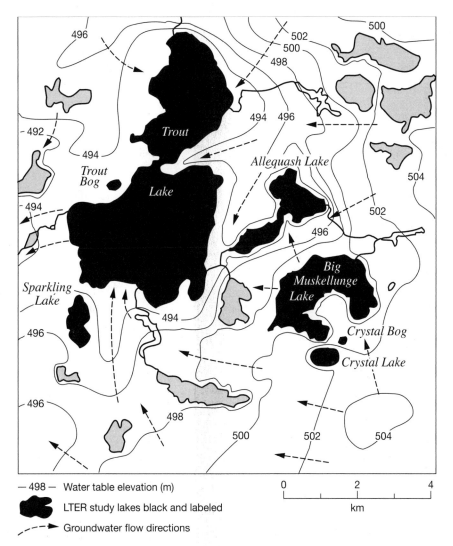

— 498 — Water table elevation (m)

LTER study lakes black and labeled

Groundwater flow directions

0   2   4
km

Figure 3.2. The seven primary LTER study lakes in northern Wisconsin on a map of water table elevations computed in 1982 and 1983 using sound refraction data (method, Fig. 14.6h, top right) and groundwater flow directions. Redrawn and modified from Okwueze (1983), in Kratz et al. (1997b) with permission of E. Okwueze and Blackwell Publishing.

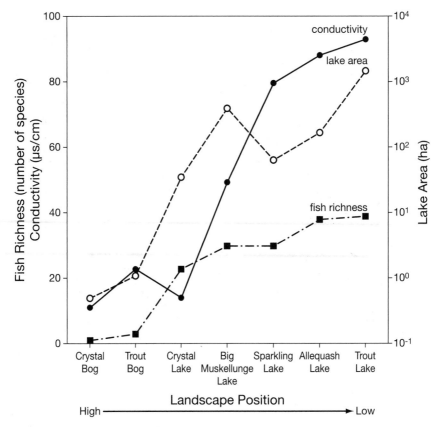

Figure 3.3. Lake surface area, conductivity, and fish species richness as a function of land-scape position for the seven primary LTER lakes in northern Wisconsin.

position. Characteristics of the seven primary study lakes vary systematically as a function of their position in the flow system. In particular, lakes higher in the flow system are lower in concentrations of major ions, are smaller in surface area, and have fewer fish species than lakes lower in the flow system (Fig. 3.3, Kratz et al. 1997b). The mechanism leading to the observed pattern for the major ions seems relatively straightforward. Precipitation and groundwater have very different chemistry. As groundwater flows through the sandy, noncalcareous tills and outwash, chemical evolution occurs through silicate hydrolysis (Kenoyer and Bowser 1992a, 1992b) and leads to increased levels of acid neutralizing capacity, calcium, magnesium, and dissolved silica. Lakes that receive a larger percentage of their incoming waters from groundwater sources tend to have higher concentrations of these chemical constituents (Chapter 2).

But why are lake areas and fish species richnesses related to landscape position? Two possible hypotheses follow. First, because lakes lower in the landscape tend to be relatively larger, the relation between species richness and landscape position actually reflects a species-area relationship of insular systems (Chapter 4). Kratz et al. (1997b) suggested that lakes higher in the landscape are restricted to being smaller in area, because additional water would spill over to the next lower basin. Lakes lower in the landscape often have complex bathymetries with many individual basins separated by shallow sills (Robertson and Ragotzkie 1990b). These lakes tend to be larger because there are no basins at lower elevations to which water can spill. Lake area is related, in turn, to species richness for many groups of organisms following the principles of island biogeography (Barbour and Brown 1974, Magnuson 1976, Magnuson et al. 1998, Tonn et al. 1990, Lewis and Magnuson 2000) (Chapter 4). Hypotheses for the pattern of species richness in relation to landscape position, independent of lake area, are that, regardless of size, lakes low in the landscape are more productive, have less stressful chemistries, and are more connected to sources of immigrants through connecting streams (Chapter 4).

Patterns in the Northern Highlands

The relationship between lake characteristics and landscape position suggests that the legacy set at the time of glacial retreat is an ultimate cause of spatial patterning of these lake ecosystems in the Northern Highlands. However, the results presented earlier were based on data from relatively few lakes. To test the generality of these results, Riera et al. (2000) expanded the analysis to include lakes throughout the entire Northern Highlands. This effort raised an immediate conceptual issue. How do we quantify a lake's landscape position? One possibility is to use lake elevation as a measure of landscape position. However, at the scale of the entire lake district, limnological characteristics are not related to elevation (Fig. 3.4). This lack of relationship suggests that landscape position affects lake characteristics at a smaller scale than the entire lake district. For example, in one part of the lake district, a particular lake may be the highest of a local subgroup of lakes. At the same time, this entire subgroup of lakes may all be much lower in elevation than other lakes in the lake district. Because this lake is relatively high in the local landscape, the hydrological and limnological characteristics may be like those of a lake high in the landscape, even though its absolute elevation might make it relatively low in the lake district.

To capture the functional elevation of lakes in the same flow system, Riera et al. (2000) characterized lakes by their order in a local flow system, rather than by their absolute elevation above sea level. They attempted to define a metric that would (1) reflect the relative importance of precipitation, groundwater, and surface water inputs to lakes, (2) be relatively simple to assign from easily accessible topographic information, and (3) allow us to increase our predictive understanding of how lake characteristics are distributed across a lake district. As described earlier, lakes often are arrayed in a predictable way along hydrologic flow gradients in lake districts where groundwater is a large part of lake water budgets. Lakes high in the flow system are usually seepage lakes, that is, lakes with no surface water inlets or outlets that receive most of their incoming water directly from precipitation. These lakes either are perched above the local groundwater table, or for most or all of the year they lose water through groundwater infiltration. The next lakes down in the

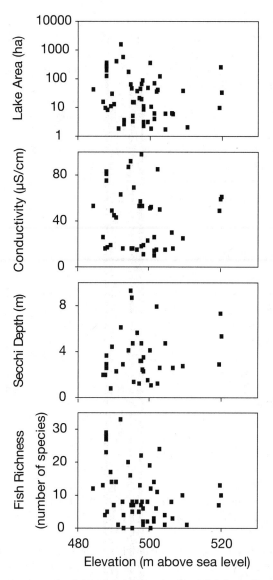

Figure 3.4. Lake area, conductivity, Secchi disc depth, and fish species richness are not related to lake elevation above sea level in the Northern Highlands Lake District.

flow system tend to be groundwater flow-through lakes, seepage lakes with both groundwater inputs and loss. These lakes may also have intermittent outlet streams. Next are lakes that have stream outlets but no inlets. In these lakes, the combination of precipitation and groundwater inflow is sufficient to support a perennial outlet stream. We know of no lakes in the Northern Highlands that have inlets but no outlets. Finally, farther down in the flow system are drainage lakes, having both inlet and outlet streams. These lakes receive input waters from precipitation, groundwater, and surface streams.

The metric of lake order attempts to quantify the position of a lake in the flow system from data that are available from topographic maps (Fig. 3.5). First, seepage and drainage lakes were separated. Seepage lakes were assigned negative lake order, and drainage lakes were assigned positive lake order. For drainage lakes, the lake order was defined as the order of the lake's outlet stream, following Strahler (1964). Headwater lakes, with a stream outlet but no inlets, were assigned a lake order of 0. Seepage lakes connected to the surface drainage network with intermittent streams were assigned a value of –1. Lakes connected to the surface drainage network by unchannelized flow through a wetland were assigned –2. Finally, lakes unconnected to the surface drainage network were assigned lake order –3.

Riera et al. (2000) related lake order to 25 physical, chemical, and biological variables for 50 to 556 lakes in the Northern Highlands Lake District (Table 3.1). Lake order explained a significant percentage of the variance of 21 of the 25 vari-

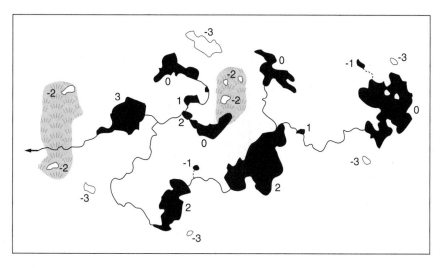

Figure 3.5. Lake order, indicated by the numbers, indicates relative position of lakes from high (negative numbers) to low (positive numbers) in the landscape, based on map analysis described in the text. Redrawn from Riera et al. 2000 with permission of Blackwell Publishing.

Table 3.1. Relation between lake attributes and landscape position for northern Wisconsin lakes based on the results of Riera et al. (2000).

Variable	Relation to Lake Order[a]	Strength of Relation[b]
Morphometry		
Lake Area	increases	strong
Lake Perimeter	increases	strong
Shoreline Development Factor	increases	strong
Maximum Depth	increases	weak
Mean Depth	increases	weak
Water Optical Properties		
Secchi Depth	decreases	weak
Turbidity	increases	weak
Water Color	increases	weak
Dissolved Organic Carbon	none	
Major Ions		
Conductivity	increases	strong
pH	increases	strong
Acid Neutralizing Capacity	increases	strong
Calcium	increases	strong
Magnesium	increases	strong
Chloride	increases	strong
Sulfate	none	
Nutrients		
Total Phosphorus	none	
Kjeldahl Nitrogen	increases	weak
Dissolved Silica	increases	strong
Biological Variables		
Chlorophyll *a*	increases	strong
Dry Weight of Plankton	none	
Crayfish Abundance	increases	strong
Fish Richness	increases	strong
Human Variables		
Boating Usage[c]	increases	Strong[c]
Density of Cottages	increases	weak
Density of Resorts	increases	weak

[a]Increases = increase in an attribute from high to low in landscape; decreases denotes the opposite.

[b]Weak = significant differences among lake orders in an ANOVA, with a tendency to increase or decrease with lake order, but multiple mean comparisons detected no significant differences among lake orders. Strong = significant differences among lake orders in ANOVA and significant differences in multiple mean comparisons.

[c]Based on Reed-Anderson et al. (2000c).

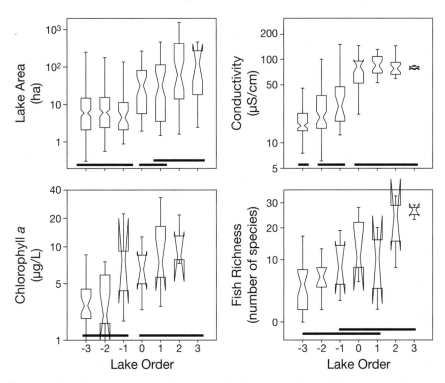

Figure 3.6. Lake area, conductivity, chlorophyll concentration, and fish species richness as a function of lake order in lakes of the Northern Highlands Lake District. The notch = median, notch edge = 95% confidence interval, top and bottom of box = quartiles, vertical lines = 1.5 times the difference between the quartiles. Dark horizontal bars indicate orders that do not differ significantly. Modified from Riera et al. (2000).

ables, for examples, lake area, conductivity, chlorophyll, and fish richness (Fig. 3.6). The variables most closely related to lake order were measures of lake size and shape; concentrations of silica, chlorophyll, and major ions other than sulfate; crayfish abundance and fish species richness; and human use. These results are consistent with the observations made on the seven primary LTER study lakes and show that a wide variety of a lake's physical, chemical, and biological characteristics are related to its position in the landscape. Equally interesting, however, are those variables that showed little relationship with lake order. Concentrations of nutrients other than silica and optical properties of the water were related poorly to lake order. Because the major source of nitrogen and phosphorus in the Northern Highlands Lake District is most likely atmospheric deposition (Cole et al. 1990, Wentz et al. 1995), we had expected that lakes with larger catchment areas would have greater concentrations of these nutrients. Because catchment area is related positively to lake order, we expected to see a positive relationship between nitrogen and phosphorus concentrations and lake order. Riera et al. (2000) speculated that uptake processes in the catchment and among-lake differences in in-lake processes could introduce variability and mask a relationship between nutrient concentration and lake order.

Lake-water optical properties, such as water color, made up the other class of variables unrelated to lake order. The fulvic and humic acid components of dissolved organic carbon are related closely to water color in northern Wisconsin lakes, where the area of riparian peatland within 25 to 100 m of the shore is correlated highly with dissolved organic carbon concentration (Gergel et al. 1999). Therefore, local factors, such as the amount of peatlands close to the lake, likely are more important influences on water color than is landscape position as measured by lake order. However, as Riera et al. (2000) argued, the current location of peatlands likely is related to the geomorphology set down by the same glacial legacy that affected the lakes.

The spatial arrangement of lakes can influence the distribution of aquatic organisms (Kratz et al. 1997b, Riera et al. 2000); however, covariation among several lake characteristics makes isolating effects of individual characteristics challenging. Pollard (2002) compared the effects of covarying lake characteristics related to lake order (benthivorous fish abundance, littoral area, stream connection, and specific conductance) on benthic invertebrate abundance and taxonomic composition. She sampled invertebrate assemblages in 32 lakes in the Northern Highlands. Benthic invertebrate abundance and diversity increased significantly with lake order. Interestingly, the dominant explanatory factors were different for invertebrate abundance and richness. Variation in invertebrate abundance was associated with specific conductance, with higher conductivity waters having higher invertebrate abundance. Hrabik et al. (2005) reasoned that specific conductance was related indirectly to the observed changes in invertebrate abundance through its direct relation with macrophyte diversity. In contrast to abundance relations, differences in invertebrate richness were associated with the presence or absence of stream connections to lakes. Although this result is not intuitive for organisms such as many aquatic insects capable of aerial dispersal, lakes with streams had higher richnesses than those without stream links. Their work provides an example of how lake characteristics summarized within the concept of lake order influence the abundance and distribution of organisms across the landscape.

Although lake order is an important determinant of the abundance and distribution of organisms (Table 3.1), the dendritic nature of surface water connections is an additional factor potentially influencing species distribution. Spatial placement of lakes was a determinant of snail species richness and community composition in lakes in the Northern Highlands (Lewis and Magnuson 2000). Snails were sampled in 18 lakes distributed among three local catchments. In each catchment, sampled lakes were arrayed along a gradient of lake order. Interestingly, species richness did not differ among catchments but was significantly related to lake order, with drainage lakes having significantly more species than seepage lakes. The authors reasoned that increased calcium concentration, the presence of stream connections, and increased amounts of macrophytes in drainage lakes likely accounted for increased snail species richness in drainage lakes.

However, relative species composition differed significantly among the three catchments. They argued that historical events such as particular species making their way into one catchment but not another may have led to current differences in species composition among catchments. Thus, colonization and dispersal routes, both functions of lake spatial placement, increase our understanding of explicit spatial variability among lakes.

The landscape position concept appears to provide useful explanatory power for peatland features, as well. Water sources and movement are critical to understanding spatial patterns of vegetation composition and water chemistry of interstitial waters in large tracts of peatlands, such as those that occur in northern Minnesota, parts of Canada, and northern Europe (Heinselman 1970, Glaser et al. 1990, Glaser et al. 1997). In a series of small, kettle-hole peatlands in the Northern Highlands, landscape position reflected the importance of groundwater input (Kratz and Medland 1989). The 10 studied peatlands ranged from high to low in the landscape on the basis of topographic maps and the groundwater flow map (Fig. 3.2). Silica concentrations in surface interstitial waters provided a tracer of groundwater input, because groundwater concentrations of silica are several orders of magnitude greater than that of precipitation and because silica might be expected to be more conservative in peatlands than major cations such as calcium or magnesium that engage in ion exchange processes with *Sphagnum* moss. In general, peatlands low in the landscape had higher concentrations of silica and appeared to have larger amounts of groundwater input than peatlands higher in the landscape. However, in addition to this regional influence of landscape position, local effects were important. Peatlands directly abutting larger lakes had silica values similar to the lake, and a peatland surrounded by steep mineral ridges had more silica than expected, probably owing to increased localized surface runoff.

Dynamics

So far, we have considered status or mean properties of lakes. However, the dynamics of limnological variables, as measured by temporal trends or temporal variability, can differ among lakes, as well. We also investigated the relation between lake dynamics and landscape position.

Kratz et al. (1991b) studied the relationship between interannual variability of limnological variables and lake landscape position. They posited that temporal variability might be related to landscape position because the way lakes respond to wet or dry years may depend on the lakes' landscape positions. Because precipitation dominates hydrologic inputs to lakes high in the landscape and groundwater dominates inputs to lakes low in the landscape, the hydrologic balance might be more dynamic across years in lakes high in the flow system than in lakes low in the flow system. Differences in the variability of the precipitation:groundwater ratio across lakes might in turn cause differences in variability of water chemistry variables associated with groundwater input. They compiled data on the seven primary LTER study lakes for 5 years from 1982 to 1986. A total of 68 parameters were analyzed, including various seasonal means for water temperature, dissolved oxygen, pH, specific conductance, nitrate, ammonia, silica, chlorophyll a, and abundances of *Chaoborus* and *Leptodora*. In general, the coefficient of variation of these parameters was related strongly to lake landscape position, with coefficients of variation tending to be greater in lakes higher in the landscape. The patterns were most pronounced for a subset of edaphic data most closely related to groundwater inputs. Although five years seemed long term at the time, the authors were not certain that the patterns would hold over longer periods. Analysis with more than the 20 years of data currently available could prove fruitful.

An opportunity to test some of these ideas about differential response to climate drivers was provided by a natural experiment, a sustained and severe 4-year drought in northern Wisconsin and most of the Upper Midwest. Using the LTER long-term data and landscape framework, Webster et al. (1996) posited that lakes would respond differentially to the drought depending on their landscape position. They focused on the relatively conservative cations, calcium, and magnesium as markers of changes in lake-groundwater interaction caused by drought. This prediction was based on the rapid loss of cations observed during drought in Nevins Lake, a seepage lake in the Upper Peninsula of Michigan (Webster et al. 1990). Webster et al. (1996) reasoned that reduced groundwater inputs during drought would cause similar decreases in cations in LTER lakes located high in the landscape. These highland lakes tend to have more transient connections with groundwater and were thus expected to be more chemically responsive to hydrologic shifts during drought compared to lakes located low in the flow system. These lowland lakes typically have stronger connections with groundwater following longer, more regional flowpaths (Chapter 2).

Webster et al. (1996) analyzed chemical data collected from 1984 to 1993 for seven lakes with groundwater input ranging from 0% to 35%. This time period included three years prior to the drought from 1984 to 1986, four drought years from 1987 to 1990, and three postdrought years from 1991 to 1993. Unlike Nevins Lake, where cation concentrations decreased, all lakes experienced increases in cation concentrations during the drought. However, much of this change in concentration was an artifact of decreased lake volume and water level owing to the concentration of ions through evaporation. Indeed, after accounting for lake level fluctuations, patterns in the total ion content, or mass, of calcium and magnesium supported the hypothesized differential response to drought. In the group of four lakes high in the landscape most hydrologically similar to Nevins Lake, calcium and magnesium mass either stayed the same or decreased during drought. The two lakes that showed no change included a lake that receives virtually no groundwater even during nondrought periods, and a second lake whose 13-year water residence time may have precluded a measurable chemical response during the 4-year drought. In contrast, calcium and magnesium mass increased during drought in the second group of lakes, all located low in the flow system (Fig. 3.7). This increase in cation content may have been caused by decreased importance of groundwater inputs from shorter, local flowpaths. These flows are more transient and the inputs to lakes less rich in ions than those inputs from longer, deeper and more stable regional flowpaths that supply water to lakes low in the flow system (Chapter 2). Clearly, these chemi-

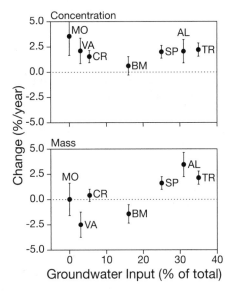

Figure 3.7. Differences among lakes during the drought in the annual percent change in calcium plus magnesium concentration (top) and mass (bottom) as a function of the percentage of lake water input from groundwater. MO = Morgan Lake. VA = Vandercook Lake. CR = Crystal Lake. BM = Big Muskellunge Lake. SP = Sparkling Lake. AL = Allequash Lake. TR = Trout Lake. Redrawn from Webster et al. (1996) with permission of the American Society of Limnology and Oceanography.

cal markers suggest that a lake's dynamic response to drought is related to its landscape position, a result that has implications for the supply of other solutes provided by groundwater such as silica as well as for predictions of differential response to prolonged global warming (Kratz et al. 1997b).

Landscape position appears to be a determinant of temporal coherence, that is, whether lakes vary synchronously. Lakes more similar in landscape position tended to be more temporally coherent than lakes that were dissimilar in landscape position (Chapter 5) (Baines et al. 2000).

Generalization to Other Regions

We next turn our attention to whether the relationships between landscape position and lake characteristics in the Northern Highlands Lake District are general for other lake districts. In particular we were interested in whether the pattern of greater concentrations of solutes in lakes lower in the landscape held in other lake districts. This pattern is expected because lakes lower in the landscape should, in general, have larger watershed-to-lake-area ratios than lakes higher in the landscape.

One of the best data sets for comparison with the North Temperate Lakes LTER site comes from the Arctic LTER site on the north slope of Alaska. Arctic LTER researchers have measured spatial and temporal patterns of 21 chemical and biological variables in 14 streams and 10 lakes several times each year since 1991 (Kling et al. 2000). Nine of the 10 lakes are connected in a direct lake chain by surface water. Despite large differences between Alaska and Wisconsin in climate, geologic setting, permafrost, soils, and vegetation, many of the relationships between landscape position of the lakes and lake characteristics were similar to those in northern Wisconsin. Conductivity, calcium, magnesium, acid neutralizing capacity, dissolved inorganic carbon, and pH all increased from high to low in the lake chain in the Alaska study. These increases were attributed partly to the increased catchment area of lakes farther down-slope and partly to the processing of materials in lakes and stream segments along the lake chain. Patterns for each of these lake variables in the northern Wisconsin lakes (Riera et al. 2000) were similar to those in Alaska.

In an exhaustive analysis of spatial patterning in 86 southcentral Ontario lakes occurring on a combination of bedrock and glacial deposits (Quinlan et al. 2003), lake order explained significant amounts of interlake variation in major ion chemistry, physical and catchment characteristics, hypolimnetic oxygen, and community composition in algal and invertebrate assemblages. Interestingly, nutrients other than silica did not have significant relationships with lake order. These results are in striking agreement with those for the northern Wisconsin lakes reported earlier (Riera et al. 2000).

However, landscape position appears to be relatively unimportant in determining lake characteristics and dynamics in at least some bedrock-dominated landscapes. Webster et al. (2000) showed that lakes in the surface-water dominated Experimental Lakes Area in western Ontario had attributes that did not vary according to a lake's landscape position. All of the lakes responded similarly, for example, to a several-year-long drought. They posited that in surface-water dominated systems, where bedrock is exposed or close to the land surface, lakes respond in a relatively simple way to changes in precipitation, whereas in groundwater-dominated systems the relationship between precipitation and solute loading is more complex (Webster et al. 2000).

To examine further the generality of the influence of landscape position on lakes, Soranno et al. (1999) compiled data from nine lake chains representing seven lake districts throughout North America. Each of the lake chains consisted of 3 to 13 water bodies organized in a flow system by groundwater or surface water connections. The lake districts included the North Temperate Lakes LTER site in northern and southern Wisconsin, the Arctic LTER site in Alaska, the Qu'Appelle Valley lakes in Saskatchewan, the Adirondack lakes in New York, Rocky Mountain lakes in Colorado, and Tennessee River Valley reservoirs. Weathering variables such as acid neutralizing capacity, specific conductance, and calcium generally, but not always, were greater in lakes farther downstream in the lake chains (Fig. 3.8). Silica had a similar pattern, suggesting that this important nutrient behaves like a weathering variable in the landscape. Interestingly, although patterns between either algal biomass or total phosphorus and lake chain position were not strong, water clarity as measured by

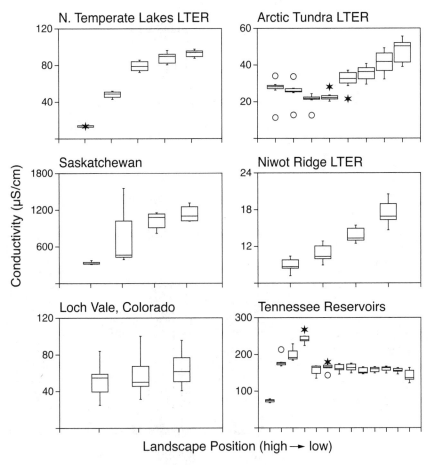

Figure 3.8. The conductivity of lakes and reservoirs as a function of position of the lakes from high to low in the landscape or in a lake chain from six areas of North America. Modified from Soranno et al. (1999).

Secchi disc generally decreased from high to low in the lake chains, perhaps suggesting a relation between lake position and turbidity or dissolved organic carbon. In general, the relations between landscape position and a series of chemical variables in lakes throughout North America were surprisingly robust (Soranno et al. 1999).

Summary

We provide examples of how a landscape perspective enriches our understanding of lake condition and dynamics (Fig. 3.9). The spatial distribution of lakes that have the various physical, chemical, and biological attributes is not random across the landscape of Northern Highlands Lake District. Neighboring lakes differ in basic features, such as surface area, ionic strength, and species richness, in systematic

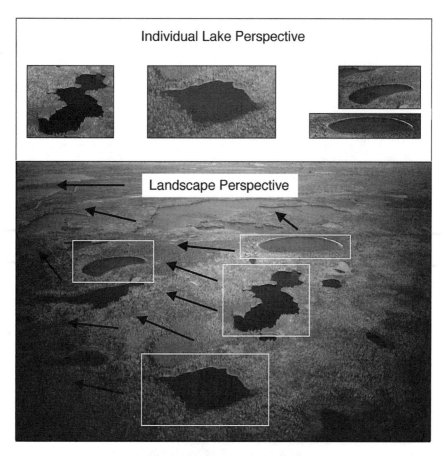

Figure 3.9. A landscape perspective provides a richer, spatially explicit understanding of land-water interactions. Lakes can be considered as individual entities (top) or as part of a larger landscape (bottom) where characteristics of lakes are dependent on their location within the landscape. Arrows indicate general direction of groundwater flow.

ways associated with the lakes' positions in the surface water/groundwater flow system. Lakes higher in the landscape are, in general, smaller, more dilute chemically, clearer, with fewer species, and less used by humans than lakes lower in the flow system. The landscape position of lakes also influenced their interannual dynamics. For example, we have evidence that, while lakes respond differently to climatic shifts such as drought, the nature of the response is coherent and related to position in the flow system. Many of these results appear to be general across diverse types of lake districts, suggesting that a landscape-scale approach to understanding the structure and dynamics of lake districts is a necessary and powerful perspective.

4

Lakes as Islands:
Biodiversity, Invasion, and Extinction

Shelley E. Arnott
John J. Magnuson
Stanley I. Dodson
Alison C. C. Colby

In the science of biogeography, the island is the first unit that the mind can pick out and begin to comprehend. By studying clusters of islands, biologists view a simpler microcosm of the seemingly infinite complexity of continental and oceanic biogeography.
—Robert H. MacArthur and Edward O. Wilson (1967, p. 3)

L akes are like islands (Fig. 4.1), that is, they can be thought of as islands of water surrounded by large areas of inhospitable land (Barbour and Brown 1974, Keddy 1976, Magnuson 1976). Because of the island nature of lakes, theories that have been developed for islands and island-like habitats may also apply to lakes (MacArthur and Wilson 1967, Gilpin and Hanski 1991, Gotelli and Kelley 1993). Given the relative isolation of islands, these theories emphasize the importance of regional population processes, suggesting that the number of species that occupy a habitat is a function of processes that control immigration as well as extinction of local populations. On the basis of these theories, we expect species composition in lake-islands to be dynamic in nature, that is, the species present at any given time will be influenced by the immigration of new species and the extinction of some species previously established in the lake.

Immigration depends on factors that control isolation, such as the distance to the nearest source of colonists or the presence of stream connections, as well as the intensity of interactions with established species. Isolation influences the distribution of species by defining the degree of movement of individuals to and from lakes. Both the physical nature of the landscape and the dispersal ability of the taxa under consideration determine isolation. Fish dispersal is typically limited by stream connections and human-aided introductions (Magnuson 1976), but aerial transport is possible (Bajkov 1949, Dennis 1993). Plankton dispersal is less constrained, be-

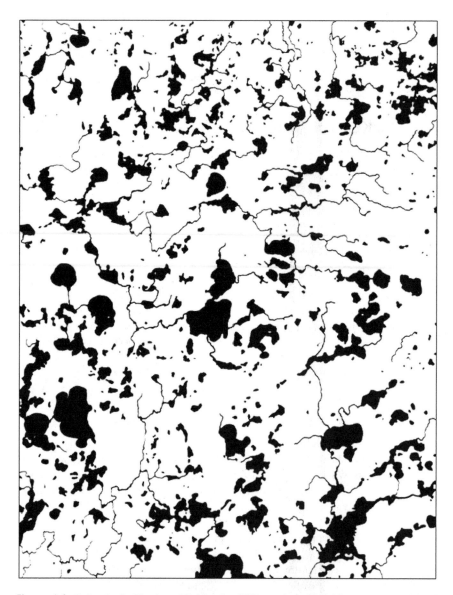

Figure 4.1. Lakes in the Northern Highlands of Wisconsin and Michigan appear as islands with and without connecting streams that may serve as transportation corridors for organisms. The map is centered at 46°3' N and 89°41' W. The vertical dimension is about 50 km. Redrawn from Magnuson et al. (1998) with permission of the Ecological Society of America.

cause wind, birds, and mammals commonly serve as vectors (Maguire 1963, Proctor and Malone 1965, Jenkins 1995, Cáceres and Soluk 2002). In addition, plankton can undergo "dispersal through time" by emerging from resting stages historically deposited in the lake sediments (Hairston 1996). In this chapter, we consider the emergence of plankton from resting stages to be a form of immigration into the water column.

Extinction probabilities on islands and in lakes are influenced by a particular system's characteristics. Lake size may be important in determining species richness because habitat heterogeneity increases with lake size (Tonn and Magnuson 1982, Eadie and Keast 1984). Larger lakes tend to have a greater diversity of habitats and therefore can provide suitable environments for a larger number of species. In addition, larger lakes with more resources enable populations to obtain larger sizes, and, therefore, extinction probability is reduced. Other local-scale characteristics such as resource levels, habitat complexity, and biotic interactions also influence the number of species that can inhabit a lake. Ultimately, biotic structure is controlled by both regional-scale mechanisms that determine which species can arrive at a particular lake and local-scale mechanisms that influence the survival of species once they are in the lake (Jackson and Harvey 1989, Tonn et al. 1990). Therefore, studies of factors that control species richness, immigration, and extinction processes should consider both local and regional drivers.

In our studies of north temperate lakes, we expected to see a relationship between extinction factors such as lake size and the number of species present. We also expected that isolation factors, such as distance to other lakes, would be important in determining species composition. And, we expected lakes to be dynamic in their species composition, with species appearing and disappearing through time and with an equilibrium number of species generated by the interplay between immigration and extinction rates. This island-biogeography framework has proven to be a useful conceptual tool for assessing the dynamic nature of species composition in lakes.

Specifically, we have addressed several key questions. How many species are present in a lake? What factors influence the species richness of lakes? Does species composition fluctuate? How well can we evaluate whether an invasion or extinction has occurred? What is the relative importance of invasion versus extinction to the assemblage richness? A related issue concerning the influence of exotic species invasions on lake ecology is treated in Chapter 8 as an external driver of lake dynamics. We focus on the Wisconsin lakes but relate cooperative intersite research (Chapter 14) on the lakes in Ontario at Dorset and at the Experimental Lakes Area (ELA) and with other international colleagues.

Species Richness

The number of species in a habitat can be estimated using a variety of metrics. One approach is to enumerate the number of species found in a single sample (or several pooled samples taken over a short period of time). Annual richness is the total number of species found after pooling samples taken throughout a single year.

Richness estimates based on a single year provide a relatively short-term view of the total species pool, particularly for communities with high species turnover rates. Another approach is to estimate the number of species observed over a long period of time, such as a decade. Cumulative richness is the total number of species observed during the time period of study. Asymptotic richness is an estimate of the total number of species based on a Walford plot, the cumulative richness for t years versus the cumulative richness for t + 1 years (Ricker 1975). The asymptotic richness is estimated from the intersection of the Walford plot and the 1:1 line, that is, where the cumulative number of species reaches an asymptote. The cumulative and asymptotic approaches provide estimates of the total species pool that can live in a particular habitat over a period of time.

The assessment of species richness can be influenced by high turnover rates of species, whether attributed to actual gains and losses of species or to sampling error. For fishes, crustacean zooplankton, and phytoplankton, the long-term species pool (based on the asymptotic richness) was, on average, two times greater than the mean annual richness (Fig. 4.2). The estimated richness increases with the number of years sampled (Magnuson et al. 1994, Arnott et al. 1998) because additional species are detected through time. This is problematic because species richness as determined by short-term assessments is not necessarily correlated with the species richness as determined by long-term estimates. For example, long-term and short-term richness for zooplankton for a set of lakes were not related because zooplankton in each lake had different annual species turnover rates (Arnott et al. 1998). Much of this discrepancy was the result of episodic species that tended to appear in lakes for only a few years. For example, in the Dorset lakes of Ontario, Heney Lake had the fewest species when considering the mean annual richness. However, because apparent turnover rate was high at approximately 25% per year and the appearance of episodic species was frequent, Heney Lake had the highest cumulative species richness.

Discrepancies between short-term and long-term richness influence the relationship between richness and explanatory variables such as lake area, lake depth, pH, and nutrients. The importance of each explanatory variable depends on whether the richness estimate is a mean annual estimate or a long-term cumulative estimate (Magnuson et al. 1994, Arnott et al. 1998). For example, mean lake depth was the most important variable predicting mean annual zooplankton species richness, explaining 61% of the variation for the Dorset lakes in Ontario. In contrast, total phosphorus concentration was the most important variable predicting long-term cumulative species richness, explaining 45% of the variation.

The discrepancy between short-term and long-term richness rankings may be resolved by considering that each estimate represents a different property of the lake community. Short-term richness is a measure of the number of potentially interacting species, whereas long-term richness is a measure of the potential species pool and may be related to the ability of a system to respond to environmental change. Because the number of species detected is dependent on the extent of the sampling and the method of calculating richness, caution must be exercised when searching for environmental correlates in multilake and multiyear comparisons. Ideally, richness estimates for comparison should be based on the same intensity

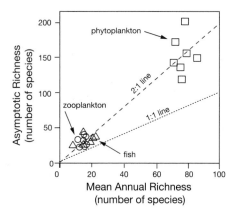

Figure 4.2. Asymptotic species richness plotted against mean annual species richness for phytoplankton, crustacean zooplankton, and fishes. The dotted line indicates where asymptotic richness equals mean annual richness. The dashed line indicates where the asymptotic richness is twice the mean annual richness. Zooplankton data for the Dorset Area (N. D. Yan, Ontario Ministry of the Environment, Canada Department of Fisheries and Oceans, and York University). Phytoplankton data from the Experimental Lakes Area (D. L. Findlay, Canada Department of Fisheries and Oceans). Fish data from North Temperate Lakes LTER. Figure redrawn from Arnott (1998) with permission.

of sampling within a year and the same number of years sampled for all lakes in the assessment. Finally, short-term richness greatly underestimates the total species pool. The greater the species turnover rate within a lake, the greater the discrepancy between long- and short-term richness estimates.

Determinants of Species Richness

The equilibrium theory of island biogeography suggests that the number of species present in a lake island is a function of immigration and extinction rates (MacArthur and Wilson 1967). On the basis of this theory, we would predict that the probability of extinction is higher in smaller than in larger lakes and the probability of colonization is lower in more isolated than in less isolated lakes. Small lakes are expected to have high extinction rates because they are less heterogeneous habitats (Tonn and Magnuson 1982, Eadie and Keast 1984, Rahel 1984) and are more likely to be severe environments for many fish species; they can be acidic, go anoxic in winter, or be more influenced by a predator than larger lakes (Magnuson et al. 1989). Small, isolated habitats would be expected to have lower colonization rates than large lakes because they are smaller targets for dispersal by wind and birds and because they are less frequented by humans (Magnuson 1976, Magnuson 1988). Also, they are typically headwater lakes high in the landscape with fewer connecting waterways (Chapter 2, 3).

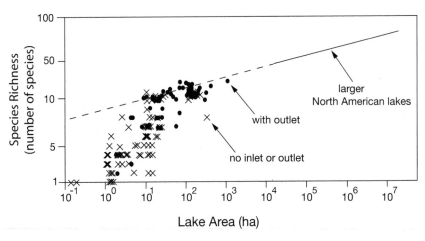

Figure 4.3. Species richness versus lake area for 155 Wisconsin lakes (Rahel 1986, Magnuson et al. 1989) overlaid with an extrapolated species-area relation for larger North American lakes (Barbour and Brown 1974). Redrawn from Magnuson et al. (1989) with permission from the editors, R. Sharitz and J. Gibbons.

Species-Area Relations

On the basis of this reasoning, we analyzed an extensive fish database from Wisconsin and Finland (Rahel 1986, Magnuson et al. 1998) for the following four predictions. The correlation between species richness and lake area should be positive and strong. Groups of lakes that include smaller lakes should have steeper species-area regressions than groups that do not. Groups of lakes that include more isolated lakes should have steeper species-area regressions than groups of less isolated lakes. More isolated lakes should have fewer species than similarly size lakes that are less isolated.

Species-area regressions from 269 forested lakes in Wisconsin and Finland were compared in total and after splitting the database by isolation and lake size. When all lakes were combined, the resulting slope was 0.42. Finland versus Wisconsin was not a statistically significant source of variation, nor was the interaction between lake area and region. The database was divided into lakes smaller or larger than 30 ha because previous research indicated different species-area relationships for lakes smaller and larger than 30 ha (Fig. 4.3). In the combined Wisconsin and Finland database, the subset of smaller lakes had a significantly steeper slope, 0.45, than did the subset of larger lakes, 0.26 (Table 4.1). This comparison suggests that factors that affect invasion and extinction change more rapidly for smaller lakes than for larger lakes as lake area declines. These results were similar to inferences from more detailed analyses of species-area relations for Finnish and Wisconsin lakes and from the literature, based on even larger lakes from a large geographic range. Species-area slopes decreased progressively from 0.45 for the smallest forest lakes (<30 ha) to 0.14 for sets of lakes that included some of the largest lakes in

Table 4.1. Regressions [(log S +1) = a + b log(A +1)] of fish species richness (S) on lake area (A), split by isolation (stream or no stream connections) and lake size (<30 ha and >30 ha) for small forest lakes in Finland and northern Wisconsin and for larger lakes from the literature.

Database	Lakes	N	Arithmetic Mean (Range)		a intercept	b (95% C.I.) slope	r^2 adjusted
			Area (ha)	Richness (number species)			
Wisconsin & Finland	All	269	57.5 (0.2–1566)	7.5 (0–33)	0.31	0.42 (0.38–0.45)	0.69
	Stream	137	83.2 (0.4–1566)	9.8 (1–33)	0.34	0.41 (0.37–0.45)	0.72
	No Stream	132	30.8 (0.2–582)	5.1 (0–19)	0.31	0.39 (0.33–0.45)	0.58
	<30 ha	193	9.0 (0.2–29.1)	4.3 (0–17)	0.28	0.45 (0.38–0.52)	0.45
	>30 ha	76	180.6 (31.2–1556)	15.5 (3–33)	0.64	0.26 (0.17–0.36)	0.28
Other Studies	Browne (1981)	12	382,411 (260–2,571,900)	40.8 (10–113)	0.58	0.23 (0.13–0.33)	0.72
	Eadie et al. (1986)	82	(0.1–8,200,000)			0.23	
	Barbour and Brown (1974)	70	2,242,876 (80–43,600,000)	41.7 (5–245)	0.77	0.14 (0.09–0.19)	0.29

the world (Barbour and Brown 1974). The slope for the small-forested lakes is similar to the slopes for distant archipelagoes, more isolated islands or less vagile species (Brown 1971, Lomolino 1984). The slope of the larger lakes, 30 ha to 1,566 ha, is within the range typically observed for islands (0.20 to 0.39; MacArthur and Wilson 1967, Abbott 1983). The slope for the largest lakes and inland seas (0.14; Barbour and Brown 1974) lies within the range observed for mainland populations where immigration, in contrast to that of islands, is affected little by isolation (0.12 to 0.17; MacArthur and Wilson 1967).

More isolated lakes were expected to have steeper species-area regressions and should have fewer species than less isolated lakes of the same area, resulting in a lower intercept of the species-area relation. For the Wisconsin and Finland lake data, regressions did not significantly differ between groups of lakes with or without a stream connection, although intercepts for lakes with streams were consistently four to six species higher than for lakes that lack streams. Given that these lakes contained only 9 to 15 species, the increase, although statistically not significant, could be biologically significant in many cases.

Differences in slopes of the species-area regressions for fish in small forest lakes (Table 4.1) do appear consistent with the expectations of island biogeography theory. However, the independent effects of extinction and isolation on richness are difficult to separate. Populations in smaller lakes are likely to suffer higher extinction rates than those in larger lakes owing to less heterogeneous habitats and more severe environments. However, the steeper species-area regressions for smaller lakes compared with those for larger lakes could reflect, in part, the fact that the former are more isolated from possible colonists. Not only are small lakes less likely targets for dispersal by wind and birds, they are less likely to receive intentional and accidental introductions by humans (Magnuson 1976, 1988). In addition, smaller lakes in the Wisconsin and Finland databases were less likely than larger lakes to have a stream connection to another lake. Other studies show strong positive species-area regressions for fishes (Barbour and Brown 1974, Browne 1981, Tonn and Magnuson 1982, Eadie et al. 1986, Rahel 1986, Minns 1989, Magnuson et al. 1994), zooplankton (Dodson 1992, Dodson et al. 2000), macrophytes (Dodson et al. 2000), and molluscs (Lassen 1975, Browne 1981). The relationship between species richness and lake size, however, often is influenced by other factors. The slope of the species-area curves differs depending on the composition of assemblages (Tonn and Magnuson 1982), the alkalinity of lakes in the dataset (Rahel 1986), and the method of estimating richness (Magnuson et al. 1994).

Limnological and Geomorphic Factors

Analyses of the environmental factors related to species richness and composition were under way at the Northern Highlands Lake District when our LTER research began in 1980 and became incorporated in the way we thought about lake biodiversity in the LTER research. John Magnuson, Frank Rahel, and Bill Tonn determined that fish species richness and composition in small forest lakes (<1 to 1600 ha in area) were influenced by physical attributes such as vegetation diversity, isolation, winter oxygen levels, and pH (Tonn and Magnuson 1982, Rahel and Magnuson 1983, Tonn et al. 1983, Rahel 1984, 1986). For *Umbra*-cyprinid assemblages (minnows), lake surface area and winter oxygen levels accounted for most of the variation in species richness across lakes (Tonn and Magnuson 1982). Rahel (1986) expanded the relationship between fish richness and physical lake attributes by including pH for lakes along a successional gradient. Three distinct fish assemblages (the centrarchid assemblage, the cyprinid assemblage, and the *Umbra-Perca* assemblage) were identified that reflected a sequential loss of species along a gradient from oligotrophic seepage lakes to bog lakes. Fish species richness is correlated with lake size and habitat complexity, both of which tend to be correlated with productivity, pH, and winter oxygen concentration. Richness decreases as the chemical environment becomes harsher and near-shore habitat is simplified. Thus, local processes that determine the harshness of the chemical conditions and individual species tolerances to the abiotic and biotic environment control fish species richness and composition.

Primary productivity is an important determinant of species richness in lake ecosystems. Species richness for phytoplankton, rotifers, cladocerans, copepods, macrophytes, and fishes exhibits a hump-shaped pattern with increasing primary productivity (Dodson et al. 2000) (Fig. 4.4). Although species richness increases with lake surface area for rotifers, cladocerans, macrophytes, and fish, lake surface area had no effect on the relation between species richness and productivity for most taxa, except for fishes and phytoplankton. For these taxa, the species richness-productivity relationship changed across a gradient of lake size (Fig. 4.5). Phytoplankton richness in smaller lakes, <1000 hectares, was highest at intermediate productivity. In larger lakes the relationship switched such that the highest richness occurred at low and high productivity levels. For fish in smaller lakes, <100 hectares, the relationship ranged from being nearly linear to having highest fish richness at low and high productivity levels. In larger lakes, >100 hectares, richness was greatest at intermediate productivity levels.

When the richness-productivity relationship from an extensive spatial survey was compared with lakes where productivity had been experimentally manipulated through nutrient additions, a variety of richness-productivity relationships emerged (Dodson et al. 2000). Two types of experimental manipulations were considered: (1) long-term nutrient additions to individual lakes at the Experimental Lakes Area (ELA) in Ontario (Malley et al. 1988), and (2) short-term nutrient additions of lakes with contrasting food webs at the University of Notre Dame's Environmental Research Center, in the Upper Penninsula of Michigan (Schindler et al. 1997). In both types of experimental manipulations, differences in the richness-productivity relation among the experimental lakes and survey lakes may have been the result of transient effects, lags in response, and possible shifts to new ecosystem states as nutrient levels and thus primary productivity were altered. Therefore, lakes that experience environmental changes may not conform to predictions based on long-term data across a broad spatial scale. For example, in L226, the site of whole-lake nutrient additions at the Experimental Lakes Area, phytoplankton richness initially increased as productivity increased, but when nutrient additions stopped and productivity declined, richness remained high. New immigrants tended to remain in

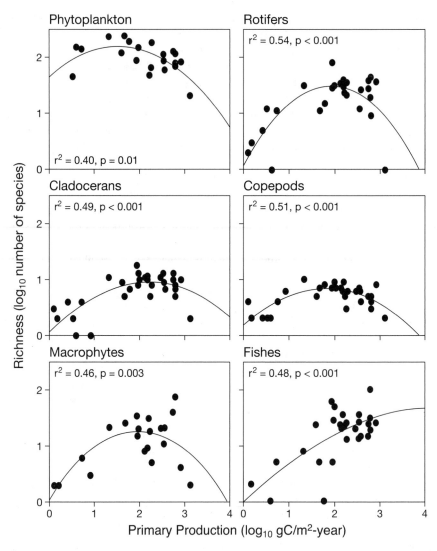

Figure 4.4. Relation between species richness and primary production for phytoplankton, rotifers, cladocerans, copepods, macrophytes, and fishes, based on values for lakes around the world. Regression lines are fitted from a quadratic model for log species richness and log primary production without including any relation between richness and area. Redrawn from Dodson et al. (2000) with permission of the Ecological Society of America.

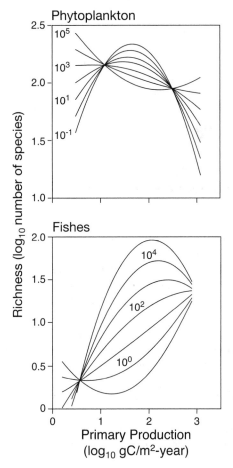

Figure 4.5. Response surfaces for the interactive effects of lake area and primary productivity on the species richness of phytoplankton (top) and fishes (bottom). Each curve indicates the relation between richness and primary productivity for a particular lake area in hectares. Curves are drawn only for the ranges of primary production associated with species richness data. Redrawn from Dodson et al. (2000) with permission of the Ecological Society of America.

the species pool, even after nutrient additions were stopped. Whether these species will continue to persist or eventually disappear is uncertain.

Individual lake characteristics also appear to influence the relationship between productivity and crustacean zooplankton. In the manipulated lakes at the Experimental Lakes Area and at the Notre Dame site, crustacean zooplankton richness decreased with increasing primary productivity (Dodson et al. 2000). This decrease likely resulted from a combination of factors, including competitive interactions, food limitation as phytoplankton species composition changed to more inedible forms, and changes in the chemical environment (low oxygen concentrations and high pH) as productivity increased. For example, in the short-term, whole-lake nutrient manipulations conducted at the Notre Dame site, shifts in food size toward larger phytoplankton probably enabled the large herbivore *Daphnia* to outcompete smaller zooplankton, thus reducing species richness at high levels of productivity. In a long-term experiment at the Experimental Lakes Area, high pH resulting from the nutrient addition and low oxygen resulting from decomposition of an increased phytoplankton biomass eliminated sensitive zooplankton species, resulting in low richness at high levels of primary production.

Local versus Regional Factors

Strong relationships between species richness and characteristics of individual lakes suggest that local biotic and abiotic conditions are more influential than regional processes. Thus, species that do not occur in a lake but that are common in a region may not exist in a lake because they cannot survive the local biotic and abiotic conditions, not because they are unable to cross dispersal barriers. Lukaszewski et al. (1999) experimentally tested this hypothesis using mesocosms. Several 300-L enclosures were suspended in Little Rock Lake and filled with filtered lake water and zooplankton from the lake. Additional colonists from other LTER lakes were added to some of the enclosures to simulate different colonization levels. At the end of the experiment, species richness of zooplankton in the experimental enclosures that received new colonists was not significantly higher than that for control enclosures that did not receive colonists. This result suggests that local biotic and abiotic conditions determine the number of species able to occur in each lake. Lakes are probably not limited by dispersal of zooplankton propagules; most species are probably capable of arriving at most lakes, although not all species are capable of surviving and reproducing once they arrive.

The relative importance of local and regional factors in determining species richness may ultimately depend on the scale of observation. Comparisons of fish assemblages in Finland and in northern Wisconsin suggest that the composition of fish assemblages in these two regions is shaped by a series of filters that operate on several temporal and spatial scales (Fig. 4.6) (Tonn et al. 1990). At the coarsest level, regional mechanisms such as Pleistocene events, dispersal barriers, climatic differences, and geomorphology determine the regional species pool. Species that pass through this first filter are then subjected to the next filter, which is characterized by habitat attributes of the lake-type. Finally, fine-scale local processes such as lake size, isolation, habitat complexity, and abiotic and biotic conditions determine fish assemblage composition and therefore species richness within individual lakes. Clearly, both local and regional processes play an influential role in determining richness and biotic structure.

Dynamics of Species Composition

Long-term annual data on the occurrence of aquatic organisms provide a unique opportunity to examine the dynamics of species composition in lakes. A key ques-

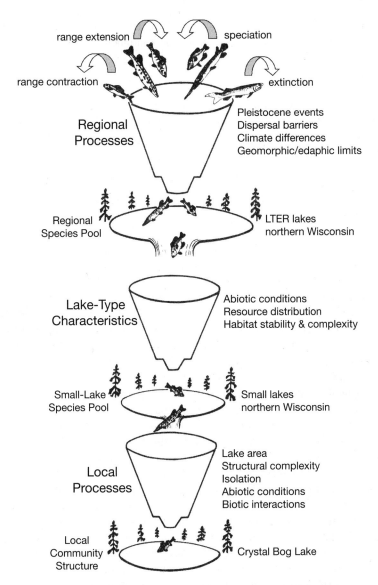

range extension   speciation

range contraction   extinction

Regional
Processes
     Pleistocene events
     Dispersal barriers
     Climate differences
     Geomorphic/edaphic limits

Regional
Species Pool
     LTER lakes
     northern Wisconsin

Lake-Type
Characteristics
     Abiotic conditions
     Resource distribution
     Habitat stability & complexity

Small-Lake
Species Pool
     Small lakes
     northern Wisconsin

Local
Processes
     Lake area
     Structural complexity
     Isolation
     Abiotic conditions
     Biotic interactions

Local
Community
Structure
     Crystal Bog Lake

Figure 4.6. Conceptual model of the origin and maintenance of fish species composition in small forest lakes illustrated with a set of filters that remove or allow passage of a species with particular features over regional to local spatial scales. Redrawn from Tonn et al. (1990) with permission of the University of Chicago Press.

tion is: do lakes contain stable assemblages of species or is turnover of species in ecological time through invasion and extinction continuous and apparent?

Species Turnover

Temporal change in community composition or species turnover for aquatic organisms in several lakes was assessed (Magnuson and Lathrop 1992, Magnuson et al. 1994, Arnott 1998, Arnott et al. 1999). The species turnover rate (T) between two sampling periods was calculated as:

$$T = 100 \left[(I + E) / (S_1 + S_2) \right] t^{-1}$$

where I = the number of species gained, E = the number of species lost, S_1 and S_2 = the number of species present in each sample, and t = the time interval between samples (Diamond 1969).

Species turnover rates for Lake Mendota, a highly managed system in an agricultural and urban watershed (Chapter 12) were calculated from historical fish records and recent fish surveys by the Wisconsin Department of Natural Resources (Magnuson and Lathrop 1992). From 1900 to 1989, the fish species turnover rate was approximately two fish species per decade. Much of this species turnover, however, occurred during two periods of intense human impact. During the 1920s, small-bodied littoral fishes declined, probably from a combination of habitat degradation associated with increased eutrophication and widespread use of herbicides and weed cutters to control macrophyte beds. Additional changes in the fish community during this time resulted from intentional fish stocking and from the release of fish by the Mississippi River Fish Rescue Program. In the 1970s and 1980s, the decline of small-bodied fishes probably resulted from increased predation on them. Stocking rates of piscivorous sport fish during this time were the highest they had been in the history of the lake (Chapter 12). Historical records for Lake Mendota indicate that species richness and community composition have changed dramatically during the past century. Much of this change was human-induced.

Species turnover rates were also high for lakes in forested landscapes with low human impact relative to the southern Wisconsin lakes. These forested sites are the Northern Highlands Lake District in Wisconsin, the Muskoka region (Dorset) in southern Ontario, and the Experimental Lakes Area in northwestern Ontario. All three regions are used primarily for recreation and have experienced relatively little human impact, at least over the past 80 years. Long-term records on fish, zooplankton, and phytoplankton species composition from these areas were well suited for addressing issues of community stability because sampling methods have been consistent since the early 1980s. Species turnover was high; on average, 22% of the fishes (Magnuson et al. 1994), 16% of the zooplankton (Arnott et al. 1999), and 18% of the phytoplankton (Arnott 1998) appeared or disappeared each year. These high rates likely include significant sampling error, however. Despite extensive sampling, rare species were sometimes missed even though they were present in the lake.

To adjust for sampling error, we calculated conservative estimates of species turnover rates by subtracting turnover rates that could be attributed to sampling error.

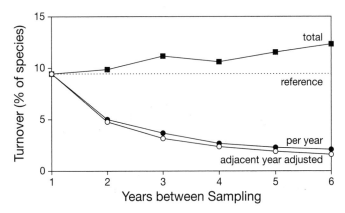

Figure 4.7. Influence of the number of years between sampling on estimates of species turnover for fishes in the LTER lakes in northern Wisconsin. The dashed horizontal line indicates the mean turnover between adjacent years. The solid squares indicate turnover between time intervals, ranging from 1 to 6 years. The solid circles indicate turnover between time intervals divided by the number of years between samples. The open circles indicate the expected turnover if all turnover resulted from sampling errors as estimated between adjacent years. Results are averaged across seven lakes from 1981 to 1991 after rare species were removed from the analysis. Redrawn from Magnuson et al. (1994) with permission of Society for Comparative and Integrative Biology.

Two approaches were used to estimate sampling error; sampling error was (1) estimated to be the turnover that occurred between adjacent years, or (2) estimated using Monte Carlo simulations based on species abundances and the probability of detecting at least one individual of a particular species in a sample.

In the analyses for fishes, Magnuson et al. (1994) removed rare species from the analysis when fewer than two individuals were encountered during the study period and subtracted the average turnover between each adjacent year from the estimate of turnover across all years (Fig. 4.7). The rationale was that turnover calculated between adjacent years probably resulted from missing rare species because the time interval was too short for changes resulting from extinction and invasion. When these adjustments were made, fish species turnover averaged 0.44% per year for all lakes, considerably less than 22% calculated from the raw data. The 0.44% turnover rate is the difference between the plotted points for per year and for adjacent year adjusted. The rate corrected for rare taxa is equivalent to the extinction or immigration of one to two species during a decade (Magnuson et al. 1994).

Because zooplankton and phytoplankton have shorter life cycles than fishes, changes in community composition as a result of invasion (either from other lakes or from resting stages in the sediments) and extinction seem more likely on yearly time scales. To account for turnover attributable to sampling error, Arnott et al. (1999) developed a Monte Carlo simulation model where the probability of detecting each species was calculated on the basis of the average number of individuals

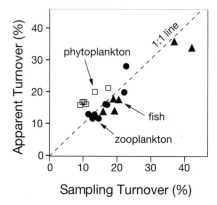

Figure 4.8. Relation between apparent turnover rates and sampling turnover rates for phytoplankton from the Experimental Lakes Area in Ontario (D. L. Findlay), crustacean zooplankton from the Dorset Lakes in Ontario (N. D. Yan), and fishes from North Temperate Lakes LTER in northern Wisconsin. Each point represents the mean annual turnover as a percent change in species composition per year for a lake. The dashed line indicates where apparent turnover equals turnover owing to sampling error. For points above the line, the difference between apparent and sampling turnover provides a conservative estimate of species turnover attributable to actual immigration and extinction. Points below the line indicate lakes where sampling error was too high to provide an estimate of real turnover. Redrawn from Arnott (1998) with permission.

caught each year. On the basis of these probabilities, the amount of turnover that could potentially result from sampling error was determined (Fig. 4.8). When potential sampling error was compared with apparent turnover estimates, that is, the turnover calculated from annual changes in the species composition of the samples, zooplankton species turnover could be detected reliably in only two of the eight Dorset study lakes. Zooplankton species turnover, corrected for sampling error, averaged 0.7% per year. Turnover rates were higher for phytoplankton in six ELA reference lakes. Phytoplankton species turnover, corrected for sampling error, averaged 6.2% per year. We suspect that our estimates of species turnover are underestimates because some of the turnover that we attributed to sampling error may, in fact, have been the result of immigration or extinction of rare species. Even so, we were able to demonstrate that, for these insular habitats with relatively little human impact, dispersal of organisms and local extinction can be important in determining local community composition and dynamics.

Detecting Invasions and Extinctions

One of the largest obstacles in identifying species that have invaded or gone extinct is that rare species are not always detected in regular sampling procedures. This is particularly a problem for detecting cryptic invaders, that is, species that are native to the region but new to a particular lake. A rare species that is occasionally detected is difficult to distinguish from a recent invader. To deal with this problem, a multicriteria approach for detecting invasions and extinctions of fish species was developed for five of the LTER lakes in northern Wisconsin (Cisneros 1993). In addition to assessing presence and absence of species, Cisneros (1993) also considered changes in abundance, dispersion, and mean body size as indicators of invasion and extinction.

Using his multicriteria approach, Cisneros (1993) estimated turnover rates of fishes in the LTER lakes from 1981 to 1991. Overall, 11% of the fish species changed per decade. Annual turnover rates ranged from 0 to 1.1% among lakes: Allequash = 0.1%; Sparkling Lake = 0.4%; Trout Lake = 0.5%; Big Muskellunge = 0.5%; and Crystal Lake = 1.1%. These estimates are slightly higher than those obtained by Magnuson et al. (1994) using a method based on annual changes in richness. Because separating out rare species from cryptic invasions of short duration is virtually impossible and because the rare species are most likely to be invaders or to go extinct, turnover rates of Cisneros (1993) and Magnuson et al. (1994) likely are underestimated. Nevertheless, these results indicate that invasion and extinction are important and common processes in determining species dynamics in lakes.

Here we expand Cisneros's (1993) analysis to include 20 years of data for one of the lakes with a known smelt invasion, Sparkling Lake (Hrabik et al. 1998). Each species was classified into six different invasion/extinction categories; apparent invasion, possible invasion, apparent extinction, possible extinction, no evidence, and rare species (Table 4.2), using invasion and extinction plots (Fig. 4.9). An invasion plot is the number of years each fish species was detected versus the first year it was detected. An extinction plot is similar to an invasion plot except that the number of years each species was detected is plotted against the last year it was detected. Some species, for example, mudminnow and muskellunge, could be placed into more than one invasion-extinction category if they were detected after several years of not being detected (possible invasion) and then disappeared again (possible extinction).

Individual species were assigned categories on the basis of their position on the invasion/extinction graphs (Fig. 4.9). On each graph, the diagonal line represents the maximum number of years that a species could be present, given the year it first appeared or when it was last observed. A point close to the diagonal line corresponds to a species with high persistence. Species that appear early or late in the record or that have few occurrences provide either weak or no convincing evidence. Classifications based on these figures were coupled with expected patterns of temporal change in abundance, mean body length, and within-lake dispersion to assess the likelihood of an invasion or extinction (Table 4.2). Positive trends in abundance and within-lake dispersion over time suggest an invasion, whereas negative trends indicate an extinction or a potential extinction. A positive trend in mean body length, an indication of an aging population and low reproduction, supports evidence of extinction. A decrease in mean body length suggests successful recruitment, indicating an invasion. We used the Mann-Kendall trend test, adjusted for autocorrelation (Hamed and Rao 1998), to evaluate the statistical significance of trends in abundance, length, and dispersion.

Using Cisneros's (1993) multicriteria approach, we found evidence for three possible extinctions (cisco, logperch, and yellow perch) (Fig. 4.9, Table 4.2). In

Table 4.2. Trends for Sparkling Lake fishes in abundance (catch/unit effort), mean body length, and dispersion along the shoreline for Invasion—Extinction categories.[a]

Invasion/Extinction Category	Species	Abundance	Mean Body Length	Dispersion
Apparent Invasions	blacknose shiner	+	−	
	bluegill	−	+	−
	largemouth bass	−	−	−
Possible Invasions	mimic shiner	+	+	+
	pumpkinseed	+	+	+
	mudminnow	+	−	+
	muskellunge	−	+	+
	smelt	+	−	+
Possible Extinctions	cisco	−*	+*	−
	logperch	−	+	−*
	iowa darter	+	+	+
	mottled sculpin	−	+	−
	mudminnow	+	−	+
	muskellunge	−	+	+
	white sucker	0	+	−
No Evidence[b]	bluntnose minnow			
	johnny darter			
	rock bass			
	walleye			
	smallmouth bass			+*
	yellow perch	−*	+	−*

[a]Direction of trends are indicated by + or −.

[b]Fishes in the No Evidence category were present in fewer than 5 years, or in the case of yellow perch, were present for all years.

*Indicates statistical significance of trends ($p < 0.05$).

each of these cases, at least two of the four criteria were met. While the immigration and extinction detection graphs provided no evidence of extinction of yellow perch as it was still present at low abundance in 2000, trends in abundance and dispersion were negative and statistically significant. This evidence may provide an early warning of a pending extinction. We found little evidence for the known smelt invasion. Trends matched the expected invasion pattern, with increasing abundance and dispersion and decreasing mean body length, but the trends were not statistically significant owing to high interannual variation in abundance. A longer period of sampling before the invasion took place (Chapter 8) could have provided stronger evidence of the smelt invasion.

Our analysis of Sparkling Lake fish data for a 20-year period suggests that fish composition is dynamic. Detecting immigration and extinction of species is difficult, however, because species abundance fluctuates over time, and some species are not detected in some years because abundances are low. A multicriteria approach, combining evidence based on abundance, distribution within the lake, and mean body size, provides a more powerful method of detecting species changes than assessments based strictly on presence or absences of species.

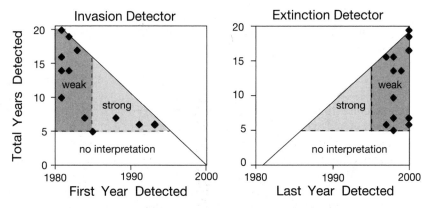

Figure 4.9. Graphical approach to detect possible invasions or extinctions in a long-term record of species presence-absence data. The example is for fishes in Sparkling Lake in northern Wisconsin from 1981 to 2000. Left: Invasions detector based on the persistence of a species and the first year it was seen. Each point represents the number of years the species was found versus the first year it was found. Right: Extinction detector is the same as the invasion graph but is based on the last year the species was found. The diagonal line in each graph represents the maximum number of years that a species could be present given the year it first appeared (left) or was last observed (right). A species present less than 5 years was considered too uncommon to interpret and is left off the graph. Points in the light gray areas suggest an invasion (left) or an extinction (right). Points in the dark gray areas are for fishes where evidence for invasions (left) or extinctions (right) is weak because the possible event occurred too close to the beginning or the end of the time series. Based on approach of Cisneros (1993).

Relative Importance of Invasion and Extinction

As islands in a sea of land, the assemblages of species in lakes are controlled by immigration and extinction events. Researchers have argued that extinction variables are most important in structuring fish assemblages based on broad-scale patterns and correlations (Magnuson 1988, Tonn et al. 1990). This hypothesis was more recently tested by comparing the relative importance of extinction and isolation variables in determining fish species richness in two lake districts, the Northern Highland Lake District in northern Wisconsin and the Upper Peninsula of Michigan and lakes in Finland (Magnuson et al. 1998). The extinction variables considered were related to variables important in structuring fish assemblages in northern Wisconsin lakes: (1) lake size, (2) conductivity (a surrogate for productivity), (3) pH, and (4) maximum lake depth (an indication of winter oxygen concentration). Several isolation variables were considered that were expected to influence the movement of fishes across the lake landscape: (1) the distance over land from a lake to the nearest surface water connection, (2) the vertical distance over land between lakes (summing up and down over ridges and depressions), (3) the distance along a water course from the study lake to the nearest downstream lake, (4) the stream gradient (the average slope along a watercourse from a lake to the next down-gradient lake), (5) the area of the nearest lake that would serve as a source pool for new species, and (6) the distance to the nearest road.

Extinction variables usually predicted fish species richness more effectively than did isolation variables in both Wisconsin and Finnish lakes (Magnuson et al. 1998). This does not suggest that isolation is unimportant in determining community structure. Rather, the result suggests that differences in the probabilities of immigration and extinction events determine their relative importance in statistical analyses of the present condition. Invasion or recolonization events for fish probably happen at an extremely low frequency, especially in small, isolated, forest lakes with relatively little human activity. In the Wisconsin database, several lakes were fishless, despite current physical and chemical properties that suggested that they could support fishes. Either fishes have not colonized some lakes during the 10,000 years after their formation or historical extinctions have not been followed by recolonization events. Extinction factors appear to operate over shorter time scales than colonization, ranging from months to decades. Magnuson et al. (1998) suggest that "the greater the lag in arrival after extinction vs. the lag in extinction after an arrival, the more important extinction variables will be identified relative to isolation variables" (p. 2953). Thus, in isolated lakes, extinction factors appear more important than isolation factors because these extinction factors were the last to effectively filter what is there today.

Aspects of the geomorphology of the two regions suggest that isolation is important in determining richness and composition of fish communities (Fig. 4.10) (Magnuson et al. 1998). Although both regions are forested and have high concentrations of lakes, the Wisconsin region has lower topographical relief, and therefore lakes tend to be connected by groundwater flows and have few stream connections (Chapter 2, 3). In contrast, the Finnish lakes are situated on a landscape with more topographical relief and are highly connected via streams. Thus, richness and community composition in the Finnish lakes were best predicted by stream gradient and area of the closest connected lake (related to size of the source species pool), whereas, in Wisconsin, horizontal distances between lakes by land or water were better predictors.

Lake surface area was related to richness and composition in both regions (Fig. 4.10). In northern Wisconsin lakes, pH was important, whereas conductivity was important in Finnish lakes. The geomorphic setting of each area may influence these differences. All four of the extinction variables in Wisconsin lakes are related to the position of the lake in the landscape (Chapter 3). Lakes high in the landscape tend to have few fish species, small surface areas, and low conductivity and pH (Chapter 3) (Kratz et al. 1997b, Riera et al. 2000).

Summary

In our studies of lakes as islands, we have emphasized the dynamic nature of aquatic communities. Temporal changes in species composition of lakes appear to be linked to within-lake drivers that determine extinction probabilities and to

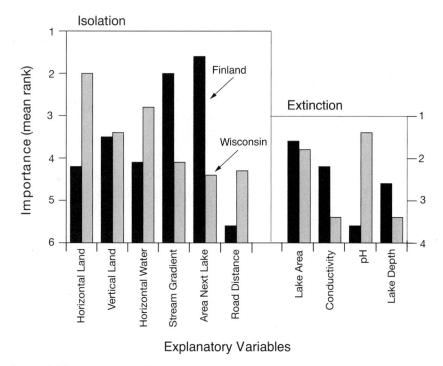

Figure 4.10. Importance of isolation variables (left) and extinction variables (right) in predicting the fish species richness and composition in lakes of Finland and northern Wisconsin. Each plotted value is the average of five different analyses with multivariate statistical models (Classification and Regression Trees and Linear Discriminate Analyses for richness and composition). Redrawn from Magnuson et al. (1998) with permission of the Ecological Society of America.

landscape-level factors that influence the movement of organisms across the lake district. Some of the local factors that we have studied and found important in determining species richness for aquatic biota include lake surface area, primary productivity, pH, and winter oxygen concentrations. For fish, predictable assemblage-types occur along environmental gradients determined by these factors. Landscape-level drivers that influence species richness through dispersal and immigration include land barriers, long streams, steep streams with waterfalls, distance to the nearest road, and the area of the next lake. Extinction factors appear to operate over shorter time scales than immigration, and their influences may be more readily detectable at a given point in time. However, detecting changes in species composition resulting from immigration and extinction has proven difficult because species can be missed when communities are sampled. We have developed analytical methods that enhance the detection of invasions and extinctions against the background of sampling problems. Despite fluctuations in apparent species composition resulting from sampling error, we have used our analytical methods to estimate realistic species turnover rates in plankton and fish communities. Our analyses suggest that the presence and absence of species is dynamic, with new species arriving and others going extinct over observable time intervals. By viewing lakes as islands surrounded by large areas of inhospitable land, we have begun to understand species dynamics in lake ecosystems as a function of both regional processes that influence the immigration of species and local processes that influence extinction probabilities.

5

Coherent Dynamics among Lakes

John J. Magnuson
Timothy K. Kratz
Barbara J. Benson
Katherine E. Webster

Everything Is Connected to Everything Else
—Barry Commoner (1971, p. 33)

Historically, limnologists have analyzed and been concerned with lake status or condition; for example, classification by trophic status into eutrophic, oligotrophic, and dystrophic has allowed us to group lakes with similar properties and to study the causes and ramifications of those differences and similarities. However, in this era of awareness about global change, we have come to focus more on regional dynamics, particularly those that act at interannual or longer time scales. By combining spatial concepts that underlie differences between lakes with the time dimension, we can gain a better understanding of the determinants of lake dynamics that allow us to make more powerful predictions of trajectories into the future. Here we focus on the interannual time scale and explore how comparing similarities and differences in dynamics among years across lakes can enrich our understanding of lake dynamics and the processes that cause lakes to change.

Many questions confront us in the quest to first describe and then understand long-term dynamics of suites of lakes. How should we characterize the coherent or shared dynamics of lakes? To what degree are interannual dynamics or long-term changes consistent among lakes? What determines how similarly or dissimilarly different lakes behave among years? Do adjacent lakes behave more similarly than more distant lakes? Do lakes of the same type or status, such as oligotrophic lakes, behave more similarly through time than different kinds of lakes, such as oligotrophic versus eutrophic? Are there common drivers of consistent dynamics among lakes such as dynamics in climate, land use, landform, or water flows? Are some types of lake parameters inherently more coherent than others? Are there internal drivers that cause individual lakes or at least certain parameters of these lakes to behave independently? Can the study of the degree of shared dynamics among lakes guide us to more sophisticated analyses and approaches in our attempts to understand long-term dynamics of ecological systems? Can the dynamics of a lake be predicted from the dynamics of other lakes?

Coherent Behavior among Lakes

Coherent behavior among lakes would be expected to occur if an external driver imposed a common dynamic to lakes in an area or region. The external driver can be thought of as a signal and the lakes as responders (Magnuson et al. 1990, Magnuson et al. 2004). But a lake and its catchment area can serve as a filter or transformer of the external signal, and thus the resulting dynamics in the lake can differ from those of the external driver. One or more lake filters may attenuate or amplify the signal, produce different time lags, or reverberate through the lake system; the effects of the signal may be hidden by interference with other signals of different frequencies and strengths (Fig. 5.1) (Magnuson et al. 2004). Several levels of filters are apparent. For example, the catchments can differ in the ratio of infiltration to runoff and thus modify the signal reaching the lake from a rain event. For another example, filters exist at the boundary of the lake for climatic factors; lakes differ in the degree of exposure to climate forcing from wind and solar radiation because lakes differ in surface area and depth. For yet another example, filters exist within lakes, and in-lake processes, such as the temperature-dependent rate of prey consumption by warmwater versus coolwater fishes, may alter the responsiveness to a climate signal that already has warmed or cooled the lake. Filters and their effectiveness would be expected to differ among lake types.

We define coherence or synchrony as the degree to which different locations, lakes in our case, behave similarly through time (Magnuson et al. 1990, Kratz et al. 1998, Magnuson and Kratz 2000). Coherence and synchrony are used interchangeably in the literature; we have used the term "coherence" throughout this chapter. We estimate coherence by the statistical correlation between time series of the same parameter on pairs of lakes. We have used data aggregated at an annual time scale and various measures of correlation between a pair of lakes, namely the Pearson product-moment correlation coefficient, r, the square of this coefficient, r^2, which is the proportion of common variance or the shared variance, and Spearman rank correlation coefficient, rho (Magnuson et al. 1990, Wynne et al. 1996, Kratz et al. 1998, Baines et al. 2000, Benson et al. 2000b). These measures can be calculated from a lake-by-year matrix for each parameter (Table 5.1), with each coherence value indicating the degree of similarity of dynamics of an individual pair of lakes for a particular variable of interest. Individual coherence values can be averaged across multiple pairs of lakes or across multiple variables of interest. Because each lake is compared with each of the other lakes and is represented more than once in all summary statistics for a suite of lakes, coherence estimates are not statistically independent. As with other integrative indices such as biological diversity (some high-quality ecosystems have low diversity), a lake pair or lake parameter with high coherence should not be judged as better or more desirable than a lake pair or parameter with low coherence.

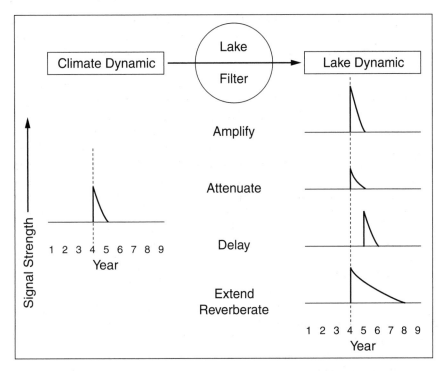

Figure 5.1. Diagram of a climate signal that can be altered by lake filters that amplify, attenuate, delay, or extend a response in a lake's dynamics. We show a pulse or event type of signal; more gradual or oscillatory signals could be envisioned in the same way. With multiple signals of various forms, interference and other interactions could further complicate the response and reduce coherence in among-lake responses. Redrawn from Magnuson et al. (2004).

Table 5.1. Matrices of lakes and years for individual lake variables arranged to calculate measures of coherence across all lake pairs or all variables or both all variables and all lake pairs.

	Summer epilimnetic temperature (°C)			
Year	$Lake_1$	$Lake_2$	$\rightarrow$	$Lake_n$
$Year_1$	22	20		25
$Year_2$	20	19		21
" "				
$Year_m$	25	23		27

Coherence contains information about the degree of shared dynamics of lakes across the landscape and about the limnological functioning of lakes from a landscape perspective. Analysis of coherence is an approach to comparing the dynamics of lakes in a spatially explicit way (Magnuson et al. 1990, Magnuson and Kratz 2000) that may contribute toward an understanding not only of individual lakes but also of dynamics of lake districts or regional groupings of lakes. One would expect adjacent lakes to vary together for a given parameter, that is, be more coherent, if the dynamics of that parameter are determined largely by interannual variability in a regional process, such as climatic variation (Chapter 7) or acid precipitation (Chapter 9). In contrast, lakes should behave more independently for a given parameter, that is, be less coherent, if the dynamics of that parameter are determined largely by internal drivers such as internal nutrient recycling or species interactions (Chapter 10). In actuality, as can be seen by what follows, the coherence among lake behaviors is determined by more complex relationships (Magnuson and Kratz 2000).

We began to use analyses of coherence to gain understanding of lake dynamics when the North Temperate Lakes LTER program had accumulated only seven years of data, namely from 1982 to 1988. This period of record was insufficient for applying simple linear regression or multivariate regression or for meeting the requirements of time series analyses. We used statistics such as r^2 as indices of coherence and compared the measures among lake pairs and parameters; we did not use them as inferential statistics. We grouped the strength of coherence into five categories from very low to very high on the basis of equal divisions of r^2, that is, the percentage of shared variability between two time series (Table 5.2). One of the advantages of this approach was that we could explore dynamics early in our site's history. We continue to find this simple approach useful for investigating questions about lake dynamics as described below.

Coherence within a Lake District

The overall coherence among northern Wisconsin LTER lakes, even between adjacent lakes, was surprisingly low when a rich set of physical, chemical, and bio-

Table 5.2. Categories of coherence based on the Pearson product moment correlation coefficient, r, and the percentage of shared variance, r^2 (coherence categories by equal units of r^2).

Coherence Category	Correlation (r)	Shared variance (r^2)
Very Low	0.00–0.44	0.00–0.19
Low	0.45–0.62	0.20–0.39
Moderate	0.63–0.78	0.40–0.59
High	0.79–0.88	0.60–0.79
Very High	0.89–1.00	0.80–1.00

logical variables was considered together. In the Northern Highlands, the seven LTER lakes had an overall average coherence of only $r^2 = 0.10$ among all pairings of the seven lakes with 15 variables over 7 years (Magnuson et al. 1990) and $r^2 = 0.07$ with 61 variables over 13 years (Kratz et al. 1998). Ninety percent or more of the overall interyear variation apparently was uniquely lake dependent.

Even though the overall average coherence is very low for the LTER lakes in the Northern Highlands of Wisconsin, it is positive and statistically greater than zero. The observed correlation values exceed those obtained when the data for each variable are rearranged randomly among years and the coherences between lake pairs are recalculated (Kratz et al. 1998). A weak regional signal of coherent interyear lake dynamics is apparent even at this level of data aggregation of a combined set of physical, chemical, and biological variables.

The very low coherence surprised us because, owing to their close proximity, the seven lakes are exposed to the same climate and weather; they lie within a circle of 10-km radius (Fig. 1.4) in the same hydrologic flow system; and the shores of some lakes are separated by only 0.1 to 0.4 km of land. In addition, the lakes are the same age, ca. 10,000 years, and lie in a rather homogeneous till of glacial sand and gravel left behind as the glacier receded (Attig 1985). Supporting our expectation of higher coherence were anecdotes and technical papers about the year of the fine French wine or the year of broad-scale recruitment of perch and walleye (Koonce et al. 1977), suggesting that the weather in a given year can impart pattern over large regions. We do know that climate drivers have a large regional footprint (Chapter 7). So, why are the dynamics of the lakes apparently so different from one another?

A necessary initial condition for coherence is that the variable must have enough interyear dynamics to rise above sampling error or noise. The length of record plays a role because increasingly long records are more likely to include more extreme events or signals such as droughts or El Niño years, or change from climate warming. As a consequence, coherence increases as the length of record increases (Fig. 5.2). Shorter time series would underestimate the shared variance between lakes or, in some cases, be overly influenced by an extreme year. The observed increase in shared variance with record length occurs for variables with very high coherence such as ice-off date and for variables with less coherence, such as acid neutralizing capacity (ANC) of the water. In some cases, the increase was marked, such as for the coherence of ANC between Sparkling and Trout Lakes; with 3 to 4 years of data, coherence was low, while with 16 to 18 years, it was very high. Coherence for some biological variables, such as animal abundance, that likely have more measurement error than physical variables are more likely to be masked by noise (Kratz et al. 1995). Aggregation of the data may change the signal to noise ratio, as well. Generally, coherence calculated from more integrated values is intermediate between the lowest and the highest coherence of individual components (Table 2 in Kratz et al. 1998, Fig. 3 in Rusak et al., unpublished manuscript). However, zooplankton were more coherent with less integrated measures, such as those for individual species or for subsets of lakes, than with zooplankton overall across all lakes (Rusak et al. 1999).

On the basis of his experience with lakes in the Northern Highlands, one of our predecessors in Wisconsin limnology, C. Juday (1871–1944), wrote, "According

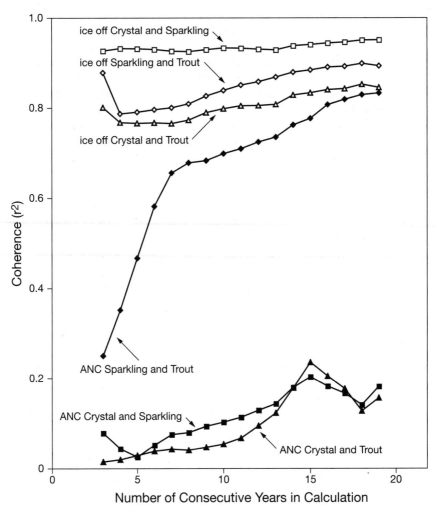

Figure 5.2. Sensitivity of coherence estimates, r^2, to the number of years of observation for ice-off date and acid neutralizing capacity (ANC) of the water for pairings of Crystal, Sparkling, and Trout Lakes. All consecutive sets of years from 3 to 19 years were used to calculate coherence. The number of sets averaged for coherence ranged from 17 for 3-year sets to 2 for 18-year sets; the 19-year set had one value of coherence. Redrawn from Magnuson et al. (2004).

to our experiences, lakes are 'rank individualists' and are *very stubborn* about fit-ting into mathematical formulae and artificial schemes proposed by man." This statement is from Juday's review recommending rejection of Raymond Lindeman's classic article (Lindeman 1942) on trophic dynamics of Cedar Bog Lake in Minne-sota (Cook 1977, p. 23). Does the very low coherence simply give evidence to this individuality of lakes and perhaps to the absence of underlying models or funda-mental relations?

Certainly, our LTER lakes in the Northern Highlands do differ from one another in many ways (Table 1.1, Table 2.1). Hypothesized explanations for the very low overall coherence include the idea that the lakes differ greatly in size, solute con-centrations and pH, shoreline structure, and biota and that these differences impart very different within-lake dynamics in response to an external driver like climatic variability. These differences among the lakes are set, in part, by the context of the position in the landscape (Chapter 3). The smaller lakes, with lower solute concen-trations and fewer fish species, are higher in the landscape, and the larger lakes, with higher solute concentration and more fish species, are lower in the landscape.

Yet, at the level of aggregation of all lake pairs and all variables, much of the interannual variability within sets of adjacent lakes appears to be lake specific. In-dividualistic dynamics may result, in part, from the island nature of lakes with in-dividualistic species assemblages (Chapter 4) and from differences in status, such as differences in position in the landscape (Chapter 3). This individualistic behav-ior is not characteristic for all limnological variables or for all subsets of lakes, as we discuss in the following section.

Differences in Coherence among Limnological Variables

While the average overall coherence discussed in the preceding section was very low, interesting patterns emerged when we considered individual variables or types of variables averaged for the seven lakes. For example, coherences for variables such as epilimnetic temperatures and date of ice breakup that are linked tightly to cli-matic forcing were very high, with $r^2 = 0.83$ to 0.88 (Kratz et al. 1998). Other vari-ables, such as crayfish density and zooplankton biomass, had very low coherences, with $r^2 < 0.01$. Other coherences were even negative, such as sedimentation rates of phosphorus, with $r = -0.33$ ($r^2 = 0.11$), and carbon, with $r = -0.29$ ($r^2 = 0.08$).

Physical variables were generally more coherent than chemical variables, which in turn were more coherent than biological variables (Fig. 5.3, top) (Magnuson et al. 1990, Kratz et al. 1998, Baines et al. 2000, George et al. 2000). However, marked differences in coherence exist even within thermal variables driven largely by cli-matic factors (Chapter 7). For example, coherence for water temperature was very high in the near surface epilimnion but very low deep in the hypolimnion (Kratz et al. 1998). Such differences in coherences are robust and were apparent in three of four sets of lakes in the Laurentian Great Lakes region (Benson et al. 2000b).

Some physical variables, in contrast to the ice phenologies, can have very low coherences if they have a significant biological component (Fig. 5.3, bottom) (Magnuson et al. 1990, Kratz et al. 1998). A good example is water clarity, as

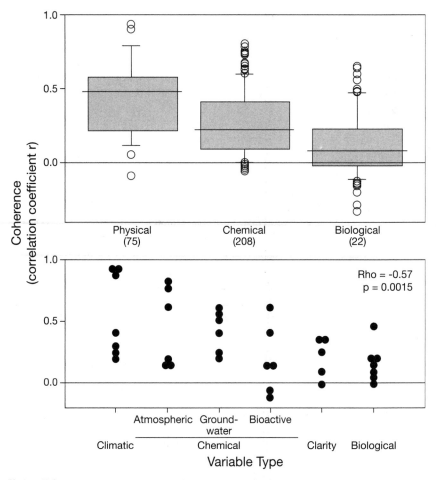

Figure 5.3. Temporal coherence, r, of physical, chemical, and biological variables for northern Wisconsin LTER lakes. Top: Coherence of physical variables, chemical variables, and biological variables. The medians, quartiles, 90th percentiles, and extreme values are shown. Redrawn from Fig. 4 in Kratz et al. (1998) with the kind permission of Kluwer Academic Publishers. Bottom: Coherence between climatic variables; chemical variables influenced by atmospheric, hydrologic, or biologic processes; water clarity variables; and biological variables. Error bars are one standard deviation. Redrawn from Magnuson et al. (1990) with permission of Blackwell Publishing.

measured by Secchi depth, $r^2 = 0.03$ to 0.05, and light extinction, $r^2 < 0.01$. Water clarity is influenced by chlorophyll (phytoplankton) densities and humic fractions of dissolved organic carbon; the coherences for chlorophyll and for dissolved organic carbon among adjacent northern Wisconsin LTER lakes differ by trophic status (Sanderson 1998) and extent of adjacent wetlands that are a source of dissolved organic carbon (Gergel et al. 1999).

Even finer differences in patterns of coherence are apparent among the chemical variables depending on their relative responsiveness to climatic or biological processes (Fig. 5.3, bottom). Chemical variables like calcium and potassium, which are less biologically active and are supplied to lakes primarily by weathering of minerals in aquifers and catchments (Chapter 2), are moderately coherent. In contrast, coherences for biologically active algal nutrients, such as total dissolved phosphorus and dissolved reactive silica, are very low (Kratz et al. 1998). This pattern of coherence strength for chemical variables is robust for different lake districts (Magnuson et al. 1990, Soranno et al. 1999, Baron and Caine 2000, George et al. 2000, Webster et al. 2000).

Patterns of Coherence among Lakes

While lakes within a lake district exhibit very low coherence when a wide variety of lake types and limnological variables is treated in aggregate, subsets of lakes can show high coherence even for biological variables (Rusak et al. 1999, Baines et al. 2000, Rusak et al., unpublished manuscript). For some variables, greater coherence has been observed between lakes that are in similar positions in the landscape or receive similar amounts of groundwater (Baines et al. 2000, Webster et al. 2000, Järvinen et al. 2002, Rusak et al., unpublished manuscript) or that are similar in area or exposure to climatic forcing (Magnuson et al. 1990, Järvinen et al. 2002), in water color or clarity (Baines et al. 2000, Järvinen et al. 2002), or in thermal stratification (Baines et al. 2000). The occurrence of high to very high coherences for some species of zooplankton in subsets of lakes at Dorset, Ontario (Rusak et al. 1999), and for some zooplankton species in lakes of the northern Great Plains (Rusak et al., unpublished manuscript), suggests that regional drivers may be important to some species.

Position in the landscape, as determined by position in the hydrologic flow system, translates to a gradient in important features that could respond to or filter the interyear variation in climatic drivers (Chapter 2, 3). For example, in the Northern Highlands Lake District, lakes high in the landscape, compared with those low in the landscape (Riera et al. 2000), tend to have smaller areas, more circular shorelines, lower concentrations of dissolved solids, lower chlorophyll concentrations, lower pH, fewer fish species, lower densities of crayfish, lower densities of cottages and resorts, and fewer or no connections via surface streams. With all these differences, we expected that lakes at different positions in the landscape might respond differently to regional drivers.

Watershed-related differences in coherence are apparent for chemical variables for the northern LTER lakes in the open-water season (Webster et al. 2000) and for

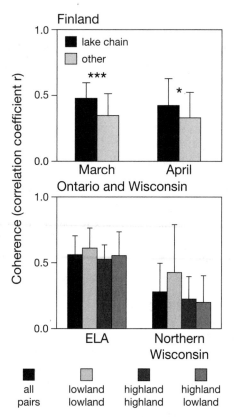

Figure 5.4. Top: The influence of hydrologic connectivity on coherence between lakes. Average temporal coherence of chemical variables in Finland between lakes with stream connections versus between unconnected lakes (*** = p < 0.001, * = p < 0.05). Modified from Järvinen et al. (2002). Bottom: Coherence between lakes at the Experimental Lakes Area (ELA) in western Ontario and northern Wisconsin LTER lakes depending on whether coherence is calculated between two lowland lakes or two highland lakes or between a highland and a lowland lake. Redrawn from Webster et al. (2000) with permission of Blackwell Publishing. Error bars are one standard deviation.

lakes near Lammi, Finland, in late winter (Järvinen et al. 2002). What both of these research groups discovered was that stream-connected or lowland lakes were more coherent than upland lakes or lakes not connected by a stream. The conclusions in Finland were based on analyses of lakes with and without a direct stream connection to the other lake in the pairing (Fig. 5.4, top); in Wisconsin, the conclusions were based on the greater coherence between lowland lakes than between upland lakes or between an upland lake and a lowland lake (Fig. 5.4, bottom right).

Spatial structure for temporal variability within a lake district as developed by Webster et al. (2000) can follow three patterns: spatially unstructured, spatially structured, and spatially uniform (Fig. 5.5). The spatial patterns described earlier

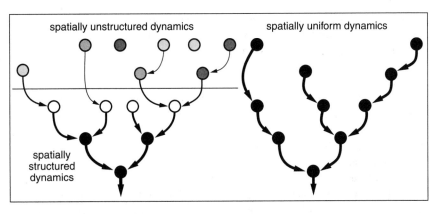

Figure 5.5. Diagram of spatial structure of temporal coherence in lake flow systems; hydrologic connections between lakes can be either via groundwater (thin lines) or surface streams (intermediate and thick lines). Lakes with the same shading have coherent dynamics. Redrawn from Magnuson et al. (2004).

for northern Wisconsin and southern Finnish lakes correspond to a variant of the spatially structured lake district, but with the set of lowland or stream-connected lakes being more spatially uniform and the set of upland lakes being spatially unstructured. An example of a more spatially unstructured lake district comes from analysis of a 15-year record at an Ontario site on the Canadian Shield (Fig. 5.4, bottom left) (Webster et al. 2000). Here coherence was similar between upland lakes, between lowland lakes, and between upland and lowland lakes. Coherence was generally higher for these Ontario lakes than for our northern Wisconsin LTER lakes, but even in these more spatially uniform lake districts most of the temporal variation was still lake-specific, suggesting that external drivers and lake-specific filters of external drivers interact to determine lake chemical dynamics.

These spatial structures occur for other variables, as well, and vary in ways that imply complex determinants of interannual variation (Magnuson et al. 2004, Rusak et al., unpublished manuscript). Uniform spatial structure in lake districts is the rule for a number of physical limnological variables, such as dates of ice-on and ice-off and surface temperature, but not for all physical limnological variables (Chapter 7) (Magnuson et al. 2004). Bottom temperatures, for example, revealed no structure in relation to landscape position. Different species of zooplankton in lakes of the northern Great Plains exhibited several structural patterns (Rusak et al., unpublished manuscript). Coherences of *Leptodiaptomus siciloides*, *Daphnia galeata mendotae*, and total crustacean density were low to moderate and spatially uniform, while *Diacyclops thomasi* exhibited very low coherences and was structured. Rusak et al. (unpublished manuscript) observed that coherences decreased with stream distance between lake pairs and that lake pairs higher in the flow system were less coherent than lake pairs lower in the flow system. Coherences for other zooplankters (*Bosmina*, *D. retrocurva*, and *Leptodora*) were very low and unstructured. Thus, within the zooplankton, populations of some species apparently respond to regional drivers, others are influenced by the gradients related to landscape position, and still others vary in lake specific fashions.

Differences in hydrologic connectivity among lakes and between the lakes and the land can affect the spatial extent of response to climate signals and thus the coherence between adjacent lakes. The northern Wisconsin LTER lakes generally have very low average coherence across all lakes and variables (Fig. 5.4, bottom), while those in the Experimental Lakes Area in Ontario are much more coherent (Webster et al. 2000). The Ontario lakes lie on Canadian Shield bedrock and are connected by surface streams. In this relatively uniform hydrogeologic landscape, responses of lakes to changes in climatic factors are transmitted rapidly from the land and to downstream lakes, regardless of position in the landscape. In contrast, the northern Wisconsin lakes are embedded in a groundwater flow system, with little overland flow of water. The response of these lakes to changes in climatic factors typically is transmitted slowly over many years from the land to the lakes via groundwater flow (Fig. 2.6) that results in a delay, as pictured in Figure 5.1. The length of the delay is greater for longer flow paths and shorter for shorter flow paths. Interlake water flow is also dependent on groundwater, especially for the upland seepage lakes. Another example of the role of the connectivity of the lakes and their watersheds comes from comparison of the LTER lakes in northern and southern Wisconsin. Nutrients are much more coherent, $r^2 = 0.4$, for the Madison Area lakes than for the Northern Highland lakes, $r^2 = 0.05$ (Kratz et al. 1998); in southern Wisconsin, the agricultural and urban watersheds are connected by drain tiles, ditches, and streams that quickly transmit nutrient inputs from the land to the lakes and from upstream to downstream lakes.

The influence of connectivity has implications for the coherence of biological variables, as well. For example, lakes connected by surface waters have fewer barriers to invasion by new species and thus may show more biological coherence compared to lakes without connecting streams (Chapter 4). Temporal coherence of variables influenced by the invading species would be greater for those connected lakes invaded at roughly the same time, while pairings of connected and isolated lakes would be less coherent. The invasion of species is augmented by hydrologic connections, such as streams, and by anthropogenic corridors, such as roads (Chapter 4) (Magnuson et al. 1998).

Differences in the spatial extent of the external driver can be important in the spatial pattern of coherence. For example, does the external driver act over a spatial scale that includes both lakes being compared? Later in this chapter, we describe regional coherence in climatic drivers that was transmitted to coherent lake dynamics in four rather distant lake districts (Benson et al. 2000b). At some time scales, coherence occurs even globally for lake and river ice phenologies (Chapter 7) (Magnuson et al. 2000b, 2004). Interestingly, coherence in ice breakup dates ranged from $r = +1.00$ to $r = -0.88$ for 62 lakes over a large area of central North America (Wynne et al. 1996). This strong positive and strong negative coherence implies that ice breakup is linked tightly to climatic variability, but for this large region, some lake pairs are being exposed to interyear climatic dynamics that are out of phase with each other. More locally, the spatial extent of events and processes can result in local differences in coherence. A severe storm with associated mixing and rainfall may pass over only some of the lakes in a lake district, or, for large lakes, only parts of a lake. Local and regional governance can have spatial extents that include some of the lakes but not others. Policies that can influence land use, nutrients, fish stocking, and restoration programs may result in higher coherence for lakes under the same jurisdiction or management, while comparison of lakes in different jurisdictions and under different management may have lower coherence.

External versus Internal Drivers

Strong external drivers that act across a lake district should produce some coherent behaviors among lake pairs, while strong internally generated drivers should promote incoherent behaviors. Evidence for these opposing causations are apparent in examples with climate drivers (Chapter 7) and with population biology of fishes (Chapter 10).

Northern Wisconsin LTER lakes that were more similar in exposure to climatic factors, indexed as the ratio of a lake's surface area to mean depth, were more coherent than were lakes that differed more in exposure (Fig. 5.6) (Magnuson et al. 1990, Kratz et al. 1998). Because the ratio of area to mean depth is greater for lowland than for highland lakes, exposure to climate, like many other features of these lakes, is related to the geomorphic legacy of position in the landscape (Chapter 3). Consequently, the upland lakes in the Northern Highlands were small in area rela-

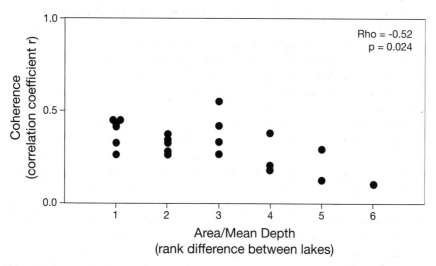

Figure 5.6. Relation of overall temporal coherence, r, between lakes and their rank difference in climate exposure as indexed by the ratio of lake area to mean depth for the northern Wisconsin LTER lakes. Redrawn from Magnuson et al. (1990) with permission of Blackwell Publishing.

tive to their depth and often colored, while the lowland lakes were large in area relative to their depth and often clear. So, size comparisons such as those for climate exposure are aliased by lake color; upland colored lakes are less coherent than the lowland clear-water lakes (Kratz et al. 1998). Analyses of winter chemical variables in Finland complicate rather than resolve the individual influence of lake variables related to landscape position. In contrast to the summer analyses for northern Wisconsin, Finland winter coherence was not related strongly to lake color or to similarity in climate exposure. Instead, lake pairs with the greatest difference in lake area and catchment area (Lake Pääjärvi was one of the pair) had the lowest coherence (Järvinen et al. 2002).

One might expect that climate-driven coherence would be greater during open water than during ice-covered periods, because the waters are isolated somewhat from direct climatic forcing by ice and snow cover. Yet, lakes do exhibit coherent behavior in winter (Järvinen et al. 2002). Coherence for 10 chemical properties and water temperature averaged $r = 0.37$ beneath the ice during late winter for 28 adjacent Finnish lakes over 11 years. These brown-water lakes were almost all small, <0.2 km^2, and most, but not all, were connected by streams. Interestingly, water temperature and chemistry in March were correlated to autumn weather and not to winter weather, while lake conditions in April were influenced by late winter weather. The very low coherence of lake conditions between lakes in winter was similar to coherences in our northern Wisconsin LTER lakes in the ice-free seasons. However, some seasonal differences were apparent. Coherence of water temperatures in March was very low across the Finnish lakes, but at our site during spring, summer, and autumn, coherence ranged from low to very high (Kratz et al. 1998). By April, coherence for water temperature had increased to high in the Finnish lakes, even though the lakes remained ice covered. By April, the snow had disappeared from the top of the ice, and solar radiation and snowmelt were entering the waters.

However, climate forcing was important in Finland; winter chemistry and water temperature were related to climate variables and to the strength of the North Atlantic Oscillation (NAO). Positive NAO values relate to mild, wet winters, and, not surprisingly, warmer water temperatures, more colored waters, lower calcium concentrations, and lower alkalinity (Chapter 7).

The dominant role of climate in determining coherence in physical variables is most apparent in the ice phenologies of lakes (Chapter 7) and in water temperature dynamics within and between four lake districts located throughout the upper Great Lakes region (Benson et al. 2000b). The four lake districts are the LTER lakes in the Northern Highlands and in the Madison area of Wisconsin, the Experimental Lakes Area in western Ontario, and the Dorset lakes in central Ontario. The maximum coherence one could expect for limnological variables, influenced strongly by climatic drivers, would be the coherence in the climate drivers themselves. The coherence in air temperatures ranged from $r = 0.78$ to 0.90 and in solar radiation from $r = 0.45$ to 0.67. These relatively high coherences occurred even though the locations of the four lake districts ranged from 43° to 49°N latitude and 78° to 93°W longitude. Coherence between lakes from different lake districts was highest for epilimnetic temperatures, $r = 0.73$ to 0.81, intermediate for thermocline depth, $r =$

0.30 to 0.63, and lowest for hypolimnetic temperatures, r = -0.02 to 0.46. The coherence for epilimnetic temperatures was only slightly less than that for air temperatures, suggesting that epilimnetic temperatures strongly reflect air temperatures. As expected, the lowest coherence occurred for hypolimnetic temperatures as the influence of climatic drivers on these temperatures is filtered strongly and differentially among lakes by the size and depth of individual lakes (Magnuson et al. 1990).

We should point out that the same climatic event can produce different responses in different lakes and contribute to incoherent behaviors between lakes. For example, a windy year can result in greater mixing for a large lake and deepen the thermocline but at the same time have little or no effect on mixing depth in a tiny meromictic, that is, permanently stratified lake. Many such possibilities can be imagined. Consider the responses of two adjacent lakes to an unusually hot and cloudless summer. Suppose that one is too shallow to thermally stratify and the other does stratify, or that one lake usually has an anoxic hypolimnion and the other an oxygen-rich hypolimnion, or that one is clear enough for photosynthesis in the metalimnion and hypolimnion and the other is not. Each of the lakes will have a different response in dissolved oxygen levels in deep water, in recycling of nutrients from the sediments, and in the habitat suitability for different thermal guilds of fishes.

Biological and biologically influenced variables generally have very low coherence, but not always (Magnuson et al. 1990, Kratz et al. 1998, Rusak et al. 1999, Baines et al. 2000, George et al. 2000, Straile 2002, Rusak et al., unpublished manuscript). Several papers on coherence and biological limnology reveal that coherences range from low to very high for subsets of lakes and species of zooplankton at Dorset (Rusak et al. 1999) and from very low to moderate in zooplankton in northern Great Plains lakes (Rusak et al., unpublished manuscript). Coherence is high for chlorophyll between certain pairs of lakes in northern Wisconsin (Baines et al. 2000). Low coherence might be expected because species compositions differ among the lakes, and internal drivers can impose unique dynamics in individual lakes, not only to the biota but also to some physical and chemical components (Chapter 10). These dynamics may result because different species or trophic levels respond differently to the same regional signal or because the individualistic biota in a lake generate dynamics that are independent of regional signals, leading to low coherence. Strong relationships exist between the timing of food web interactions that cause the spring clear-water period in European lakes and the North Atlantic Oscillations; this external driver produces coherence in the timing of the clear-water period (Straile 2002, Straile et al. 2003). Thus, large-scale climatic drivers do influence biological coherence in a number of examples.

Predicting Lake Dynamics

This chapter has presented one approach to understanding interannual dynamics between lakes that may help predict interannual dynamics of a given lake or groups of lakes. For some variables and for some sets of lakes, the higher r^2 values and percentages of common variability they represent suggest that predictions can be

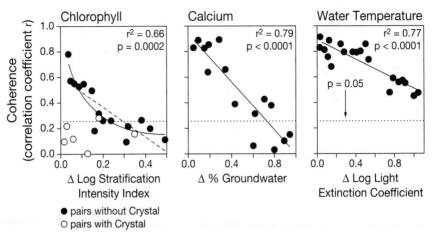

Figure 5.7. Temporal coherence, r, of chlorophyll, calcium, and epilimnetic temperature related to differences between lakes in variables related to vertical thermal stratification, groundwater input, and water clarity for the northern Wisconsin LTER lakes. Correlation coefficients above the dotted line are significant at $p = 0.05$. Redrawn from Baines et al. (2000) with permission of the Ecological Society of America.

made from simple linear regression. Clearly, this approach has little utility for lakes and lake variables with lower coherence, because these have little shared variability. Occasionally, the inclusion of a second lake as an explanatory variable in the regression model has increased the ability to predict the year-to-year variation in a third lake (Baines et al. 2000). Regardless, for most variables and lakes these statistical relationships have low predictive power because our present level of understanding of what drives differences and similarities in dynamics of adjacent lakes is inadequate and because individualistic lake behavior remains a strong component in interannual dynamics, especially for most chemical and biological variables. Prediction of the interannual dynamics of an unmonitored lake is likely to be possible at this time for some physical variables for most lakes, for some chemical variables for lakes that are connected by streams, and for a landscape where precipitation flows rapidly from the land to the lakes and among lakes. For biological variables, examples of coherent behavior are appearing in the literature and hold promise for prediction for certain subsets of lakes and variables.

Because the coherence of many physical, some chemical, and even a few biological variables is related to the similarity in the status of the lakes (Fig. 5.7) (Baines et al. 2000), prediction is more likely for subsets of similar lakes than for entire lake districts. For summer epilimnetic chlorophyll, the most coherent lake pairs were similar in their intensity of thermal stratification. For epilimnetic calcium concentrations during the ice-free season, coherence was higher for lakes that received similar percentages of groundwater in their total water input. For summer epilimnetic temperatures, lake pairs with more similar light extinction coefficients were more coherent than pairs with large differences in light extinction.

Extending the Concept

Limnologists (Chapter 1) have traditions of intensively studying one lake for a long time (Edmondson 1991, Goldman 2000) or surveying multiple lakes (Rawson 1939, and many other limnologists) to explore differences in trophic status based on climate, lake morphometry, edaphic conditions, and anthropogenic influences. Others have used comparative limnology to explain species richness and composition (e.g., Tonn et al. 1990) and fish production (e.g., Ryder 1982). In these cases, the entity is the lake and the objective is to estimate, understand, and predict a lake's status. Gaining a predictive understanding of differences among lakes in status is a worthy mission.

We posit that analyses of coherence will help us obtain a predictive understanding of lake dynamics in a context of regional or landscape ecology. Coherence makes more explicit the patterns of dynamics across the landscape and suggests reasons for those dynamics. Knowledge of coherence is useful to management and stewardship of lakes. Without an awareness of the spatial extent of both the driver and the coherent response of lakes, we will continue to confuse the local with the regional nature of lake stewardship and management problems. Analyses of coherence can help reduce the mismatch between the spatial scale of influence of responsible governmental units and the spatial scale of the drivers and of the lake's responses.

While we have treated lakes here, the approach has utility for other kinds of ecosystems, both wet and dry. Others have used the general approach of coherent dynamics in analyses of other ecological systems, for examples, timing of flowering in Norway (Post 2003), global fish populations (Kawasaki 1993), forest productivity in the Pacific Northwest (Graumlich et al. 1989), and lynx populations in Canada (Stenseth et al. 1999).

Summary

Shared dynamics between lakes occur at interannual, interdecadal, and century time scales. Climate dynamics serve as one important signal resulting in shared regional dynamics, but other large-scale regional drivers related to human settlement and land-use change exist. The coherence or shared variance, r^2, in the dynamics of adjacent lakes is low when a diverse set of limnological variables and a diverse set of lakes are examined. However, coherence is greater for lakes that are more similar in climate exposure, as indicated by their ratios of area to mean depth. Climatic signals can be attenuated or delayed, or they can reverberate through an aquatic ecosystem; such alterations in a signal are progressively greater for physical, chemical, and biological variables. Consistently, coherence is higher for climate-driven physical variables, such as ice-on dates and surface water temperatures; intermediate for chemical variables, such as pH and calcium concentrations; and lower for biological variables, such as invertebrate and fish abundances. However, coherence can be high, even for some biological variables among some lake pairs. Coherence values increase when the length of record is increased, because a greater range of the external drivers occurs—a drought, for example. Patterns of interlake coherence, from high to low, in the landscape differ among lake districts and among limnological variables. Some lake districts with strong surface water connections tend to be spatially uniform in the coherence of chemical variables. Other lake districts that are dominated by groundwater and thus have weak surface-water connections tend to be spatially structured with higher coherences lower in the landscape and low coherences higher in the landscape. Some limnological variables, such as ice-on date and surface water temperature, exhibit uniform dynamics across the landscape while, many biological variables are spatially unstructured. Coherent patterns of dynamics may lead to an ability to predict the dynamics for some variables for some lakes. In an analysis of coherence, the entity of interest is a set of lakes, rather than a single lake ecosystem; the goal is to understand and predict lake dynamics, rather than lake status or average condition. We posit that analysis of coherent behavior is a useful early step in helping obtain a predictive understanding of lake dynamics across the landscape.

6

Generalization from Intersite Research

Joan L. Riera
Timothy K. Kratz
John J. Magnuson

The scientific mind does not so much provide the right answers as ask the right questions.
—Claude Lévi-Strauss (1964), in Weightman and Weightman (1969, p.7)

It has been said that the primary function of schools is to impart enough facts to make children stop asking questions. Some, with whom the schools do not succeed, become scientists.
—Knut Schmidt-Nielsen (1998, p. 3)

A hallmark of the LTER network is that it provides the data, the expertise, and the infrastructure to encourage scientists studying diverse ecosystems to seek ecological principles. In attempting to understand similarities and differences among disparate ecosystems (forests, grasslands, deserts, lakes, estuaries, and so on), we are led to ask new and unorthodox questions. Given the scope of LTER data sets, these questions necessarily place emphasis on temporal and spatial variability, address issues of scale, and are informed by nonequilibrium concepts. If we ask the right questions, the reward is ecological synthesis through greater generality, greater integration, and new levels of understanding. The challenges are to transgress disciplinary boundaries, migrate concepts and theories, and compare patterns and trends associated with disparate ecosystems and landscapes (Allen and Hoekstra 1992, Picket et al. 1994).

Comparative ecology has many virtues, from the generation of new hypotheses, to the detection of emergent properties manifested at large scales, to theory testing and prediction (Cole et al. 1991). The approach is facilitated by ensuring accessibility of comparable data, enhancing opportunities for funding research and education in support of studies at regional and long-term scales, and providing incentives to evolve new knowledge for policies to manage ecological resources. Comparative ecology demands creativity in using our analytical toolbox and, perhaps most important, ingenuity in devising appropriate questions.

Recognizing these challenges, the Long-Term Ecological Research (LTER) program, funded by the National Science Foundation, included from the onset the difficult goal (Chapter 14) of conducting synthetic and theoretical research (Callahan 1984). To aid these efforts, all sites are required to articulate and address broadly defined common goals that provide understanding of common processes on the basis of data sets that facilitate intersite comparisons.

Data availability and accessibility are essential for intersite synthetic efforts (Chapter 13), but data alone are not sufficient. Participating scientists must share common visions and interests and must be motivated to step into the realms of other disciplines, unfamiliar ecosystems, multiple scales of analysis, and different levels of organization. At site and intersite levels, the strengths of the LTER program are that scientists are brought together who share common interests and that synthetic efforts are facilitated through workshops, funding opportunities, and educational incentives.

Here, we describe our efforts at synthesis across ecosystems and landscapes. These endeavors include both theoretical exercises and data-rich intersite comparisons. These examples of synthesis across systems provide a more generalized ecological understanding of the processes than could be achieved by studying a single ecological system. The dynamics of lakes in the landscape, the focus of the North Temperate Lakes LTER program, provides the template for expanding our ecological understanding.

Geography for Limnologists

Comparisons between ecological systems often start as theoretical exercises in which inference by analogy plays a central role. If two ecosystems are analogous in structure, then they may be similar in function. If two ecosystems are functionally analogous, then a theory developed for one may be valid for the other. This is the thinking behind the application of the theory of island biogeography (MacArthur and Wilson 1963, 1967) to lake community ecology (Tonn and Magnuson 1982, Rahel 1984, Tonn et al. 1990, Magnuson et al. 1998) (Chapter 4). Accordingly, lake communities emerge from the net result of invasion and extinction processes because the formation of the lakes started with the recession of the Wisconsin glaciation about 10,000 years ago.

Taking this analogy one step further, if a lake is an island analogue, then a lake district is an archipelago analogue. Comparisons between islands and lakes and between oceans and continents where connectivity and size arose as key concepts, suggested that lakes are more like islands and oceans are more like continents than lakes are like oceans and islands are like continents (Magnuson 1976, 1988, Magnuson et al. 1991). Analogies are useful, but, when tensed, they eventually break, and the breaking point can be richly informative. How similar is the structure of lake districts to island archipelagos? Lake districts may differ from island archipelagos in their degree of connectivity. Lakes are rarely isolated by land from other aquatic systems because streams or wetlands, permanent or temporary, often connect them. Yet, compared to oceans, lake districts are connected poorly, prompting Magnuson (1988) to propose that, as far as population dispersal processes go, the difference between oceans and lakes resides not so much in size per se as in the closed, isolated nature of lakes versus the openness and connectedness of an ocean system.

The benefits of studying lakes from a landscape perspective (Johnson and Gage 1996) have been considered in Chapter 3. Exercises of analogy and contrast, the application of theories from other disciplines, and a constant quest for new, innovative perspectives and approaches to the analysis of waterscapes pervades the research at our site.

A Fundamental Question

The most engaging comparative analyses in ecology stem from questions so simple and fundamental that they appeal to diverse ecological communities. One of these simple questions is the characterization of ecological variability (Duarte 1991). What constitutes the so-called normal variability? Are some ecosystems more variable than others? Are some measures of ecosystem structure or function more variable than others? How do temporal and spatial variability between ecosystems compare? Do variance components vary with scale? While these questions are of general interest, and indeed are addressed by many ecologists who study particular ecosystems or ecosystem components, few researchers have attempted to describe the temporal and spatial variability exhibited by ecosystems in a comprehensive manner.

To investigate these questions, 12 LTER sites representing ecosystems as diverse as deserts, northern deciduous forests, north temperate lakes, and estuaries chose to participate in an intersite analysis of spatial and temporal variability of ecological attributes (Table 6.1). Data from each site were compiled, and a workshop was held at Trout Lake in 1988 at which participants were provided the tools to manipulate and analyze data to compare variability across these 12 LTER sites (Fig. 6.1).

Variability and Landscape Position

Kratz et al. (1991b) tested the hypothesis that temporal variability in ecosystems changed systematically with landscape position (see also Chapter 3). Data from four contrasting ecosystems: north-temperate lakes (North Temperate Lakes LTER site), a northern deciduous forest (Hubbard Brook Experimental Forest LTER site), an estuary (North Inlet Estuary former LTER site), and a hot desert (Jornada Desert LTER site) were used in this analysis. These sites were selected because data were available for at least three locations for at least 5 years and because the data collected from these locations at each site could be arranged along an elevation gradient.

A standardized and dimensionless metric was required to compare across ecosystems because each site did not measure the same parameters. For example, pri-

Table 6.1. Key characteristics of the LTER sites included in the study of Variability in North American Ecosystems (VARNAE).

LTER Site	Biome or Ecosystems	Ecosystem Parameters (number)	Locations within Site (number)	Years (number)
H. J. Andrews Experimental Forest	temperate coniferous forest	33	3–9	4–17
Bonanza Creek Experimental Forest	taiga	4	7–14	12–14
Cedar Creek Natural History Area	eastern deciduous forest and tallgrass prairie	17	4–18	5–6
Central Plains Experimental Range	short-grass steppe	13	4–9	4–6
Coweeta Hydrologic Laboratory	eastern deciduous forest	18	3	6–17
Hubbard Brook Experimental Forest	eastern deciduous forest	49	2–6	5–6
Illinois River	riverine and riparian ecosystems	19	5	4–5
Jornada Basin	hot desert	56	6–7	3–5
Konza Prairie	tallgrass prairie	52	6–9	5–7
Niwot Ridge	alpine tundra	28	6	5
North Inlet Estuary	atlantic coastal wetland	57	3	6
North Temperate Lakes	northern temperate lakes in forested watersheds	102	5–7	4–6

mary production is measured very differently in a forest ecosystem and in a lake ecosystem; in a similar vein, decomposition rates of logs are measured in forest ecosystems but not in open ocean ecosystems. This metric was developed by calculating the coefficient of among-year variation for each parameter at each location at every site. The coefficient of variation is a useful metric that has no units and standardizes the variation of a parameter by its mean value. Using this approach, we were able to compare variability in measures as diverse as, for example, calcium concentration and biomass or air temperature and thermocline depth. Once the coefficients of variation were computed, they were ranked by location to test the hypothesis that they changed systematically with landscape position (Kratz et al. 1991b).

This analysis revealed several general ecological patterns that applied to each site. At each site, the interannual variability of ecological parameters differed by location and, at least for a subset of the parameters, by the position of the locations in the landscape, demonstrating that a mechanism associated with landscape position may be affecting the predictability of temporal variability (Fig. 6.2). Although the mechanisms relating spatial positioning to interannual variability were specific to each site, one underlying factor, water movement across the landscape, determined the observed pattern. This LTER cross-site study demonstrated a common context for understanding variability in diverse ecosystems.

Figure 6.1. Participants at the workshop on Variability in North American Ecosystems (VARNAE) held in April 1988 at the Center for Limnology's Trout Lake Station. From left to right by row from front to back: First row, Randy A. Dahlgren (Hubbard Brook), Thomas M. Frost (North Temperate Lakes), Dennis Heisey (North Temperate Lakes); second row, Caroline S. Bledsoe (National Science Foundation), Stephen R. Carpenter (North Temperate Lakes), John J. Magnuson (North Temperate Lakes), John Yarie (Bonanza Creek); third row, Barbara J. Benson (North Temperate Lakes), James C. Halfpenny (Niwot Ridge), Elizabeth R. Blood (North Inlet), Arthur McKee (H. J. Andrews Forest); fourth row, Donald W. Kaufman (Konza Prairie), Peter B. Bayley (Illinois River), Cory W. Berish (Coweeta), Gary L. Cunningham (Jornada); fifth row, Jon Anderson (Short Grass Steppe), Timothy K. Kratz (North Temperate Lakes), Richard S. Inouye (Cedar Creek). (C. Bowser)

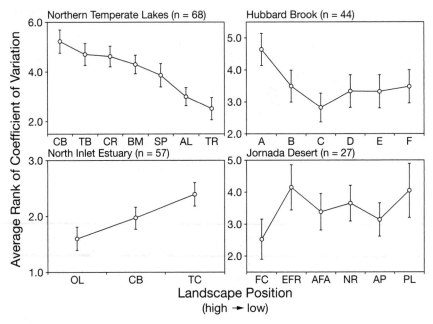

Figure 6.2. Average coefficients of variation ranked by landscape position for four LTER sites. Top left: CB = Crystal Bog, TB = Trout Bog, CR = Crystal Lake, BM = Big Muskellunge Lake, SP = Sparkling Lake, AL = Allequash Lake, TR = Trout Lake. Top right: A to F are locations 1 to 6 in Lawrence et al. (1986). Bottom left: OL = Oyster Landing, CB = Clambank, TC = Town Creek. Bottom right: FC = alluvial fan collar, EFR = erosional fan remnant, AFA = alluvial fan apron, NR = nonburied fan remnant, AP = alluvial plain, PL = playa. Higher ranks correspond to higher coefficients of variation. Redrawn from Kratz et al. (1991b) with the kind permission of The University of Chicago Press.

Expanding the Scope

In a collaborative effort among LTER scientists, Magnuson et al. (1991) and Kratz et al. (1995) described general patterns of temporal and spatial variability exhibited by ecological parameters across 12 LTER sites representing a diverse array of ecosystem types. By comparing the temporal and spatial variability of a large number of ecological parameters, they examined whether different types of parameters differed systematically in variability and how variability was partitioned into year-specific variability, location-specific variability, and variability that reflects the interaction of spatial and temporal factors.

Year-specific variability reflects the variation in an ecological parameter that occurs independent of location (within each LTER site). This variability represents another measure of temporal coherence, the tendency for locations within a landscape to behave similarly in different years (Chapter 5). Analogously, location-specific variability is a measure of spatial coherence, the tendency for a parameter to differ consistently among locations independent of the variation among years.

The third partition, the interaction between temporal and spatial variability, reflects the behavior of those parameters that display temporal variation differently in different locations.

The scope of the study was broad and encompassed disparate ecosystems, but the conceptual framework and the methodology were not new. Kratz et al. (1987a) presented them in an earlier study that partitioned zooplankton variability in lakes into their site-specific and year-specific components. The paper used historic zooplankton data from E. A. Birge and C. Juday and their colleagues gathered between 1925 and 1942 from five lakes in northern Wisconsin, including three LTER lakes. Zooplankton community data are well suited to explore questions about temporal and spatial variability because zooplankton have short life spans and zooplankton communities typically undergo one successional sequence each year. Several parameters were analyzed, with varying expectations of the extent to which variability would be year-specific or site-specific. Those parameters that were influenced by weather variability, either directly, for example, through the effect of water temperature or indirectly, for example, through the effect of precipitation on nutrient loading, tended to be year-specific. The abundance of most zooplankton species fit this pattern. For parameters that were farther removed from the influence of weather variability, site-specific factors emerged as system drivers. For example, many factors associated with the depth that zooplankton maxima occur are invariant over the years but vary consistently among lakes; accordingly, depth of zooplankton maxima showed high site-specific variability.

Translating this conceptual framework into the expanded scope of an inter-ecosystem study presented new challenges. Because none of the 448 parameters included in the intersite data set was measured at all sites, parameters were grouped so that their variability could be compared. Kratz et al. (1995) classified the parameter set into one of four groups: climatic, edaphic, plant, and animal. The rationale was that climatic parameters, such as water temperature or soil moisture, are affected directly by external climate forcing, whereas, in the biotic domain, animal parameters, such as fish biomass or diversity of small mammals, should be influenced by climate variation and biological components of variability. Thus, variability was hypothesized to be lower for climatic parameters and higher for animal parameters.

To analyze these interactions, a two-way analysis of variance with year and location as fixed effects was used to test the significance of variance components for each parameter. Because replicates were not available for each combination of year and location, the portion of variance attributable to the interaction of space and time could not be estimated independently from the variance owing to random error.

Several important lessons were learned from this analytical process. The hypothesis that variability would increase from climatic (physical) to animal (biological) parameters was supported (Fig. 6.3, top). Although climatic and edaphic parameters did not differ significantly in terms of total variance, biological parameters were more variable; among them, animal parameters were the most variable. One interpretation of these patterns is that measurement error increases from nonbiological to biological parameters, accounting for the higher observed variability in the biological parameters. But other explanations are more intriguing. For example, perhaps climate signals are amplified as they travel through ecosystem components, suggesting that variability in precipitation begets greater variability in nutrient loading, which, in turn, results in greater variability in plant and animal production.

The proportion of variance explained by each type of parameter exhibited interesting patterns (Fig. 6.3, bottom). The proportion of variance explained by location was greater than the proportion explained by year, with the exception of climatic parameters, where the year factor explained most of the variance. The variance component "other," which contains the interaction factor, was increasingly larger from climatic to edaphic to plant to animal parameters. Plant and animal parameters differed in the proportion of variance explained by location versus year and interaction (other). Most of the variability in plant parameters was explained by location, whereas for animal parameters, interaction (other) was more important. As in the analysis of zooplankton data (Kratz et al. 1987a), we may hypothesize that the mobility of organisms and their ability to rapidly respond to a changing climate explains this pattern.

The observation that, in almost all cases, spatial variability within LTER sites was larger than among-year variability suggests that an understanding of the dynamics of a single location within a landscape is insufficient to allow us to describe the full range of ecosystem dynamics within a landscape (Chapter 3). To our knowledge, the strategy of studying the long-term dynamics at multiple locations has been adopted by perhaps all LTER sites.

Despite these general trends across ecosystem types, one would expect different ecosystems to behave differently with regard to the partition of temporal and spatial variance. Kratz et al. (1995) examined this hypothesis by comparing two strikingly dissimilar ecosystems: deserts and lakes. Deserts were more variable among years than lakes, possibly because deserts are highly sensitive to variability in precipitation, whereas lakes are buffered physically from thermal change owing to the heat capacity of water. Conversely, lakes exhibited more spatial variability than deserts, perhaps because lakes in a lake district are more isolated than are locations in a desert landscape. Of course, these results should be interpreted with some caution because they are dependent on the choice of parameters and locations at the desert and lake sites.

Heterogeneity among Disparate Landscapes

In the analyses of temporal and spatial variability at LTER sites in the study described earlier (Kratz et al. 1991b, Magnuson et al. 1991, Kratz et al. 1995), not one of the 448 ecological parameters included in the intersite data set had been measured in common across sites, because many of the measurements that ecologists make are ecosystem-specific. Additionally, each measurement is associated with a specific temporal and spatial scale that differs among parameters and could integrate variability in noncomparable ways. Vande Castle et al. (1995) and Riera et al. (1998) addressed this problem by using satellite remote sensing to measure the same parameter at the same spatial scale regardless of the system. This approach

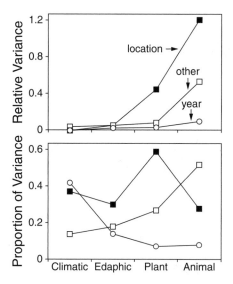

Figure 6.3. Relative variance (top) and proportion (bottom) of the variance explained by year (among years variation), location (among locations at an LTER site), and other (interaction term and measurement error) for climatic, edaphic, plant, and animal parameters measured at 12 LTER sites. Data compiled from at least five years and at least three locations per site for each LTER site in Table 6.1. Redrawn from Magnuson et al. (1991) with permission of John Wiley & Sons Limited, Sussex, England.

constrained the type of parameters available for analysis but provided new insight into cross-site spatial heterogeneity.

In an analysis of the determinants of spatial heterogeneity of landscapes, Vande Castle et al. (1995) and Riera et al. (1998) evaluated landscape heterogeneity using the standard deviation of a vegetation index, the Normalized Difference Vegetation Index (NDVI), derived from Landsat Thematic Mapper satellite images (Vande Castle et al. 1995). When open water was excluded from the analyses, landscapes with low average vegetation index, such as deserts, and landscapes with high average vegetation index, such as deciduous forests, had similarly low spatial heterogeneity relative to other landscape types (Fig. 6.4). Not surprisingly, the most highly heterogeneous landscapes had values for their average vegetation index that were midway through the observed range; no landscape with a mean vegetation index in the midrange had low spatial heterogeneity.

When aquatic ecosystems were added to the dataset for analysis, the heterogeneity of the landscapes increased significantly for images that included substantial surface waters. This result is not surprising, because surface waters absorb light and oligotrophic lakes appear as black bodies in a Landsat image owing to the terrestrial bias in the radiance range of the Landsat sensor system. The increase in heterogeneity when including water makes the point that aquatic ecosystems contribute significantly to overall heterogeneity and that waterscapes should be considered as integral components of landscapes in such analyses.

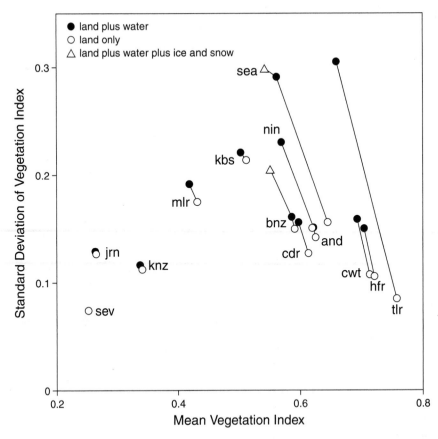

Figure 6.4. Relationship between the standard deviation and the mean of the normalized difference vegetation index (NDVI) for full scenes from Landsat Thematic Mapper. Solid lines connect values for land, land plus water, and land plus water plus ice and snow, for each of the landscapes. The landscapes represented encompass a wide range of biomes, including boreal forest (bnz = Bonanza Creek Experimental Forest); temperate rainforest (sea = Seattle Area); temperate coniferous (and = H. J. Andrews Experimental Forest) and deciduous forests (cdr = Cedar Creek Natural History Area, cwt = Coweeta Hydrologic Laboratory, hfr = Harvard Forest); grassland (knz = Konza Prairie); and desert (jrn = Jornada Experimental Range, sev = Sevilleta National Wildlife Refuge). Two of the scenes were dominated by agricultural landscapes (kbs = Kellogg Biological Station and mlr = Madison Lakes Region). Water bodies were common in northern Wisconsin (tlr = Trout Lake Region), coastal South Carolina (nin = North Inlet Estuary), and the Seattle Area. Glaciers were significant at the Alaska site (bnz). Redrawn from Riera et al. (1998) with permission of Springer-Verlag Publishers.

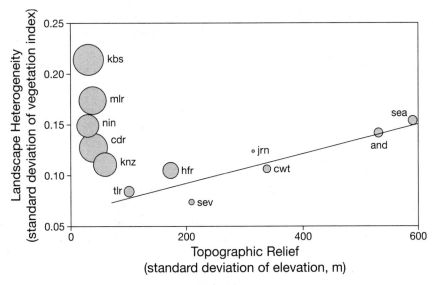

Figure 6.5. Landscape heterogeneity (standard deviation of the normalized difference vegetation index (NDVI) is related to topographic relief (standard deviation of elevation) and to agricultural intensity (percentage of agricultural land cover). Sites are coded as in Fig. 6.4. Circle area is proportional to the percentage of agricultural lands in each of the landscapes; kbs is almost 100% agriculture, and jrn is almost 0% agriculture. Agricultural landscapes present high heterogeneity and deviate from the linear relationship between heterogeneity and topographic relief; the regression line is for sites with less than 40% agricultural land cover (n = 7, r^2 = 0.8, p<0.005). Data are for 12 full Landsat scenes at their original resolution of about 30-m pixel size. Redrawn from Riera et al. (1998) with permission of Springer-Verlag Publishers.

Describing spatial variability is an essential first step toward understanding the role of spatial heterogeneity in landscape function. A useful next step is to investigate the observed differences in spatial heterogeneity. Riera et al. (1998) hypothesized that, at the scale of a full Landsat image (185 km swath with a resolution of 900 m^2; Lillesand and Kiefer 2003), the spatial variability in vegetation cover and biomass is related to topographic relief and to land cover and use, especially by agriculture. These two factors explained a large portion of the variability among the landscapes (Fig. 6.5).

Topographic relief determines differences in vegetation type and plant vigor through a combination of the effects of elevation, slope, and orientation on soil type, humidity, temperature, sun exposure, and precipitation. Thus, vegetation indices are spatially more variable in landscapes with higher topographic relief. Land use interacts strongly with topography to influence heterogeneity. Thus, agricultural lands are major contributors to spatial heterogeneity, because different crops or different phenological states of fields are associated with widely varying values of the vegetation index. Because agricultural land use is predominant in flat lands, low-relief landscapes will be likely to have high heterogeneity. Medium relief landscapes, dominated by natural or managed vegetation cover, had the lowest spatial heterogeneity.

Investigating Spatial Scale

We recognize that the scale of analysis influences the patterns observed. Every study described in this chapter explicitly takes scale into account. The components of scale include the grain (finest spatial and temporal resolution of the measurements), the extent (the spatial area considered and the temporal span of the study), and the level of aggregation of the parameters measured or estimated (Allen and Star 1982, Frost et al. 1988, Frost et al. 1992). All these components of scale depend on the object and methodology of analysis. Therefore, one of the fundamental principles in ecological research is that the scale of the analysis should always be documented and reported, and the sensitivity of the results to variations in scale should be evaluated.

In their intersite comparison of variability in North American ecosystems, Kratz et al. (1995) noted that variability in ecosystem parameters was sensitive to the temporal and spatial scales associated with their data sets and to the level of aggregation of the parameters of interest. Accordingly, they evaluated the effect of sampling unit size at the different sites and the temporal extent of the data set on the variability of the measurements and assessed to what extent the observed variability was related to the degree of aggregation of the parameters measured. The number of years used in the analysis, given the minimum of five years, was adequate to estimate year-to-year variability, and the relationship between the spatial resolution of the analysis and variability was weak at best. In contrast, total variability was related strongly to the level of aggregation for biological parameters, that is, variability relative to the mean was consistently higher for variables measured at the species level than at guild level and was the lowest for variables referring to major groups, for example, biomass and diversity. Compensatory species interactions such as predation and competition may provide an explanation; for example, while species of zooplankton in a lake may wax and wane, total zooplankton biomass may remain relatively unaltered.

The choice of sensor imposed the scale of analyses for the study of Riera et al. (1998). In this case, scale was determined by the spatial extent and the resolution or pixel size of the images and by the spectral resolution of the sensor given as the number and width of spectral bands. Spatial variability often is manifested at nested scales, for example, variability associated with forest stands at a small scale versus variability associated with forest communities at a larger scale. Thus, the spatial resolution of the analysis would be expected to influence the measure of heterogeneity. The standard deviation of the vegetation index decreased with increasing pixel sizes (Fig. 6.6, middle) in a multiscale analysis that aggregated the images to increasingly larger pixel sizes and recalculated the measure of heterogeneity at each new scale. The rate of decrease differed among landscapes. Just as landscapes could be described by their spatial heterogeneity, they could be described by the way their spatial variability responded to changes in grain. The rate of decrease was related

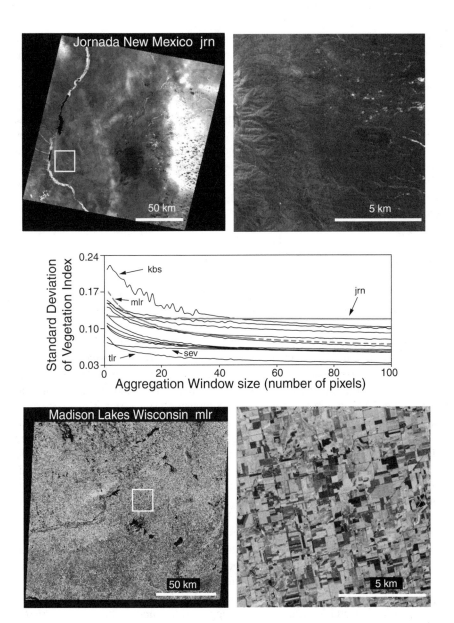

Figure 6.6. Patterns of change (middle panel) in landscape heterogeneity (standard deviation of the normalized difference vegetation index or NDVI) with image grain size (from 1 pixel, or 30 m x 30 m, to aggregation windows of 100 pixels by side, or 3 km x 3 km) for full Landsat images that include 13 LTER sites. Two extreme cases are illustrated with Landsat images: jrn (top) includes the Jornada LTER site, a desert with low small-scale heterogeneity and high large-scale heterogeneity; mlr (bottom) corresponds to the Madison Lakes Region, with high small-scale heterogeneity and low large-scale heterogeneity. Redrawn from Riera et al. (1998) with permission of Springer-Verlag Publishers.

to the scale at which pattern was manifested. On one end were desert landscapes, jrn, which, relative to other landscapes, are very heterogeneous at large scales influenced by major landforms but rather average at small scales (Fig. 6.6, top). On the other end were agricultural, low relief landscapes, mlr, where spatial heterogeneity was highest relative to other sites at the small scales determined by field size and average at large scales (Fig. 6.6, bottom). Another agricultural landscape, kbs, and another dry area with high relief, sev, are identified for comparison. Between these end members were landscapes with more complex mosaics of land cover types where heterogeneity was manifested at nested spatial scales. Owing to these differences in the source of spatial heterogeneity, from checkered agricultural fields to mountain-and-valley land forms, the ordering of sites from high to low spatial heterogeneity changed with grain size (Fig. 6.6, middle).

Summary

Comparative analyses of dissimilar ecosystems and landscapes are valuable both to generalize research and to suggest new questions, as the intersite work presented here demonstrates. We used three approaches to perform comparative analyses. First, we used analogies, such as lakes as islands, to discover commonalities among systems. These commonalities, in turn, suggested ways to borrow perspectives or theories from other disciplines or from other systems, for example, island biogeography and species equilibria, and to use them to expand, enrich, and synthesize our understanding. Second, our application of dimensionless metrics was a necessary approach to overcome the limitations to comparisons imposed by the absence of standard measurement systems across disparate ecosystems. This approach revealed fundamental patterns of temporal and spatial variability that appear to characterize all LTER sites. Finally, a common measure, the standard deviation of the normalized difference vegetation index, was identified and applied using satellite imaging to compare the spatial heterogeneity of disparate landscapes. This measure is related strongly to land use and topography and is dependent on the spatial scale of analysis in a predictable way. The last two approaches originated from our desire to discover fundamental properties that could not be observed by analyzing a single ecosystem or landscape type. The objective was to characterize similarities and differences among any and all LTER sites with robust approaches that did not require us to classify ecosystems or landscapes into homogeneous systems such as forests, lakes, or prairies. Our analyses suggested that approaches to comparisons among disparate systems have merit when applied to more homogeneous systems for which our historical approach has been to use more conventional methods of analysis.

Part II

Drivers of
Long-Term Dynamics

Understanding long-term dynamics of lakes in the landscape requires consideration of major drivers of change and variability. At the beginning of our study, we identified three external drivers, one physical, one chemical, and one biological, that we expected to be primary influences on lake dynamics. These drivers are climate (Chapter 7), acid precipitation (Chapter 9), and establishment of nonnative species (Chapter 8). Each of these drivers interacts with internal lake processes to affect lake dynamics. Understanding internal drivers and processes is essential to explaining the dynamics of lakes (Chapter 10).

When we chose our seven study lakes in northern Wisconsin, in 1981, we thought they were minimally affected by humans. However, the lakes we chose, and perhaps most lakes in the northwoods, are not reference sites immune from human influence. Our studies soon had to be broadened to understand the human drivers of dynamics, including reciprocal interactions between people and lakes. These interactions are important in both northern Wisconsin (Chapter 11), in a forested lake district used largely for recreation, and southern Wisconsin (Chapter 12), in an urban and agricultural lake district.

We planned and implemented long-term experiments, such as the Little Rock Lake acidification study (Chapter 9). We also are observing the effects of watershed restoration efforts and the continuing saga of human-caused effects and management on Lake Mendota (Chapter 12). Additionally, we took advantage of unplanned disturbances to the LTER lake ecosystems that served as natural experiments, such as the invasion and establishment of nonnative species (Chapter 8). Even for the planned manipulations, interactions between the manipulations and "natural" perturbations were important in determining the resulting lake dynamics.

7

Climate-Driven Variability and Change

John J. Magnuson
Barbara J. Benson
John D. Lenters
Dale M. Robertson

Truth, like climate, is common property.
—Elizabeth Stuart Phelps (1896, p. 259)

The dynamics of climate and the differences in how lakes respond to these dynamics are the focus of our North Temperate Lakes Long-Term Ecological Research (LTER) on climatic variability and change. We regard climate and climatic variability as external drivers, or, in a sense, signals that impart their mark on the status and dynamics of the lakes (Fig. 5.1). Status is the average conditions of the lake. Variations in solar radiation, cloudiness, air temperature, wind, and humidity cause changes in water temperature, lake mixing, stratification, and ice cover. These variations, along with changes in precipitation, cause changes in the lakes' water budgets, water levels, and geochemical cycles. The biology and chemistry of the lakes respond, in turn, to these limnological dynamics. Climatic variability affects lakes at all time and space scales. The lakes respond to climatic factors over days, seasons, years, decades, and longer time scales. These lake responses or dynamics have some patterns that are similar among lakes and some that are different among lakes. These patterns of similarities and differences extend between adjacent lakes, across a lake district, and across Wisconsin, the Great Lakes region, and the Northern Hemisphere, in part, owing to differences in climate and climate dynamics.

A series of scale-related questions emerges. How well does the climatic variation over the LTER years represent earlier periods of intensive study on the LTER lakes by E. A. Birge and C. Juday between 1925 and 1941 or, more generally, over the twentieth century? How similar or how different are the climates and climate dynamics of our northern and southern Wisconsin sites? Do adjacent lakes respond similarly to interyear climatic variability? What is the spatial extent of coherent interyear variation in climate from the perspective of lake responses to climatic factors? Are the lakes responding to large-scale drivers of climatic variability such as the El Niño Southern Oscillation or the variation in the Aleutian Low (Pacific Decadal Oscillation) or the North Atlantic Oscillation? Are long-term trends in the lakes at the century time scale and at a global spatial scale consistent with global changes in climate associated with greenhouse warming? How are the Wisconsin climates projected to change in the coming decades and century?

Our purposes here are to examine the influence of climatic drivers on lakes from years to centuries and from a lake, to adjacent lakes, to regions, and to the Northern Hemisphere. We focus on lake properties that are driven strongly by climate, such as ice cover, water temperature, and water level. Here, meteorological data from Minocqua Dam represent the climate for LTER lakes in northern Wisconsin, and data from the Dane County Regional Airport represent the LTER lakes in southern Wisconsin. We focus on the Wisconsin lakes but relate cooperative intersite research (Chapter 14) on lakes in Ontario at Dorset and at the Experimental Lakes Area (ELA) and from the lake ice studies with colleagues around the globe. We present and discuss our findings with respect to global climate change and the potential influences of such change on lakes and lake ecology.

Climate at the LTER Locations

Wisconsin has a continental climate modified somewhat by Lakes Superior and Michigan. As is typical in continental climates, the climate varies markedly among seasons and among years. Winters are cold and snowy, and summers are warm. According to the Wisconsin State Climatology office, the highest temperature ever recorded in Wisconsin was 46°C, recorded at Wisconsin Dells on July 13, 1936, and the lowest temperature on record was –48°C, reported from Couderay on February 2 and 4, 1996. Storms generally move eastward across the state. Thunderstorms average about 30 per year in northern Wisconsin and about 40 per year in southern counties and occur mostly in the summer.

Average temperatures and total precipitation are presented for the 21-year LTER period from 1981 to 2001 (Table 7.1). Annual air temperatures averaged 4.5°C in the north and 8.1°C in the south; precipitation totaled 83.2 cm/year in the north and 86 cm/year in the south. In both locations, temperatures average above 17°C in summer and below –5°C in winter. Precipitation is greatest in summer and least in winter. In all seasons except fall, when the south was dryer than the north, the south received more precipitation than the north. Northern Wisconsin has considerably more snow than southern Wisconsin. Interannual differences in snowfall were markedly greater in the north than the south.

How similar is the weather during the 21-year LTER period to the weather during the past century or, more specifically, to the weather during the years 1925–41, when our predecessors, E. A. Birge and C. Juday, intensively studied the northern Wisconsin lakes? In general (Fig. 7.1), the range in seasonal mean air temperatures and total precipitation during the LTER years from 1981 to 2001 encompass the variation observed during the non-LTER years from 1904 to 1980. With the exception of winter mean air temperature, the LTER years largely spanned the range

Table 7.1. Mean values and differences in air temperature and total precipitation between the southern (Dane County Regional Airport) and the National Oceanic and Atmospheric Administration National Weather Service Cooperative Station at the northern (Minocqua Dam) Wisconsin sites during the LTER period 1981–2001.

Seasons	Mean Air Temperature (°C season or year)			Total Precipitation (cm/season or year)		
	South	North	Difference (S − N)	South	North	Difference (S − N)
Spring	7.9	4.3	+3.6	21.7	19.7	+2.0
Summer	20.7	17.6	+3.1	32.8	30.6	+2.2
Fall	9.2	5.7	+3.5	20.6	24.2	-3.6
Winter	-5.5	-9.6	+4.1	10.9	8.7	+2.2
Total Year	8.1	4.5	+3.6	86.0	83.2	+2.8

of extreme mean air temperatures. For the winter season, the LTER years were overrepresented in the warmest winters, and they did not include winters with a mean temperature less than −12°C. LTER years span the range of variation in precipitation for spring and winter and, except for two extremely dry years, for fall. The range in summer precipitation is smaller than the range for the non-LTER years.

Climate-Driven Dynamics of Lakes

Limnological climates include not just atmospheric properties like air temperature, precipitation, and wind but also the physical properties of the lakes. In deeper lakes, climatic factors generate thermal structures through heat transfer and wind-driven mixing. In summer, lakes thermally stratify, with a warm, upper-mixed layer or epilimnion, a thermocline or metalimnion, and a lower, colder hypolimnion. In winter, ice cover eliminates wind mixing, and thermal stratification is reversed, with water near 0°C just beneath the ice and warmer water at greater depths. Water at 4°C is at maximum density, and warmer and cooler waters are less dense. Other physical features of the limnological climate include the depth of the mixed layer, duration of summer stratification, and duration of winter ice cover. All of our lakes have the classic winter thermal structure, and, except for Lake Wingra in the south and Crystal Bog and perhaps Allequash Lake in the north, all have the classic summer stratification. The exceptions are lakes shallow enough that wind events frequently mix the water from surface to bottom, eliminating any stable thermal stratification.

The summer thermal structures of a northern and a southern LTER lake in warmer and cooler summers illustrate the importance of climatic variability among years in Wisconsin (Fig. 7.2). Using the deepest LTER study lakes, Trout Lake and Lake Mendota, as examples, average temperatures in both the epilimnia and the hypolimnia were warmer in warmer summers; differences in the epilimnia were greater than

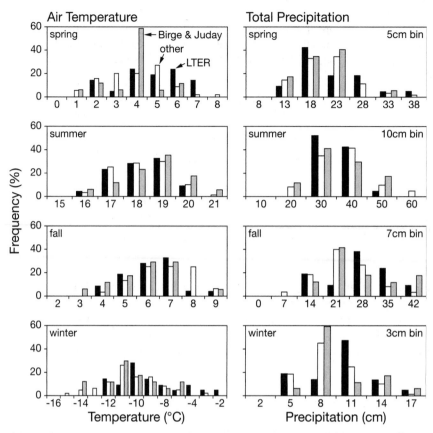

Figure 7.1. Histograms for seasonal air temperature and total precipitation that compare the LTER period 1981–2001 with the Birge and Juday period, 1925–41, and the other years since 1904, that is, 1904–24, 1942–80.

those of the hypolimnia. Thus, vertical stratification is more stable in warmer than in cooler summers because the near-surface and deeper waters differ more in temperature and therefore differ more in density.

Differences in the thermal structure of lakes (Fig. 7.2) relate as well to differences in atmospheric climate between northern and southern Wisconsin (Table 7.1). Not surprisingly, Trout Lake, in the north, and Lake Mendota, in the south, differ in their limnological character owing to the differences in atmospheric climate between the two locations. The differences in surface water temperature are the result primarily of climatic differences and are of a magnitude similar to differences in summer air temperatures. The differences in hypolimnion temperatures are influenced by differences in both climate and lake morphometry. The deeper and smaller Trout Lake experiences less influence of wind mixing than Lake Mendota, and Trout Lake therefore stratifies before the water has warmed as much as in Lake Mendota.

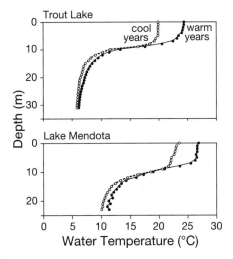

Figure 7.2. Summer water temperature profiles for Trout Lake in northern Wisconsin and Lake Mendota in southern Wisconsin for cool and warm summers. Data are averaged for three cool summers (1992, 1994, and 1996) and three warm summers (1988, 1995, and 2001) for a single sample date between August 2 and August 16.

Water temperatures of individual LTER lakes did not always respond consistently to the difference in air temperatures between the north and the south. As expected, the southern lakes had warmer surface temperatures than the northern lakes; however, summer bottom temperatures and winter temperatures were more lake-specific than location-specific, that is, north versus south. In summer, near-surface temperatures averaged across the LTER years ranged from 20.5°C in Trout Bog to 24.4°C in Allequash Lake in the north and from 24.6°C in Lake Mendota to 25.8°C in Lake Wingra in the south. In summer, near-bottom temperatures in the lakes with minimal stratification, Crystal Bog at 22.0°C in the north and Lake Wingra at 23.6°C in the south, had bottom temperatures averaging 9 to 19°C warmer than the near-bottom temperatures in the stratified lakes. Summer near-bottom temperatures of the stratified lakes ranged from 4.3°C in Trout Bog to 13.2°C for Allequash Lake. Water temperatures just below the ice ranged from 0.6°C in Lake Mendota to 3.3°C in Lake Wingra in the south and from 1.2°C in Trout Lake to 2.7°C in Crystal Bog in the north. Near-bottom temperatures in winter ranged from 1.8°C for Lake Mendota to 4.2°C for Lake Wingra in the south and from 2.2°C for Trout Lake to 4.0°C for Crystal Lake in the north. While differences in lake thermal characteristics reflected differences between locations, many differences among the lakes, both in the north and in the south, relate to lake morphometry.

Limnological climates are responsive to both temporal and spatial climatic variability, as seen in Trout Lake and Lake Mendota (Fig. 7.2). What patterns of variability exist over interdecadal and century time scales? Responses of lakes to climate

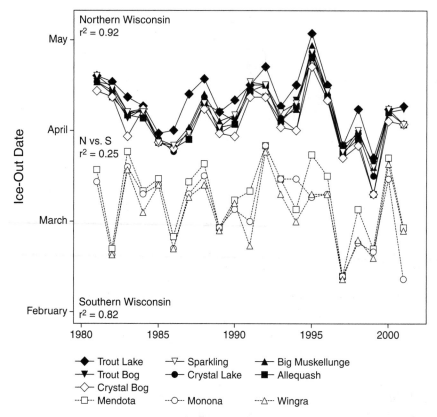

Figure 7.3. Interannual variation and coherence in ice-off time series for the LTER lakes excluding Fish Lake, 1981–2001. Coherences values, r^2, are between pairs of northern LTER lakes, pairs of southern LTER lakes, and pairs of southern and northern LTER lakes. Redrawn from Magnuson et al. (2004).

at these time scales are illustrated in our consideration of climatic-driven coherence in lake districts, regional climate footprints, and large-scale climatic drivers.

Coherent Dynamics in Lake Districts

Coherent dynamics, that is, synchronous interannual variability between lakes, are evidence for a common external driver such as climate, especially if the dynamic lake feature is one linked strongly through causal mechanisms with climatic factors. The concept and definitions of coherent dynamics are developed in Chapter 5; essentially, coherence is a measure of the shared variance between two time series. The relative strengths of coherence from very low to very high are indexed in Table 5.2.

The date of ice-off is remarkably coherent (Fig. 7.3). Interannual variation in ice-off dates in the northern LTER lakes was high for each lake, differing among years by 35 to 45 days over the LTER period, while interlake variation in the same year was low; on average, the ice broke up on all the northern LTER lakes within a span of eight days. Coherence for the 21 possible northern lake pairings was very high, with a median of $r^2 = 0.92$ (range = 0.83 to 0.98), even though the seven lakes differ morphologically by factors of 3,000 in surface area and 14 in mean depth. Coherence was also very high for the three southern lakes, $r^2 = 0.82$ (range = 0.79 to 0.86), even though they differ in morphology by factors of 28 in surface area and 5 in mean depth. The LTER lakes have similar interannual dynamics in ice-off date regardless of their size.

One might expect that all physical responses of lakes would be similarly linked to climate dynamics as demonstrated by the ice-off patterns (Fig. 7.3). Yet, even physical variables differ greatly in coherence expressed as r^2 or common variance. For a suite of eight physical variables for the northern LTER lakes, coherences ranged from a median r^2 of 0.92 for ice-off date to only 0.07 for near bottom temperatures in summer (Fig. 7.4). Magnuson et al. (1990) suggested that individual lake features filter the ways that climatic signals such as solar radiation, air temperature, and wind are expressed in a lake's behavior (Fig. 5.1). A comparison of coherences of ice-on with ice-off dates makes the point. Ice-on date depends primarily on the rate of water cooling. This rate depends on the interaction of climatic (primarily air temperature and radiation) and morphologic (mean depth) factors. With the same rates of heat loss, waters cool faster in a shallow lake than in a deep lake; shallow lakes cool more rapidly because they store less heat compared to deeper lakes. Thus, freeze dates occur earlier in shallow than in deep lakes on average. More important for coherence, shallow lakes are more responsive to a period of cold weather and more likely to freeze as a result. In addition, the larger lakes (deeper lakes are usually larger in area) are more exposed to wind mixing, which delays the formation of ice, than are smaller lakes. In contrast, ice-off date depends just on when the ice melts, not on the time to warm the entire water body; thus, ice-off date depends primarily on solar radiation and air temperature. These factors are similar for each lake regardless of size. Differences in ice thickness and lake fetch cause some, but apparently only small, differences in ice-off dates among lakes. Differences in morphometry among lakes play a larger role in determining ice-on date than ice-off date.

The thickness of the surface mixed layer, that is, the epilimnion, differs among lakes and among years owing to differences in lake-specific morphometric factors related to interactions with air temperature, wind strength, and cloudiness, which influences solar radiation reaching the lake. In warm years, a lake should have a more stable thermal structure, and more wind force will be required to deepen the thermocline as summer progresses. In less windy years, the wind does not mix the water as deeply, and a shallower thermocline should result. The same wind force contributes more to mixing in larger lakes than in smaller lakes. In less cloudy years, more solar radiation enters the lake, resulting in warmer surface waters and thus a more stable thermal structure and a thinner epilimnion. Differences in mixing in small and large lakes make the response among lakes complex. These within-lake factors result in low coherence values for the epilimnion thickness, which averages only $r^2 = 0.26$ (Fig. 7.4).

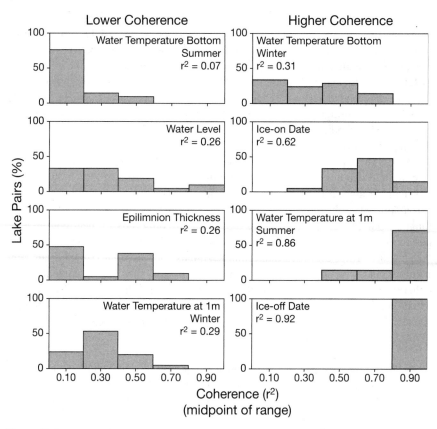

Figure 7.4. Percent frequency distributions of the 21 coherence, r^2, values between pairs of LTER lakes in northern Wisconsin for eight physical lake variables from 1981 to 2001. Panels are arranged from the variable with the lowest median coherence on the top left to the highest median coherence on the bottom right. Redrawn from Magnuson et al. (2004).

Average coherence for water level is low (Fig. 7.5), in part, because many interacting factors related to landscape position influence water levels (Chapter 2, 3) and thus coherence between lakes. Patterns of variation differ for lakes high and low in the landscape; some lakes are strongly coherent, others are not. Three of the northern LTER lakes, Crystal, Sparkling, and Big Muskellunge Lakes, have high to very high coherence. Each is moderately high in the landscape, with clear water and no surface water inlet or outlet. The highest coherence, $r^2 = 0.95$, is between Crystal and Big Muskellunge Lakes, which are separated by only a narrow spit of land and are linked through common groundwater flow. Crystal Bog, once a bay of Big Muskellunge Lake, is moderately coherent, average $r^2 = 0.49$, with each of these lakes. Three other lakes behave more independently: Allequash Lake, Trout Lake, and Trout Bog. Both Allequash and Trout Lakes are drainage lakes with surface water inlets and an outlet (Table 2.1); they are lower in the landscape than the three groundwater-dominated lakes just mentioned. Allequash Lake has a beaver dam at

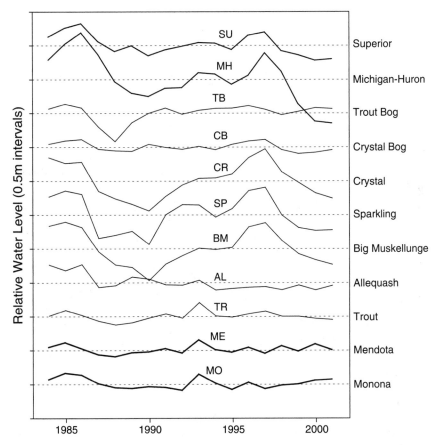

Figure 7.5. Time series of annual mean lake levels from 1984 to 2001 for two Laurentian Great lakes, the northern LTER lakes, and two southern Wisconsin LTER lakes. The Laurentian Great Lakes, Michigan-Huron and Superior, are at the top; the southern LTER lakes, Lakes Mendota and Monona, are at the bottom (http://wi.water.usgs.gov/pubs/wi-03-01/wi-03-01.pdf); and the seven northern LTER lakes are in the middle (http://lter.limnology.wisc.edu). Mean levels for each lake, designated by the dashed lines, are arbitrarily separated by an interval of 0.5 m. Annual means are calculated by averaging monthly mean data (Lakes Superior, Michigan-Huron, Mendota, and Monona) or biweekly samples. Data gaps during the winter months for the northern LTER lakes are ignored.

the outlet that has been removed from time to time as a management action by the Wisconsin Department of Natural Resources. The only lake that is at least moderately coherent with Trout Lake is Sparkling Lake; these two lakes are separated by a sandy spit, as are Crystal and Big Muskellunge Lakes, and are hydrologically related through groundwater flows. Trout Bog, while immediately adjacent to Trout Lake, is mounded about 2.3 m above the water level of Trout Lake and appears weakly connected hydrologically to any of the lakes. Its coherence with Trout Lake

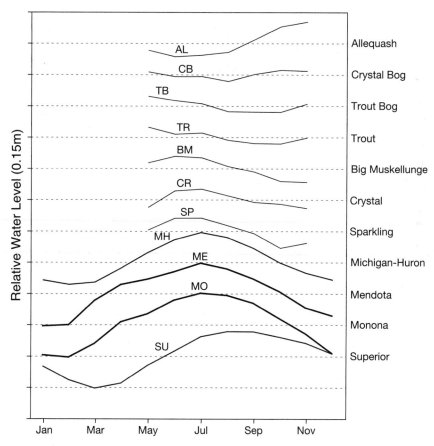

Figure 7.6. Long-term monthly mean lake levels averaged from 1984 to 2001. Annual mean values, designated by the dashed lines, are separated arbitrarily by an interval of 0.15 m. Except for large data gaps in winter months, missing data are ignored.

is barely moderate at $r^2 = 0.40$, and for all the other lakes its coherence is very low to low.

Seasonal changes in water level differed among lakes (Fig. 7.6). Four of the northern LTER lakes had highest water in late fall or early spring: Allequash Lake, Crystal Bog, Trout Bog, and Trout Lake. The other three, Big Muskellunge, Crystal, and Sparkling Lakes, had the highest levels in early summer. The two southern LTER lakes, Lakes Mendota and Monona, had highest levels in midsummer. Of the Great Lakes, Lake Michigan-Huron had the highest water levels in midsummer and Lake Superior, in early fall. These differences in seasonal lake levels reflect lake-to-lake differences in the seasonal timing of the various water budget components. For example, the northern Wisconsin LTER lakes differ in their ratios of surface water to groundwater components. Groundwater has a significantly longer lag time in influencing lake levels. The southern Wisconsin LTER lakes, on the other hand, are influenced less by spring snowmelt and more by the summer maximum in precipitation, as well as by water management of lake levels. Deeper lakes, such as Lakes Superior and Michigan-Huron, have significant seasonal lags in surface temperature and evaporation, delaying lake level maximums until midsummer or early autumn.

Patterns of interannual and seasonal variation in water levels of the LTER lakes were often, but not always, individualistic. While differences in climate between the north and the south could cause low coherence between locations, the low coherences within a location must result from lake-specific differences in their water budgets. Those lake pairs in the north that were hydrologically similar were more coherent than those that were hydrologically dissimilar, suggesting that coherence in interannual variation in water level is determined by similarities in the degree and timing of lakes' responses to variation in precipitation and evaporation. The importance of direct precipitation, snowmelt, and groundwater to these lakes and the absence of surface runoff as a component of the water budget for each of these lakes appears to be influential to their coherence.

Lake evaporation can be an important driver of lake-level variability, but accurate, long-term monitoring of evaporation is an expensive and time-consuming undertaking (Winter 1981). As a result, intensive measurements of lake evaporation have been made for only one of the LTER lakes, Sparkling Lake, in the north. Lenters et al. (2005) described these results and used an energy budget analysis to understand short- to long-term variations in evaporation for Sparkling Lake over 10 years, from 1989 to 1998. An important conclusion of the study is that lake evaporation undergoes significant changes on a wide range of time scales, from biweekly to seasonal to interannual. Primary climatic drivers of the variability depend on the time scale but include factors such as solar radiation (interannual), lake temperature (seasonal), and lake-air temperature difference and relative humidity (biweekly).

Seasonal variations in evaporation for Sparkling Lake are illustrated in Figure 7.7, top, for the five primary ice-free months of the year, along with monthly mean evaporation estimates for Lakes Michigan-Huron and Superior, based on results from Hunter and Croley (1993). On average, the highest evaporation rates are attained in July for Sparkling Lake, in September through October for Lake Michigan-Huron, and in December for Lake Superior. Maximum rates are similar among the lakes, but differences in timing reflect primarily differences in lake depth and, as noted earlier, can have a noticeable influence on seasonal water levels. For example, Sparkling Lake reaches its highest level 1 month earlier than deeper Lake Michigan-Huron and 2 to 3 months earlier than even deeper Lake Superior (Fig. 7.6). Besides evaporation, other factors such as precipitation and runoff contribute to seasonal variations in lake level. In the case of Sparkling Lake, for instance, summer precipitation rates reach a maximum of about 10 cm/month during June, July, and August, that is, to evaporation rates, but drop off significantly during other times of the year to roughly 1 to 3 cm/month during winter.

Interannual variations in late summer evaporation for Sparkling Lake, Lake Michigan-Huron, and Lake Superior exhibit varying degrees of coherence over 10

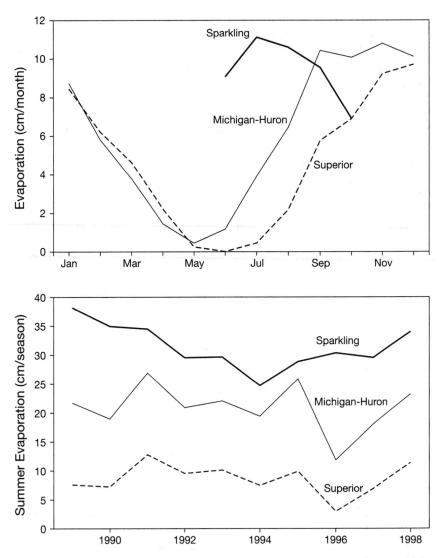

Figure 7.7. Top: Monthly mean evaporation based on Lenters et al. (2005) for Sparkling Lake and for Lakes Michigan-Huron and Superior from Hunter and Croley (1993). Means are calculated from 1989 to 1998. Sparkling Lake has missing data for June and October 1994 and for October 1996 to 1998. Bottom: Total evaporation for July through September from 1989 to 1998 for Sparkling Lake, Lake Michigan-Huron, and Lake Superior.

years (Fig. 7.7, bottom). The average evaporation for Lake Michigan-Huron from July through September of 20.9 cm is more than twice that of the colder Lake Superior at 8.6 cm, but the interannual coherence for these two large lakes is very high, $r^2 = 0.87$. On the other hand, the smaller and shallower Sparkling Lake has much higher late summer evaporation, 31.4 cm, and significantly lower coherence with the other two lakes, $r^2 = 0.04$ to 0.06. This low coherence, however, primarily reflects three years (1989, 1990, and 1996) when Sparkling Lake evaporation rates were abnormally higher than the other two lakes. For the remaining time period, particularly 1991 to 1995, the three lakes had similar interannual variations, including the drop in evaporation from 1991 to 1992 following the eruption of Mt. Pinatubo (Lenters et al. 2005). Overall, of the nine year-to-year changes in evaporation for Sparkling Lake, six are in the same direction as those for the two Great Lakes. Thus, nearby lakes of vastly different size and depth, and thus mean temperature, can still exhibit similarities in interannual variability in evaporation. Prominent exceptions to this, for example, such as occurred in 1996, suggest that local-to-regional variations in lake-climate interactions can still be significant. Furthermore, as was noted for the seasonal variability, evaporation is not the only water budget component that exhibits significant interannual variations. For example, July-to-September precipitation at Sparkling Lake averages around 30 cm, which is similar to evaporation, but has an interannual standard deviation of 9.5 cm, which is more than twice that of evaporation, 4.0 cm, over the same years from 1989 to 1998. Thus, at least in this region, precipitation is generally a more significant driver of interannual variations in lake level than is evaporation.

Regional Climatic Footprint

Coherence among lakes can exhibit a large regional footprint owing to the broad regional similarities in climatic variability among years. For example, compare the coherence in ice-off dates for the northern and southern Wisconsin LTER lakes (Fig. 7.3). While the match is not as strong between northern and southern Wisconsin as it is within each location (see the data for the early 1980s), the different locations exhibit some coherent behavior (see the data for the 1990s). The coherence in the 21 possible lake pairings between the locations averaged $r^2 = 0.25$. For a second example, coherence in water levels for some lakes extends from the northern LTER lakes to the Laurentian Great Lakes to the southern LTER lakes (Fig. 7.5). The three groundwater lakes, plus Crystal Bog at the northern LTER location, have an average coherence or shared variance of $r^2 = 0.61$ (range = 0.56 to 0.68) with Lake Superior and 0.48 (range = 0.39 to 0.56) with Lakes Michigan-Huron. The coherence with the Great Lakes may result from the Great Lakes' watersheds, especially Superior's, having a similar climate to northern Wisconsin and provides some evidence for a broad hydrologic footprint for parts of our region. Trout Lake exhibits some coherence with the southern lakes, with $r^2 = 0.53$ with Lake Mendota and 0.28 with Lake Monona. These similarities in behavior must derive from a large regional pattern of precipitation and evaporation that affects northern and southern lakes in similar ways.

This observed interannual coherence between lakes, as well as the coherence of the climatic drivers, extends beyond Wisconsin to the upper Great Lakes region

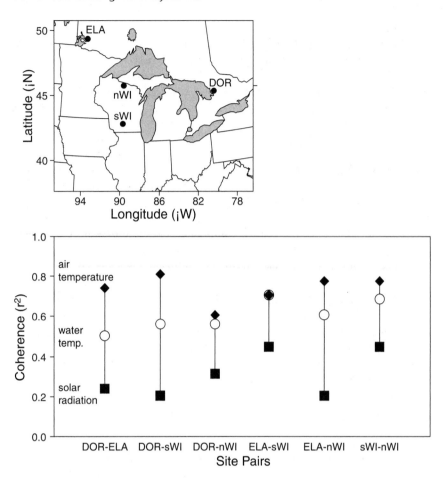

Figure 7.8. Coherence of summer mean air temperature, near-surface water temperature, and solar radiation between lake districts. The location of lake districts in the top panel are: DOR = Dorset Lakes, ELA = Experimental Lakes Area, nWI = northern Wisconsin LTER lakes, and sWI = southern Wisconsin LTER lakes. Data from Benson et al. (2000b).

(Fig. 7.8), that is, to western Ontario at the Experimental Lake Area and to central Ontario at the Dorset Lakes (Benson et al. 2000b). Interannual variation in summer mean air temperature was strongly coherent across this broad region, as evidenced by the coherence of lakes between lake districts, which ranged from $r^2 = 0.61$ to 0.81. Mean daily solar radiation during summer was less coherent than air temperature among lake districts, which ranged from $r^2 = 0.20$ to 0.45. The regional scale climatic drivers would be expected to produce coherence across the region in lake thermal response variables closely driven by air temperature. Indeed, the near-surface summer water temperature had strong coherence among lake districts, with the mean r^2 between pairs of lakes from different lake districts ranging between

0.50 and 0.71 for the lake-district pairs. Thus, broad-scale climatic signals lead to similarities in lake responses across large regions.

The regional climatic footprint can be broad for precipitation and runoff variables, as demonstrated by the drought during the late 1980s (Fig. 7.9). The first signs of this drought began showing up during the winter of 1986–87 when warm El Niño conditions led to reduced snow pack throughout the Upper Midwest. This resulted in reduced snowmelt during the spring of 1987 and associated below-normal fluxes of water to lakes, soil, and groundwater. These hydrologic responses to the regional climate variability have been simulated with a land-surface hydrologic model known as the Integrated Biosphere Simulator (Lenters et al. 2000). The simulated anomalies from the long-term mean in snowfall and related runoff display patterns that extended across the western Great Lakes region in the winter and spring of 1986–87 (Fig. 7.9). Lower-than-average snow depth was prevalent across a broad area from west central Minnesota through Wisconsin and into Ontario and Michigan. Similarly, lower-than-average spring runoff was simulated across the entire western Great Lakes region. The large-scale features of the simulation were validated against observations of river discharge, snow depth, soil moisture, and regional water balance studies (Lenters et al. 2000). Fine spatial features of these simulations should be interpreted with caution.

Large-Scale Climate Drivers

Some interannual variation in our lakes is associated with large-scale climatic drivers. These drivers include intense volcanic eruptions, the El Niño/Southern Oscillation (ENSO), the North Atlantic Oscillation (NAO), and various drivers in the North Pacific, such as variation in the strength of the Aleutian Low (Table 7.2). Because this climatic variation results from rare events such as intense volcanic eruptions or from quasi periodic oscillations of three to seven years (ENSO) or longer (NAO and the Aleutian Low), the long-term ice-on and ice-off data have been especially valuable in determining relationships. Long-term ice data around the Northern Hemisphere were acquired as part of an LTER-funded international workshop that took place in October 1996 at the Trout Lake Station (Chapter 14) (Magnuson et al. 2000a, 2000b). The database of lake and river ice dates with 165 time series longer than 50 years and 27 longer than 100 years was deposited in the National Snow and Ice Data Center (www.nsidc.org; search for lake ice). Some additional data have been deposited since.

Ice-off dates in Wisconsin and in the Great Lakes region are associated with at least four large-scale climate drivers centered far from Wisconsin in the South Pacific, the North Pacific, and the North Atlantic Oceans (Table 7.2). The importance of these drivers has changed through time. The relation between ice-off on Lake Mendota and these large-scale climate drivers is strongest for the North Atlantic Oscillation (NAO), accounting for 37% of the variation from 1879 to 1928 (Livingstone 2000). Strong to moderate-intensity El Niño events were associated with early ice-off after 1940 and late ice-off before 1940 (Robertson et al. 2000). Responses to El Niño conditions also are not consistent across Wisconsin (Anderson et al. 1996); ice-off dates in northern Wisconsin lakes are less responsive to a

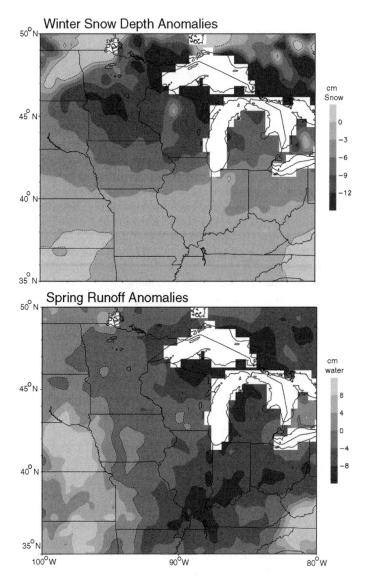

Figure 7.9. Maps of anomaly contours for winter snow depth (top) and spring runoff (bottom) for the winter and spring of 1986–87 as modeled by the Integrated Biosphere Simulator (Lenters et al. 2000). Snow depth anomalies are deviations from the December through February mean from 1965 to 1994. Spring runoff anomalies are deviations from the March through May mean from 1965 to 1994. The zero lines are shown as dashed contours.

Table 7.2. Relation of ice-off dates to large-scale climate drivers.*

Large Scale Driver[a]	Years	Location	Relation to ice off	Source
ENSO	1900–1940	Canada and Eastern U.S.	Late with El Niño	Robertson et al. 2000
ENSO	1940–1995	Canada and Eastern U.S.	Early with El Niño	Robertson et al. 2000
ENSO	1942–1991	Lake Mendota, WI	Highest correlation with SO index $r^2 = 0.25$	Livingstone 2000
NAO	1879–1928	Lake Mendota	Highest correlation with NAO index $r^2 = 0.37$	Livingstone 2000
PNA (Winter or Spring)	1969–1988	South Boreal and Temperate in North America	Correlation with index $r^2 = 0.09$ to 0.20	Benson et al. 2000b
WP (Spring)	1969–1988	South Boreal and Temperate in North America	Correlation with index $r^2 = 0.07$ to 0.18	Benson et al. 2000b
Intense Volcanic Eruption	1850s–1995	Five lakes in Northern Hemisphere including Lake Mendota	Delayed 1–4 days in 2nd and/or 3rd year after eruption	Livingstone 2000

Source: from Magnuson (2002a).

*See also Magnuson et al. (2004).

[a]ENSO = El Niño Southern Oscillation (SOI = Southern Oscillation Index); NAO = North Atlantic Oscillation; PNA = Pacific/North American Pattern; and WP = West Pacific Pattern.

strong El Niño event than are those for Lake Mendota in southern Wisconsin. The relation is stronger in southern Wisconsin because the strong El Niño events influence air temperatures near the dates when the southern lakes are breaking up, but the northern lakes break up later. The indices differ in influence even on a broader spatial scale. The influence of the Western Pacific Pattern (WP) and Pacific/North American Pattern (PNA) indices is stronger in some parts of North America than in others (Benson et al. 2000a). Differences in the strength of the relations between ice-off dates and the NAO and the Southern Oscillation Index (SOI) are noted among regions (Finland, Siberia, Switzerland, and Wisconsin) and among years from 1871 to 1920 (Livingstone 2000).

Long-Term Trends

The Climate Assessment in 2001 by the Intergovernmental Panel on Climate Change concludes that evidence for climate change is stronger than ever (Houghton et al. 2001, Trenberth 2001). An overall global warming trend has occurred, and evidence suggests that some of the increases in temperature are associated with increases in greenhouse gas emission from the burning of fossil fuel. The 1990s were the warmest decade in the twentieth century. Changes in climate are hypothesized to have occurred both over the LTER years from 1981 to 2001 and during the entire twentieth century. In the next section we examine whether trends in climate and the physical limnology of LTER lakes were evident at the northern and southern Wisconsin LTER locations.

Trends in Climate

We examined the climatic records for northern (Minocqua Dam) and southern (Madison) Wisconsin LTER locations for 21 years from 1981 to 2001, and for the twentieth century from 1904 to 2001 to determine whether long-term trends suggest that climate change has occurred. These years were chosen because we had comparable data from both north and south. Analyses were made for each of the four seasons of the year at each location for mean air temperature and total precipitation (Table 7.3). Linear least square regressions and the tau statistic for testing whether the ranking of a variable by size is related to temporal order produced similar results for almost all 16 sets of data.

For the 21 years from 1981 to 2001, only a few significant trends in air temperature and precipitation were apparent. The area near Trout Lake experienced no statistically significant trends in air temperature or precipitation, but Madison tended toward wetter summers and warmer winters. At Madison, winter temperatures increased by 1.5°C/decade ($p = 0.07$ for the slope and 0.05 for tau), and summer precipitation increased by 5.9 cm/decade ($p = 0.04$ for the slope and 0.01 for tau). For those of us who live in Madison, these trends have been apparent.

In contrast to the shorter LTER period, the longer time series from 1904 to 2001 exhibited many statistically significant trends in climate at both locations and, except for fall, in all seasons (Table 7.3). Madison air temperatures warmed significantly in spring, summer, and winter, and for the total year; summer precipitation increased, as it did during the LTER period, but at a slower rate of 7.8 cm/century. At Minocqua Dam, spring temperatures increased significantly, while summer temperatures decreased significantly; winter total precipitation also increased. Five out of six significant slopes were for warmer air temperatures, whereas only two out of four significant slopes were for wetter conditions. In general, the largest increases in air temperature occurred in spring and winter, the same period that strongly affects the timing of ice-off.

The most dramatic trends in climatic conditions were for snowfall. We have chosen not to present the quantitative snow data here, because the snow data do not seem to be reliable at least for the northern location (personal communication, K. E. Kunkel, Illinois State Water Survey). General observations of the data sets, however, suggest that snowfall increased at both the southern and the northern locations. The increase in snowfall in the north was much greater than that in the south. Total snowfall appeared similar in the north and in the south in the early 1900s but by 2001 snowfall appeared to be much greater in the north than the south. This may be an artifact of the Minocqua Dam snowfall data, although records from at least one other nearby station (Ironwood, Michigan) reveal a similar increase in snow-

Table 7.3. Rate of change (slope) from 1904 to 2001 for air temperature and precipitation of the southern and northern Wisconsin LTER sites for each season and the total year.[a]

	Rate of change (season or year)			
	Air temperature (°C/century)		Total precipitation (cm/century)	
Season	South	North	South	North
Spring	+2.0*	+1.2*	−0.9	+2.7
Summer	+0.9*	−0.8*	+7.8*	−0.5*
Fall	+0.4	−0.4	−3.4	+2.6
Winter	+1.9*	+1.3	−0.2*	+2.4*
Total Year	+1.3*	+0.3	+3.3	+7.2

[a]Northern data are from the National Oceanic and Atmospheric Administration National Weather Service Cooperative Station at Minocqua Dam and the southern data are from Dane County Regional Airport.
*Slopes significant at p ≤ 0.05.

fall over time. Interannual variation at the northern and southern locations appeared moderately coherent.

Climate records for temperature and precipitation at the Wisconsin LTER site tend to mirror the general climatic changes occurring in the upper midwest and are consistent with the expectations from greenhouse gas–induced climate change (Sousounis and Bisanz 2000, Kling et al. 2003). We can speculate that the northern location would be more buffered from climatic change owing to its proximity to Lake Superior (see Kling et al. 2003) and that the southern location would be more influenced by air temperature changes that melt the snow and thus reduce the albedo. The apparent greater snowfall in the north might be related to the proximity of Lake Superior and to increases in lake-effect snow (Leathers and Ellis 1996, Burnett et al. 2003). Ice cover on Lake Superior has decreased in the past 20 years (Assel et al. 2003), perhaps contributing to increases in lake-effect snowfall.

Trends in Lake Conditions

The LTER southern lakes have displayed long-term trends in the direction of shorter durations of ice cover since the mid-1800s, consistent with those found for other lakes with long-term records throughout the Northern Hemisphere (Magnuson et al. 2000b). The record on Lake Mendota from 1853 was used to make the point about the invisible present (Chapter 1)—as you open an observational time series from a single year to progressively longer periods, new information makes the context of the single year more apparent. The trend record for Lake Mendota has been published frequently (Robertson 1989, Robertson et al. 1992, Assel and Robertson 1995, Magnuson et al. 2000a, Magnuson 2002a, Kling et al. 2003). In this analysis, ice-on dates are occurring later, ice-off dates are occurring earlier, and the duration of ice cover is shortening. We fit Mendota's ice-duration data (Fig. 7.10) with two statistical models, a linear regression model and an intervention time series

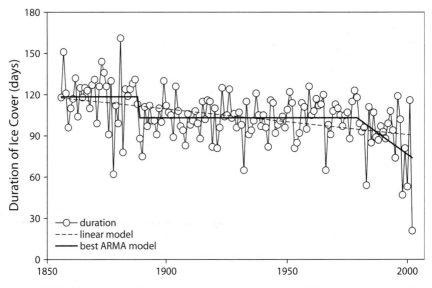

Figure 7.10. Time series of annual ice cover duration on Lake Mendota, a southern Wisconsin LTER lake, from winter 1855–56 to winter 2001–02. Two models are fit to the data, a linear model and an autoregressive-moving average (ARMA) model.

model (Box and Tiao 1975). With the linear regression, ice duration decreased by 18.6 days/century. For the time series model, a step change of 15.4 days to a shorter duration occurred about 1888 or 1889, and a recent declining trend of 12.6 days/decade occurred from 1980 to 2002. The intervention time series model explains more of the variability, $r^2 = 27$, than the linear model, $r^2 = 17$, and has implications somewhat different from those of the linear regression. Rather than a gradual warming in winter air temperatures since 1855, the climatic changes have occurred rather abruptly over two relatively discrete periods. Around 1890, fall-through-spring air temperatures warmed by approximately 1.5°C (Robertson et al. 1992), and a similar increase in fall-through-spring air temperatures of about 1.6°C occurred from about 1980 to 2002.

Lake Mendota is not unique among the LTER lakes in revealing trends in ice cover and increased rates of change in recent years. Over the past 151 years, Lake Monona in southern Wisconsin had a statistically significant, linear trend of 8.5 days/century later ice-on and 13.3 days/century earlier ice-off. The rate of change in ice cover on Lake Monona is more rapid than that for Lake Mendota, probably because a power plant that releases heated effluent into the lake was built and expanded from 1925 to 1959. The changes in ice cover on four other Wisconsin lakes with at least 90 years of data were all in the direction of warming, but not all were statistically significant (Magnuson et al. 2003). While none of the 10 LTER lakes with ice data in the north and the south had statistically significant slopes for the 21-year LTER period, they averaged 4.9 days/decade later ice-on and 3.2 days/

decade earlier ice-off. Each of the 10 lakes had a slope in the direction of later ice-on and earlier ice-off, indicating a robust pattern of reduced ice duration.

Trends for particular 20-year or 50-year windows oscillate between negative and positive slopes in the Lake Mendota record as well as in three records in New York, Finland, and Russia (Wynne 2000, Magnuson 2002a). These 20- and 50-year slopes are calculated for successive 20- and 50-year runs of the data in the same manner as one generates a set of running averages. The oscillations in the slopes occur because extreme winters have tended to occur in clusters of years along the record, and these impose reversals in the short-term slopes as the extreme short or the extreme long durations move through the 20- or 50-year windows. While these oscillations do not negate the long-term trend of decreasing ice cover that has occurred since 1853, they indicate that shorter time periods can produce trends that are opposite in direction to the longer-term 150-year trend. These oscillations also reveal that large-scale climatic drivers are important, as discussed earlier (Table 7.2).

Summer water temperatures over the LTER period, 1981 to 2001, for one southern lake and for the seven northern lakes had no consistent trends either near the surface at 1 m depth or at 1 to 2 m off the bottom. This result is not surprising, given that summer air temperatures have not changed significantly during the 21 years. Only one of the lakes, Crystal Lake in the north, had a statistically significant change; near-bottom temperatures decreased 1.2°C/decade. We have no explanation for this decrease, but perhaps a decrease in light penetration associated with a suite of biological changes that occurred in the lake increased the concentration of phytoplankton and reduced light penetration. In contrast to the ice dates, century-scale water temperature data are limited. Lake Mendota has extensive water temperature data from 1894 to 2003 but did not demonstrate any apparent trends in seasonal temperatures (Robertson 1989).

With respect to long-term trends in water level, the southern LTER lakes behave somewhat individually, each reflecting a combination of factors affecting their water budgets (Fig. 7.11). Positive and negative slopes occurred. Trends were toward higher levels in two of the three LTER lakes with long-term data, namely Lake Mendota (1916–2001) and Fish Lake (1966–2001), both in the south. Lake Mendota had a slope of 2.2 cm/decade (p <0.0001), and Fish Lake had an extreme slope of 73.3 cm/decade (p <0.0001). Fish Lake is an upland seepage lake, and Lake Mendota is a drainage lake in the Yahara River flowage. Buffalo Lake, another seepage lake, located only 7 km south of the northern LTER lakes, with a record from 1943 to 1988, also had a significant increasing trend of 3.7 cm/decade, but with high variability. Buffalo Lake is similar to Crystal Lake in clarity and surface area, but its maximum depth is 8 m versus 20 m for Crystal Lake. For the LTER years from 1984 to 2001, only two lakes had statistically significant slopes, namely Allequash Lake, with a slope of -16.5 cm/decade (p <0.0006), and Fish Lake, again with an extreme slope of 73.7 cm/decade (p <0.0001). Fish Lake has four missing years in this time period.

Explanations for the statistically significant trends (Fig. 7.5, 7.11) differ for each of the three LTER lakes: Fish Lake, Lake Mendota, and Allequash Lake. For Fish Lake, with no outlet stream, increased water levels totaled 2.6 m from 1966 to 2001, and homes have been flooded along the shoreline (Krohelski et al. 2002). Such

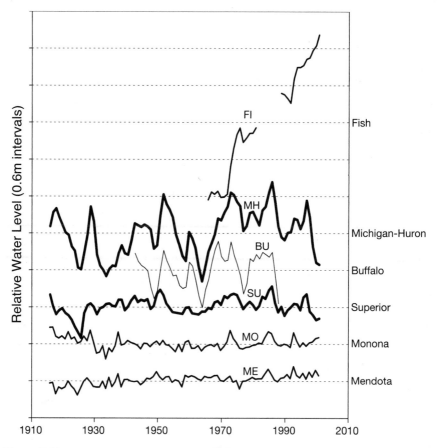

Figure 7.11. Long-term mean annual lake levels, arranged from most variable (top) to least variable (bottom) on the basis of standard deviation. Thick lines are Laurentian Great Lakes (Michigan-Huron and Superior), intermediate thickness lines are southern LTER lakes (Fish, Monona, and Mendota), and the thin line is Buffalo Lake, near the northern Wisconsin LTER lakes. Long-term means, designated by the dashed lines, are arbitrarily separated by an interval of 0.6 m. Annual means are calculated by averaging monthly mean data (Lakes Superior, Michigan-Huron, Mendota, Monona, and Fish Lake) or weekly samples (Buffalo Lake). Except for the large data gap for Fish Lake, missing data are ignored.

dramatic increases in water levels have occurred in Wisconsin for lakes without a surface outlet: Shell Lake, in the northwest, from 1935 to 1997 (Krohelski et al. 1999), Devils Lake, in the south center, from 1922 to 2002, and Buffalo Lake, near our northern LTER lakes. For these lakes without a surface outlet, the explanation for increased water levels appears to be long-term increases in precipitation and groundwater recharge as documented for Fish Lake, perhaps related to increased snowfall (Krohelski et al. 2002). For such lakes, alternatives such as relocating buildings or pumping water from the lake are being considered (Krohelski and Batten

1995, Krohelski et al. 1999, Krohelski et al. 2002). The explanation appears more complex for Lake Mendota, with both an outlet stream and several inlets. Water levels have increased by 19 cm over 85 years. As a consequence of increased development in the watershed and associated increases in impervious surfaces, the lake level has become more responsive to a standardized rain event in 1995 than it was in 1935; for example, a 15 cm rainstorm increased lake level about 15 cm in 1935 but 25 cm in 1995 (Wegener 2001). The proportion of rain falling in intense events has increased in the Great Lakes region (Kunkel et al. 1999, Kling et al 2003), further exacerbating increased runoff relative to infiltration. In addition, water management of the lock and dam at the outlet may have retained water in some seasons for a variety of reasons, such as flooding the spawning marshes for northern pike in the spring, holding water to reduce flooding around the downstream lakes, and maintaining adequate water for large pleasure boats in upstream channels. Regulatory limits are in place for high and low water levels; in the past, both limits have been exceeded in some of the years (Krug 1999). For Allequash Lake, decreases in water level have totaled about 30 cm in 17 years (Fig. 7.5). This change has resulted from the regular removal of the beaver dams at the outlet after 1990 as part of a program to lower water levels, especially in spring, to augment wild rice production in the shallow lower basin of Allequash Lake (personal communication, Ron Ekstein, Wisconsin Department of Natural Resources).

Climate Change and the Future

Future climate scenarios (Table 7.4) for the Laurentian Great Lakes region forecast significant warming at the century time scale. The air temperature and precipitation scenarios are available in the U.S. Global Change Research Program's "Great Lakes Overview" (Sousounis and Bisanz 2000) and in the Ecological Society of America and Union of Concerned Scientists' "Confronting Climate Change in the Great Lakes Region" (Kling et al. 2003). As with the historic data, possible future changes are more evident at the century scale than the decadal scale. Winter air temperature increases, according to the HadCM3 model, range from increases of 3

Table 7.4. Temperature change scenarios for the Great Lakes region in 2030 and 2090 with increases in greenhouse gases using General Circulation Models driven by possible increases in greenhouse gases given in the *Special Report on Emission Scenarios* (Nakicenovic and Swart 2000).

Climate Models	2030		2090		Source
	Summer	Winter	Summer	Winter	
HadleyCM2 & Canadian GCM1	1–3°C	unclear	2–4°C	4–7°C	Sousounis & Bisanz 2000
HadleyCM3 & Parallel Climate Model	unclear	unclear	3–8°C	4°C	Kling et al. 2003

to 5°C with a low-emission scenario and from 5 to 8°C for a high-emission scenario (Kling et al. 2003). Summer air temperature increases, according to the HadCM3 model, range from 3 to 4°C with the low-emission scenario and from 6 to 9°C with the high-emission scenario. Earlier scenarios suggested that air temperature will increase more in winter than summer (Sousounis and Bisanz 2000). Precipitation scenarios are more uncertain than those for temperatures. For the models in Sousounis and Grover (2002), precipitation is simulated to increase especially in summer. For the models in Kling et al. (2003), annual precipitation is simulated to increase by 10% to 20% by the end of the twenty-first century, but seasonal differences are notable, with summer precipitation declining by as much as 50%, while winter and spring precipitation increase. Snowfall may increase, with more winter precipitation, but snow depth may decrease owing to warmer temperatures and more frequent winter melting. Heavy precipitation, that is, the proportion falling in extreme events, is simulated to increase for the Great Lakes region (Sousounis and Grover 2002, Kling et al. 2003), consistent with historic increases in extreme rain events in the Great Lakes region (Kunkel et al. 1999, Kling et al. 2003). By 2095, the summer climate of Wisconsin, according to a scenario that uses the HadCM3 model, would be similar to the climate of Arkansas and eastern Oklahoma today (Kling et al. 2003, Magnuson et al. 2003).

The physical limnology of the LTER lakes should respond to such climate changes, according to studies by Robertson and Ragotzkie (1990a) and De Stasio et al. (1996). Lake Mendota's response to changes in air temperature were simulated using the Dynamic Reservoir Simulation Model of Imberger and Patterson. For each 1°C increase in air temperature, surface water temperatures increased about 0.4 to 0.85°C, bottom temperatures increased from 0.0 to 0.3°C, and the length of stratification increased about 3 to 4 days (Robertson and Ragotzkie 1990a). The response of four LTER lakes to climates with doubled carbon dioxide (De Stasio et al. 1996) were simulated using four 1980s vintage General Circulation Models (GCMs) and three base climates drawn from a warm, intermediate, and cool year in the LTER records. The lakes were Trout, Sparkling, and Crystal Lakes in the north and Lake Mendota in the south. A total of 48 simulations were run with four lakes, by four GCMs, by 3 base years. Increases in annual air temperatures with doubled carbon dioxide ranged from 1°C to 6.5°C among the four General Circulation Models, with the greatest warming occurring in summer (De Stasio et al. 1996). Across all lakes, General Circulation Models, and the three base climates, summer epilimnetic temperatures increased from 1°C to 7°C, maximum summer epilimnetic temperatures increased an average of 4.6°C, and maximum bottom temperatures increased an average of 2.4°C. Again across all simulations, average duration of stratification increased by 10 days, and average mixing depth decreased by 0.3 m. Only Crystal Lake simulations indicated a large change in mixing depth, averaging 1.2 m shallower.

The changes in temperature and stratification would be expected to change the size of the thermal niche for warmwater, coolwater, and coldwater fishes (Magnuson et al. 1997b). Changes in thermal structure of lakes simulated from the climate with doubled carbon dioxide substantially increased the thermal habitat for warmwater and coolwater fishes (De Stasio et al. 1996). The greatest increases were for warmwater fishes like bluegill and largemouth bass. In one-third of the simulations, the thermal niche decreased for coldwater fishes like cisco, lake trout, and lake white-fish. Cisco is, or has been, common in most Wisconsin LTER lakes and would be stressed by warming. But the simulated declines in thermal habitat underestimate the negative influence on the coldwater fishes because the simulations do not in-clude the possibility that the deep, cold waters will become hypoxic or anoxic. Lower dissolved oxygen levels would be expected to result because summer stratification would be longer and oxygen depletion would occur for a longer time and thus be more likely to reach a critical threshold or zero. Simulations by Stefan et al. (1995) included the oxygen depletion process, and their results showed a decline in coldwater habitat in Minnesota lakes. The loss of coldwater habitat was greater for shallower and more eutrophic lakes but occurred in deep oligotrophic and me-sotrophic lakes, as well. Thus, climate warming would increase the likelihood that cisco would be extirpated in eutrophic Lake Mendota, Lake Monona, and Fish Lake, in the south and in the relatively shallow Big Muskellunge Lake, in the north. Cisco already appears to have been extirpated from Sparkling Lake owing to the inva-sion of rainbow smelt (Chapter 8); climate change is a slower process than an inva-sion of smelt in extirpating the cisco. Warming temperatures would likely increase the rates of invasion by species and genotypes now found in the southern United States but not yet present in Wisconsin (Magnuson et al. 1997b, Gitay et al. 2001, Kling et al. 2003).

Hydrological impacts of a changing climate are complex because precipitation and evaporation (or evapotransporation) are competing influences. Warmer tem-peratures contribute to greater rates of evapotranspiration and increase the likeli-hood of a negative water budget and thus lower lake levels, unless precipitation increases occur at an even greater rate. Basin discharge from Lake 240 at the Ex-perimental Lakes Area in western Ontario did decline during a 20-year warming period as evapotranspiration increased (Schindler et al. 1996b). In the few scenarios projected for lake levels in the Wisconsin area, water levels have been projected to decline. A scenario for Crystal Lake at our northern location (D. P. Hamilton, per-sonal communication) simulated declines in water level of 1.0 to 1.9 m. Other simu-lations (Sousounis and Bisanz 2000, Lofgren et al. 2002) have been done for Lake Superior and Lake Michigan-Huron, which in the past have varied in a coherent way with some of the Wisconsin LTER lakes. These scenarios generally project declines for the adjacent Great Lakes but range from 0.1 m to -1.4 m by the 2090s (Kling et al. 2003). Stream baseflow, seepage-lake levels, runoff, and groundwater levels have been increasing in nonurban areas of southern Wisconsin but not in the Trout Lake area (Magnuson et al. 2003). The scenarios given in this section sug-gest that these increasing trends should reverse during the 2000s.

Past and Future Climate Dynamics

What can we say about future climate trends? If long-term historic changes and future scenarios agree in direction of change, we suggest that the future direction is relatively certain. If they disagree or the evidence for either is weak, we suggest that the future direction is more uncertain. On the basis of historic trends and future scenarios presented earlier, we provide the following expectations (Table 7.5). The climate at the North Temperate Lakes LTER site is expected to continue to warm. The warming may be greater for the southern location at Madison than for the northern location at Trout Lake, judging from past trends and the potential climate buffering by Lake Superior in the north (Fig. 10 in Kling et al. 2003). The present difference in air temperatures between the southern and the northern LTER locations in Wisconsin (Table 7.1) is smaller than the increases in temperature, according to the HadCM3 scenarios, by the end of this century (Table 7.4).

Physical variables of the lakes that are strongly coherent between lakes would be expected to track the changes in climate. Surface water temperatures would be expected to increase, ice-on would occur later, and ice-off would occur earlier. The hypolimnetic waters would be more likely to be hypoxic or anoxic. Other temperature-related changes of the lakes would occur but would be more lake-specific.

Evapotranspiration, precipitation, and the frequency of heavy rainfall events should increase. The winter season should shorten. Changes in water levels are uncertain because historical changes and changes in future scenarios move in opposite directions. Observations suggest that the climate has been getting wetter and the rain events more intense; in combination, these changes produce greater rates of runoff to infiltration. Future simulations are complex because increases in evapotranspiration driven by higher temperatures suggest that climate will become drier, while continued increases in intensity of rain events suggest more runoff and flooding during stormy periods. The indications are that water levels and baseflow will switch from the increases of the past to decreases in the future. Drought would be more likely during late summer and fall, and risk of flooding in summer will increase with the increase in heavy rainfall events. Flooding would be less likely in the north than in the south owing to the sandy till and groundwater-moderated stream flow.

Interdecadal changes should persist and will influence lake dynamics. The large-scale driver that will be most important is uncertain. Over the past century or so, the influence of these drivers has varied in magnitude and direction at least with respect to changes in lake ice cover. Future scenarios for the strengths and dynamics of these drivers are uncertain and may well be influenced by greenhouse warming (Houghton et al. 2001, Gillett et al. 2003). What is certain is that the lakes will continue to respond to these dynamics at a variety of time scales.

In conclusion, climatic and lake physical variables are expected to continue to be dynamic at interdecadal and century time scales. Trends associated with warming should continue and become more apparent at the century time scale. Decadal trends, as has been observed in the LTER years 1981–2002, may be obscured by the large interannual and interdecadal variability. Climate change and variability can be expected to drive significant processes of lake physics and watershed hydrology.

The evolving LTER data set is comprehensive and beginning to be long enough to permit the discovery of new facets in the relations between climate and lake dynamics that include not only the physics but also the dynamics of lake and watershed chemistry and biology, and perhaps human behavior. The latter aspects of lakes and climate change are largely unexplored.

Table 7.5. Long-term climate and limnological changes in northern and southern Wisconsin based on historic trends in climate and limnology and on future climate change scenarios.

Parameter	Historical Changes (1904–2001) North	Historical Changes (1904–2001) South	Future scenarios by late 2000s	Sources	Uncertainty for Wisconsin[a]
Air Temperature (1904–2001)					
Summer	decrease	increase	increase	Sousounis & Bisanz (2000), Kling et al. (2003)	low
Winter	none	increase	increase	"	moderate
Spring	increase	increase	increase	"	low
Fall	none	none	increase	Kling et al. (2003)	high
Precipitation (1904–2001)					
Summer	decrease	increase	increase / no change, decrease	Sousounis & Bisanz (2000) / Kling et al. (2003)	high
Winter	increase	none	increase	Sousounis & Bisanz (2000), Kling et al. (2003)	moderate
Snowfall	increase	increase	decrease / increase	Sousounis & Bisanz (2000) / Kling et al. (2003)	high
Intense Events	increase	increase	increase	Sousounis & Grover (2002), Kling et al. (2003)	low
	Great Lakes Region — Kunkel et al. (1999), Kling et al. 2003		Great Lakes Region		
Limnology (> 90 years)					
Surface Water Temperature	no information		increase	Robertson & Ragotzkie (1990a), DeStasio et al. (1996)	low
Bottom Water Temperature	"		increase	"	low
Duration Summer Stratification	"		increase	"	low
Depth Mixed Layer	"		decrease	"	moderate
Hypolimnetic Dissolved Oxygen			decrease	Stefan et al. (1995)	low
Ice Cover	decrease	Great Lakes & inland lakes	decrease	Lofgren et al. (2002), Hamilton et al. (2002)	low
Water Levels & Flows	various — Magnuson et al. (2003)	Great Lakes & inland lakes	decrease — Great Lakes & inland lakes	Fang and Stefan (1998) "	moderate

[a]Criteria are based on consistency between long-term historical observations and future scenarios, consistency between northern and southern Wisconsin or among scenarios, and magnitude of changes observed or simulated.

Summary

Climate is a major external driver of lake dynamics. Climatic forcing acts over years, decades, and centuries, and from individual lakes to lake districts and regions to the globe. Wisconsin has a continental climate at temperate latitudes modified somewhat by Lakes Michigan and Superior. Weather during the LTER study from 1981 to 2001 is similar to that from 1904 to 1980 except that extreme cold years are underrepresented and extreme warm years are overrepresented. Limnological climates, that is, water temperatures, thermal stratification, water levels, and ice cover, differ among LTER lakes and between northern and southern Wisconsin LTER lakes as a group. The coherence of interannual dynamics of limnological climates can be very high among lakes for ice-off dates and near-surface water temperatures in summer; moderate for ice-on dates; low for water levels, mixed layer depth, and winter water temperatures; and very low for bottom water temperatures in summer. Coherence persists between northern and southern LTER lakes but is lower than within lake-district coherence. Water levels are more coherent for lakes in similar positions in the landscape than for other pairings. LTER lakes share some of their variability with Lakes Michigan-Huron and Superior owing in part to coherent patterns of evaporation, but interannual differences in precipitation may be even more important to interannual variation in water budgets. Climate dynamics have a large regional footprint across the Great Lakes region. For example, near-surface water temperatures in summer of LTER lakes are coherent between southern and northern Wisconsin and with lakes in Ontario. Large-scale climatic drivers such as the El Niño Southern Oscillation, the North Atlantic Oscillation, the variation in the strength of the Aleutian Low, and intense volcanic eruptions are reflected in the interannual and interdecadal variation in ice dates of the LTER lakes and other lakes around the Northern Hemisphere with greater and lesser influence among regions. For example, the southern Wisconsin LTER lakes are more responsive to ENSO than are the northern Wisconsin lakes. Long-term trends are apparent in climatic and lake climate data at the century or longer time scales. Generally, air temperatures have warmed, precipitation has increased, and ice cover durations have shortened. Trends over the LTER study from 1981 to 2001 usually are not statistically significant owing to high interannual variability. The declining durations of ice cover are occurring across the Great Lakes region and the Northern Hemisphere. Long-term trends in water level have been more erratic, with both increases and decreases apparent; most extreme is Fish Lake in southern Wisconsin, with an increase of about 70 cm per decade. Climatic changes are expected to continue; scenarios that take in to account greenhouse gas increases suggest that warming will continue, ice cover will continue to decline, and precipitation events will become more extreme. Lake warming, changes in thermal stratification, and hydrology are expected, as well.

8

Ecological Change
and Exotic Invaders

Karen A. Wilson
Thomas R. Hrabik

A hundred years of faster and bigger transport has kept up and intensified this bombardment of every country by foreign species, brought accidentally or on purpose, by vessel and by air, and also overland from places that used to be isolated.
—Charles S. Elton (1958, p. 29)

Advances in invasion ecology are often hindered by a gross lack of preinvasion information. In aquatic systems, this inadequacy is exacerbated by the challenges inherent in observing underwater habitats, and, as with terrestrial invasions, many invasions are often well under way before they are discovered. The North Temperate Lakes Long-Term Ecological Research (LTER) site has provided an especially valuable research platform for the analysis of the spread and influence of exotic species by providing pre-invasion data and the early detection of invaders. Our long-term data function as an ecological context for short-term investigations into mechanisms behind species invasions. Without long-term information, observations of ephemeral interactions and time lags in response to the invader are lost.

The serendipitous discovery of two invasive species, the rainbow smelt and the rusty crayfish, in the LTER lakes in northern Wisconsin has led to a research program that has provided insights on the invasion patterns and mechanisms of these two organisms. We have been able to extrapolate our results to assess the threat represented by these organisms to other lakes in the region. Lakes, owing to their island nature (Chapter 4), provide ideal systems through which to study the invasions of exotic organisms. Here, we review the history of recent invasions, mechanisms, and long-term impacts of these species on aquatic communities in the lakes. The invasions we use as examples are ongoing in Wisconsin's Northern Highlands Lake District and elsewhere; we discuss future invasion and management scenarios in the context of the LTER findings and the role LTER may play in these scenarios.

North Temperate Lake Invaders

We distinguish between species native to a region that are expanding their range to new localities and exotic invaders that are not native to the region. In particular, we discuss two species that originated from native ranges well removed from the Northern Highlands Lakes District, the rusty crayfish (*Orconectes rusticus*), originally from the Ohio River Drainage (Fig. 8.1), and the rainbow smelt (*Osmerus mordax*), which originated along the Atlantic coast of North America (Fig. 8.2). The invasion histories of these two species are well documented, and human-mediated dispersal mechanisms are well known. However, other species have invaded these lakes since the arrival of European settlers. Game and food fish have been stocked in northern Wisconsin during the twentieth century, potentially resulting in large unrecorded impacts on lake food webs (Vander Zanden et al. 1999). These historical introductions suggest that, even in our relatively long data set, the conditions before invasion likely represent the ghosts of human-mediated invasions of the past in the sense used by Connell (1980).

Rusty Crayfish

Early inventories of the benthic biota from northern Wisconsin lakes paint a very different picture from what we observe today. In the early 1930s, *Orconectes virilis* was the only fully aquatic crayfish species found in the Northern Highlands Lake District (Creaser 1932). By 1970, however, Capelli (1975) observed four other crayfish species (three of them *Orconectes* species) inhabiting these waters. *O. propinquus*, originally present only in southern and eastern Wisconsin (Creaser 1932), invaded (or reinvaded) the region after 1932, becoming the dominant crayfish species by 1970. Rusty crayfish likely was introduced in the 1960s, and its range and abundance have increased over the past three decades (Fig. 8.3).

Rusty crayfish replace other orconectid species through a variety of mechanisms (Lodge and Hill 1994). During daylight hours, all crayfish respond behaviorally to the presence of predatory fish by seeking shelter, primarily in complex cobble substrate (Stein and Magnuson 1976, Hill 1994). In the presence of other crayfish species, rusty crayfish successfully preclude other species from accessing shelter (Garvey et al. 1994, Hill and Lodge 1994), potentially increasing the vulnerability of other species to fish predation. In addition, *O. propinquus* generally are smaller than rusty crayfish and thus experience greater predation by size-selective fish predators (DiDonato and Lodge 1993, Roth 2001). High growth rates enable rusty crayfish to obtain a size refuge from fish predators more quickly than other crayfishes present (Hill and Lodge 1999). Another mechanism of species replacement is hybridization. Rusty crayfish readily hybridize with *O. propinquus*, and hybrids are fertile (Perry 2001a), such that most populations of rusty crayfish likely include many hybrid individuals.

Several factors influence the distribution of rusty crayfish. All crayfish are limited by the concentration of calcium, requiring a minimum of approximately 2.5 ppm (Capelli and Magnuson 1983). In addition, unlike *O. virilis* and *O. propinquus*, the distribution of rusty crayfish among suitable lakes appears to be related to human

Figure 8.1. Rusty crayfish (*Orconectes rusticus*) from Trout Lake in 1998. Rust-colored spots on either side of the carapace and rust coloration on the tail superficially distinguish the rusty crayfish from other crayfishes in northern Wisconsin, although coloration varies from water body to water body. This individual is approximately 45 mm in length from the tip of the rostrum to the base of the tail. (K. Wilson)

Figure 8.2. Rainbow smelt (*Osmerus mordax*) from Lake Superior that range in age from young of year to reproductive adults. (J. Magnuson)

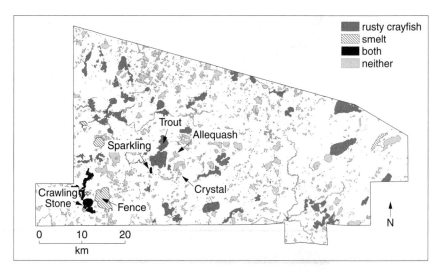

Figure 8.3. Known distribution of rusty crayfish and rainbow smelt in Vilas County, Wisconsin, in 2002. Lakes invaded only by rusty crayfish (74 named lakes) are gray; lakes invaded only by rainbow smelt (11 named lakes) are lined. Only a few lakes (Sparkling, Pokegama, Flambeau, Long Interlaken, and Crawling Stone) contain both species and are black. Lakes mentioned in the text are identified. Records are based on Capelli (1982), Capelli and Magnuson (1983), Olsen et al. (1991), Hrabik (1999), Perry et al. (2002), records of the Wisconsin Department of Natural Resources, and personal observations of the authors. (Map prepared by J. Maxted and K. Lord, UW-Madison Center for Limnology)

activity and distance from major roads (Capelli and Magnuson 1983). These patterns of invasive crayfish distribution suggest that humans have increased greatly the probability of rusty crayfish introductions. Recreational anglers using crayfish for bait (Ludwig and Leitch 1996) and lakeshore owners interested in controlling aquatic plants (Magnuson et al. 1975) are considered the most likely vectors of rusty crayfish invasions, although information is sparse on the number of lakes with early introductions of rusty crayfish. In 1982, Wisconsin outlawed the use of crayfish as live bait and the introduction of live crayfish to inland waters (Wisconsin Administrative Code 2002). However, rusty crayfish are common throughout the region today and continue to spread (Capelli and Magnuson 1983, Olsen et al. 1991), suggesting that these species are capable of interlake dispersal through natural and human-mediated means.

In 1979, just before the North Temperate Lakes LTER program began, rusty crayfish were first detected near a heavily used boat landing in Trout Lake (Lodge et al. 1986). This set the stage for rusty crayfish research at the site. Early work provided pre-invasion information on conditions in Trout Lake and suggested that rusty crayfish were likely to have large impacts on the littoral zone community (Capelli 1975, Magnuson et al. 1975, Stein 1977, Lorman 1980). Initial expectations were that rusty crayfish would reduce populations of existing crayfish, macrophytes, macroinvertebrates associated with macrophytes, and the abundance of some littoral fish species. As a result of these changes in the littoral zone, littoral

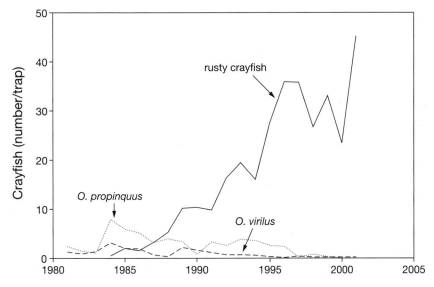

Figure 8.4. Change in abundances of rusty crayfish, *Orconectes propinquus*, and *O. virilis* in Trout Lake, northern Wisconsin, from 1980 to 2001. Crayfish abundances are from the the annual LTER summer survey.

nutrients likely would be assimilated into the trophic structure through phytoplankton, fueling greater pelagic productivity. To test these predictions, LTER researchers chose permanent sampling sites at various distances from the point of first rusty crayfish detection. Over time, these sampling sites have been invaded sequentially, representing multiple opportunities to observe community change within one lake.

In contrast to the rapid invasions observed by species with pelagic life stages, rusty crayfish took over 18 years to completely encompass Trout Lake (~ 1979–97) at a rate of 1.68 km/yr, suggesting a relatively slow, natural dispersal (Perry et al. 2001b). Detecting rusty crayfish populations in other lakes has required long time series, as well (Capelli 1982, Olsen et al. 1991), although comparisons with these more infrequently sampled invasions suggest that in Trout Lake the rate of spread is slow. In Trout Lake, rusty crayfish have reached and maintained abundances up to 20 times the abundance of native crayfish alone (Fig. 8.4). As predicted in laboratory studies and observations of other lakes in the region (Capelli 1982, Olsen et al. 1991), over time rusty crayfish have displaced *O. virilis* and *O. propinquus* (Fig. 8.4). Climate may have played an important role in facilitating rusty crayfish increases through its effects on the Trout Lake fish community. The number of ice-free days and summer water temperature varied from well above to well below average through the late 1980s and early 1990s. These temperature fluctuations may have temporarily suppressed recruitment of predators and competitors of rusty crayfish, allowing the population to establish and grow (Willis 2003).

Changes in the abundance and species composition of crayfish in Trout Lake have been accompanied by sometimes dramatic transformations of the littoral zone community. The most obvious change that accompanied rusty crayfish invasion in Trout

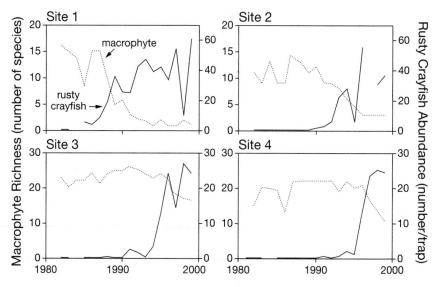

Figure 8.5. Changes in macrophyte species richness related to changes in rusty crayfish abundance at four sites in the littoral zone of Trout Lake, Wisconsin. Redrawn from Wilson (2002) with permission.

Lake has been a dramatic loss of submersed aquatic macrophyte species and biomass. Submersed macrophytes are important habitat for fishes and invertebrates (Sagova and Adams 1993a), are food for waterfowl and other herbivores (Fassett 1957), and play a significant role in nutrient cycling between the sediment and the water column (Carpenter and Lodge 1986). In Trout Lake, as many as 14 of 15 macrophyte species have been lost at sites occupied by rusty crayfish for many years (Fig. 8.5). Over the course of the rusty crayfish invasion, macrophyte communities in Trout Lake have shifted from diverse communities to communities dominated by low biomass of *Potamogeton amplifolius* and *P. robbinsii,* two evergreen, tough-leaved species (Wilson 2002). Loss of macrophytes in the presence of rusty crayfish has been observed in several natural and experimental systems (Magnuson et al. 1975, Lorman and Magnuson 1978), but long-term records of macrophyte change in other lakes are rare.

Long-term observations and experiments suggest that crayfish have had both direct and indirect negative effects on macrophyte establishment and survival. Crayfish directly consume macrophytes in feeding trials (Cronin et al. 2002) and in the field (personal observation K. A. Wilson). In early crayfish enclosure experiments, rusty crayfish clipped single-stemmed plants at the sediment surface and either consumed them or let them float free (Lodge and Lorman 1987, Olsen et al. 1991). More recently, experiments using open-topped crayfish exclosures suggest that rusty crayfish decrease establishment of submersed macrophytes by consuming seedlings (Fig. 8.6, top). However, most macrophytes in Trout Lake reproduce primarily with vegetative fragments (personal observation K. A. Wilson) that can be dislodged by rusty crayfish. Thus, over the long-term, rusty crayfish prevent the establishment of rooting plant fragments by consumption, dislodging the fragments from the substrate,

Figure 8.6. Effects of crayfish on plants in Trout Lake in 1998. Top: Crayfish exclosure at 2.5 m depth in Trout Lake. The annuals *Najas flexilis* and *Chara* cover almost 100% of the substrate inside the exclosure. The bare sand outside the exclosure is pitted with crayfish tracks. Walls are approximately 14 cm high. For scale, the white shell is approximately 4 cm long. (K. Wilson) Bottom: *Potamogeton amplifolius* (musky weed or cabbage weed) in Trout Lake after rusty crayfish were abundant. (K. Wilson)

and reducing the extant macrophytes that produce new fragments and function to anchor fragments (Wilson 2002). Feedbacks between the loss of mature macrophytes and establishment failure of rooting fragments may have accelerated the loss of macrophytes in Trout Lake and likely will reduce recovery rates if rusty crayfish populations decrease in the future (Wilson 2002).

Rusty crayfish are selective grazers (Lorman 1980, Lodge 1991). In Trout Lake, small, delicate macrophyte species (*Chara*, *Najas*, and thin-leaved *Potamogeton* species) have been lost as rusty crayfish became established (Wilson 2002). In many cases, these species were most desired by crayfish in feeding trials. However, with increases in rusty crayfish abundance, the macrophyte community has shifted to communities dominated by *Potamogeton amplifolius* (Fig. 8.6, bottom) (Wilson 2002). This species is preferred highly by crayfish in feeding trials conducted with plant shoots from Trout Lake and from nearby Sparkling Lake (Lodge 1991). But, unlike many of the other species of macrophytes in Trout Lake, *P. amplifolius* is evergreen, with large leaves, thick stems, and sturdy shoots that arise from underground rhizomes. Together, these traits should assist in the long-term persistence of this species through reduced herbivory and rapid re-establishment of individual plants.

Snails are highly susceptible to crayfish predation and rarely are found at high densities in the presence of crayfish (Lodge et al. 1998). Snails at densities of 25,000/ m^2 occurred in Trout Lake soon after rusty crayfish first appeared in the southeast bay in 1984 (Lodge and Lorman 1987), but by 2000 had declined precipitously to 4.6 snails/m^2 in the same area (Wilson et al. 2005). The loss of snails, a major grazer of attached algae on macrophytes, may have negative consequences for surviving macrophytes. Although crayfish consume attached algae (Lodge et al. 1999), they are not as efficient grazers as snails (Luttenton et al. 1998). Thus, replacement of snail grazing by crayfish grazing potentially could result in an increased biomass of attached algae that may reduce the productivity of macrophytes by shading (Brönmark 1985).

Fishes are the longest-lived animals in Trout Lake, with high mobility and, presumably, the ability to use areas of the lake with few or no rusty crayfish. Thus, we expected to see slow responses of the fish community to rusty crayfish invasion. Changes in the Trout Lake fish community became apparent in the late 1980s. Rusty crayfish appeared to negatively affect benthivorous fishes, including sunfish (bluegill, *Lepomis macrochirus*, and pumpkinseed, *Lepomis gibbosus*) and mottled sculpins (*Cottus bairdi*), which became less prevalent components of the Trout Lake fish community (Willis 2003). In addition, two small-bodied cyprinid species experienced population explosions in the mid-1990s, possibly as compensation for reduced competition and predation from declining sunfish and mottled sculpin populations (Willis 2003).

Currently, only adult bluegill and pumpkinseed sunfish have declined significantly in abundance with increases in rusty crayfish, while abundances of several fish species that readily consume smaller rusty crayfish have remained similar over time (Wilson et al. 2005). Long-term records suggest that bluegill and pumpkinseed have experienced reductions in breeding success and survival of young of the year as compared to populations in a nearby lake where rusty crayfish are not present. Direct consumption of eggs by crayfish, loss of macrophyte habitat, and competition for invertebrate prey with crayfish may have contributed to a long-term decline in the sunfish species (Dorn and Mittelbach 1999).

Overall, the invasion of rusty crayfish in Trout Lake was associated with a loss of many species of macrophytes, decreases in invertebrate abundance, and a gradual decline in some fish species. We expect that changes to the littoral food web are large (Fig. 8.7), with large rusty crayfish being an energetic dead-end because they are both abundant and inaccessible to predatory fishes. Before rusty crayfish invasion, aquatic macrophytes provided habitat and a food source for a diversity of macroinvertebrates, including native crayfish that were consumed by predators such

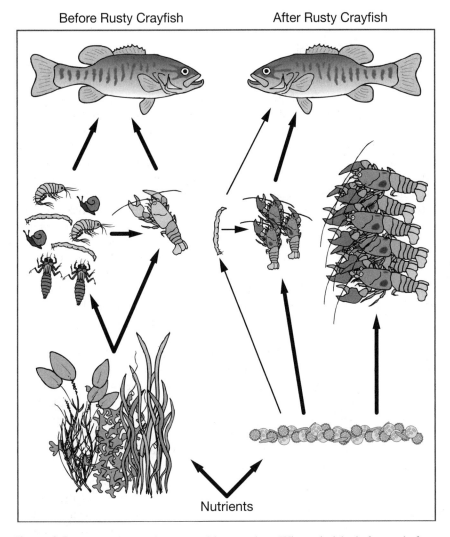

Before Rusty Crayfish   After Rusty Crayfish

Nutrients

Figure 8.7. Littoral food webs expected in a northern Wisconsin lake before and after a rusty crayfish invasion. Arrow thickness suggests the importance of a link.

as smallmouth bass. At high abundances of rusty crayfish, benthic algae dominates littoral primary productivity. Macroinvertebrates are rare through loss of habitat and direct predation by crayfish. Even though small rusty crayfish are consumed in great numbers, rusty crayfish quickly reach a size refuge from fish predation, possibly leading to an energetic dead end in the littoral food web. Since 1997, when rusty crayfish were first present around the entire littoral zone of Trout Lake, their abundances have continued to rise in some areas but stabilized in other areas. Over time we expect to see additional changes associated with rusty crayfish, especially for long-lived fish species.

Rainbow Smelt

Like the rusty crayfish, rainbow smelt (*Osmerus mordax*), an anadromous fish species that is native to the Atlantic coastal areas of North America, have expanded their range during the 1900s through intentional and accidental introductions (Becker 1983, Nellbring 1989). Rainbow smelt exist across a wide range of environmental conditions. Their ability to exhibit different life history strategies under various conditions make them well adapted for invasion into new environments. Smelt are now reported throughout North America, and they have colonized many different types of habitats, including reservoirs, as well as small and large lakes (Loftus and Hulsman 1986, Franzin et al. 1994, Jones et al. 1994, Hrabik and Magnuson 1999). Smelt in the upper Great Lakes region are thought to have originated from an intentional introduction in 1912 into Crystal Lake, Benzie County, Michigan. The exotic smelt from Crystal Lake in Michigan is believed to have invaded Lake Michigan via a connecting stream (Creaser 1927); by the 1940s, smelt were abundant in all of the upper Great Lakes (Dymond 1944, Van Oosten 1947, Christie 1974). Populations of smelt in the Great Lakes have served as sources for subsequent invasions into inland waters. As a result, the inland distribution of smelt continued to expand in the 1980s and 1990s (Evans and Loftus 1987, Mayden et al. 1987, McLain 1991, Franzin et al. 1994, Hrabik et al. 1998, Hrabik and Magnuson 1999). Several small lakes in northern Wisconsin have smelt populations (Fig. 8.3), and we continue to find additional lakes with smelt populations each year. The migratory behavior of adult smelt defines their anadromous behavior in marine systems; this behavior is retained in freshwater systems and likely facilitates dispersal through lakes with connecting streams (Hrabik and Magnuson 1999). The recent invasion of smelt into several isolated lakes in Wisconsin suggests that humans are introducing smelt (Hrabik and Magnuson 1999).

Smelt invasions into small lakes in the upper Great Lakes region likely are facilitated through stocking by individuals and management agencies. Possible mechanisms for many of the introductions not initiated by management agencies include (1) intentional introduction of live smelt by members of the public, (2) the escape of live smelt used as bait by anglers, (3) unintentional introductions of fertilized gametes by smelt fisherman cleaning ripe smelt and discarding smelt viscera, and (4) the cleaning of nets used to catch smelt during spawning runs at other lakes.

After smelt become established in new ecosystems, substantial changes in the local fish communities often occur through a variety of mechanisms. Negative effects of the smelt invasions on other fish species are clear from long-term data from two LTER lakes, Sparkling and Crystal Lakes. The long-term data sets have been instrumental in allowing researchers to identify the mechanisms of interspecific interaction, rates of decline, and extinctions of native species. For example, predation by smelt has led to the apparent extinction of native cisco (*Coregonus artedii*) in Sparkling Lake over two decades (Fig. 8.8, top). In Sparkling Lake, spatial overlap between adult smelt and young-of-year cisco led to predation-induced extirpation of a native species (McLain 1991, Hrabik et al. 1998). In spring, young ciscos inhabit the cool epilimnion of small lakes; then, by early summer, as the epilim-

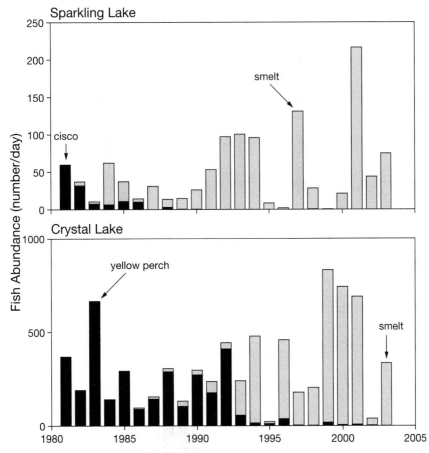

Figure 8.8. Abundance of exotic rainbow smelt and native cisco in Sparkling Lake and exotic rainbow smelt and native yellow perch in Crystal Lake from 1981 to 2003. Catch per day are from the annual LTER surveys. Updated from Hrabik et al. (1998).

nion warms, they migrate to the cool hypolimnetic waters. During the period of several cisco year-class failures in Sparkling Lake, adult smelt were caught only in the hypolimnetic waters during the summer, and a substantial proportion ate small fish. Because young-of-year cisco was the only small fish species located within the hypolimnion during the summer, adult smelt were probably feeding on them. These findings suggest that adult smelt caused the precipitous decline in cisco through predation-induced recruitment failure.

Competitive interactions between smelt and native fishes can lead to significant declines in native fish populations (Fig. 8.8, bottom). In Crystal Lake, the negative effect of smelt on a historically dominant yellow perch (*Perca flavescens*) population has been evident in long-term observations. Yellow perch now represent a small and declining component of the fish community. Long-term records of temperature, zooplankton, and fish size were combined with short-term seasonal studies of body

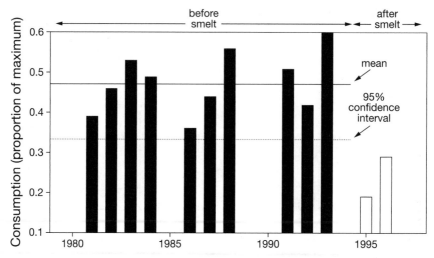

Figure 8.9. Proportion of maximum consumption realized by young-of-year yellow perch in Crystal Lake estimated from 1981 through 1996. A value of one indicates that fish are feeding at the maximum consumption rate. The solid horizontal line = the mean proportion in years when young-of-year perch survived to August. The dashed horizontal line = the lower 95% confidence interval for the same period. In 1995 and 1996, smelt were abundant, and young-of-year perch ate less and did not survive to August. Redrawn from Hrabik et al. (2001) with permission of the National Research Council of Canada.

size, mouth gape, and feeding to estimate consumption rates of young-of-year perch and smelt to determine the ecological effects of the smelt invasion. Hrabik et al. (2001) used long-term lake temperature records and fish growth rates as inputs to bioenergetic models that estimated consumption rates and feeding success of young-of-year perch in years with low smelt reproduction and in years with large year classes of smelt. In these analyses, young-of-year perch fed at a much lower rate in years when young-of-year smelt were abundant. Low feeding rates were linked to year-class failures and to declines in abundance of perch. Both smelt and perch fed selectively on *Diaptomus minutus*, a small calanoid zooplankter. The consumption rates by young-of-year perch were correlated positively with the abundance of *Diaptomus*, indicating the importance of *Diaptomus* as prey. Other variables, including densities of six potential zooplankton prey species and two potential invertebrate competitors, were not related to consumption rates of young-of-year perch. A large year class of smelt in 1996 consumed a substantial proportion of the *Diaptomus* standing stock on each day, suggesting that competition with smelt was related to the decrease in perch feeding rates (Fig. 8.9). This evidence suggests that the presence of smelt has shifted the plankton community available to perch in Crystal Lake, resulting in decreases in perch reproductive success and population size.

The information contained in the long-term records was essential for the identification of mechanisms of negative interaction among exotic smelt and native fish species in these northern lakes. Short-term studies, 2 to 5 years long, would not have been able to identify the shifts in abundance and performance of the native fish community that were the result of invasions of exotic rainbow smelt. Long-term information on spatial overlap between species first suggested the possibilities of diet overlap and predation; short-term research was able to confirm the mechanisms involved, as diagrammed in Fig. 8.10.

Scaling Up to Regional Impacts

Conclusions based on long-term data on five of the LTER lakes in northern Wisconsin were tested at broader spatial scales using a larger set of lakes in the Northern Highlands of Wisconsin. We examined the confounding effects of water chemistry and surface water connections on the composition of fauna in the primary LTER lakes plus an additional 24 nearby lakes from 1998 to 2000. In this larger set of lakes, macrophyte species richness was lower in lakes with high densities of rusty crayfish for lakes with similar chemical properties, that is,

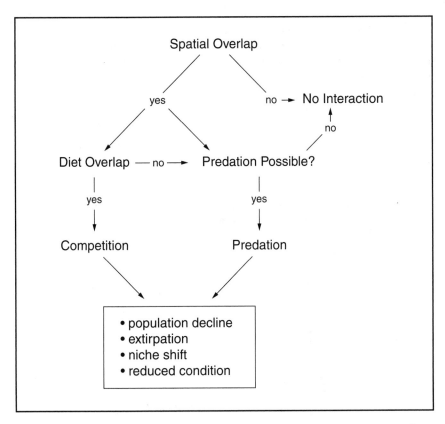

Figure 8.10. Potential interactions diagrammed between rainbow smelt and young-of-year and adult yellow perch. Observable consequences of competition or predation or both include the decline or loss of fish populations, niche shifts by surviving species, and reduced body condition of surviving species.

conductivity (Fig. 8.11). Surface-water inlets and outlets were not important predictors of macrophyte species richness. We conclude that rusty crayfish pose a significant threat to lakes throughout the region, and, on the basis of the long-term changes we observed in Trout Lake, we conclude that these changes may be dramatic, with negative consequences for recreational fisheries.

This broader-scale sampling identified new lakes with rainbow smelt. Several of these lakes have fish community characteristics consistent with the negative interactions observed in the LTER lakes, namely that cisco are generally rare in lakes with high smelt abundance. Again, our long-term studies on LTER lakes have relevance for other lakes across the Northern Highlands Lake District.

Management of Exotics

Rusty Crayfish

Rusty crayfish were recognized as a problematic species in the 1970s (Magnuson et al. 1975) in northern Wisconsin. In 1982, the State of Wisconsin outlawed the use of live crayfish as bait and the introduction of live crayfish to inland waters of Wisconsin (Wisconsin Administrative Code 1983) in an attempt to minimize the spread of rusty crayfish throughout the state. Commercial exploitation of rusty crayfish was explored (Arora and Wik 1988), although only a few commercial trappers remain active in northern Wisconsin. More recently, some proactive lake owners associations have voluntarily encouraged catch-and-release fisheries of smallmouth bass, a major predator of rusty crayfish, to reduce abundances of rusty crayfish (Fig. 8.12) although the impacts of these strategies are unknown as of yet.

Rainbow Smelt

Management actions aimed at predatory fishes may influence interactions between rainbow smelt and native cisco populations. In the early 1990s, the Lac du Flam-

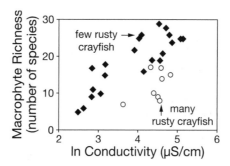

Figure 8.11. Number of macrophyte species in 29 lakes located in the Northern Highlands Lake District, Wisconsin. Conductivity alone explained 34% of variation in species richness (p = 0.001); the inclusion of rusty crayfish abundance explained an additional 30% of the variation in species richness (p<0.001). Redrawn from Wilson (2002) with permission.

Figure 8.12. A sign at a boat landing advocates catch and release fishing on crayfish predators in a lake near the LTER lakes in northern Wisconsin. In summer 1998, rusty crayfish had a mean catch rate of 7 per trap in Boulder Lake. (K. Wilson)

beau Tribal Natural Resource Department decreased tribal spearing and restricted sport fishery harvest by altering the bag limits and minimum size limits for the native predatory walleye (*Sander vitreum*) in Fence and Crawling Stone Lakes, which have been studied intermittently by researchers at the Center for Limnology, University of Wisconsin-Madison. Stocking efforts were increased to restore historical levels of walleye. The result was a substantial increase in the abundance of walleye, an associated reduction in the number and average body size of adult smelt, and a reduction in predation on young cisco by adult smelt (Krueger 2003). This management of walleye resulted in dramatic increases in the native cisco, owing to selective predation on smelt by adult walleye in late summer (Krueger 2003). Despite the high abundance of young cisco in both Fence and Crawling Stone Lakes, no young cisco occurred in walleye stomachs we sampled, suggesting that cisco may have an ability to avoid predation by walleye, whereas smelt do not.

Our observations on Sparkling Lake, however, indicate less influence of walleye on the smelt population than in Fence and Crawling Stone Lakes. Reasons for this may include the lower densities of walleye in Sparkling Lake resulting from more liberal fishing bag limits and lower stocking rates. In addition, smelt fishing during the spring spawning runs in Fence and Crawling Stone Lakes may have contributed to lower smelt populations. The increase in native cisco populations in Fence and Crawling Stone Lakes probably resulted from smelt population decreases that were direct results of spring smelt fishing and management actions that increased walleye predation on smelt.

Future Scenarios

Even with our relatively long time series, we cannot yet determine whether rusty crayfish will persist in invaded lakes at high densities or will boom and bust. Population declines after invasions have not been reported in published accounts, but populations did decline in at least one shallow lake (Kentuck Lake, Forest County, Wisconsin: personal communication Wisconsin Department of Natural Resources). Currently, Sparkling Lake, a LTER primary study lake, is being biomanipulated to reduce or

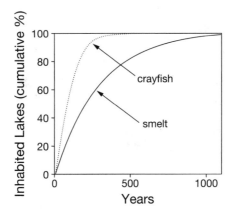

Figure 8.13. Scenarios for the cumulative percentage of lakes invaded by rainbow smelt (Hrabik and Magnuson 1999) and rusty crayfish in the Bear River watershed over 1000 years. The crayfish scenario was projected theoretically from differences between rusty crayfish and smelt in life history, dispersal abilities, and habitat requirements. Modified from Hrabik and Magnuson (1999).

eliminate well-established populations of both rusty crayfish and smelt (Martinez 2002, Turner et al. 2003). The hope is that invaders can be controlled by extensive trapping and by increased predation from large, predatory fish. Lessons learned may be applied to other invaded lakes.

The prediction of factors that influence the spread of exotic species will continue to be among the most important aspects of conservation ecology in the coming decades. Many studies focus on how exotics influence the native biota or ecosystem dynamics. However, conservation and management must focus on reducing the spread of harmful exotics. Hrabik and Magnuson (1999) constructed a spatially explicit population model to predict the dispersal of smelt within a watershed in Wisconsin. The model predicted that, at the current rate of introductions by humans, more than 50% of lakes habitable by smelt will be invaded within 200 years (Fig. 8.13). Crayfish have a broader set of habitat requirements than do smelt and likely have higher natural dispersal rates because they can survive warm temperatures and exist out of water for several hours. As a result, rusty crayfish will likely saturate watersheds more rapidly than smelt. Dispersal models highlight lakes that are less likely to be invaded and thus, given appropriate conditions, are prime candidates for conservation. However, introductions by humans are clearly the most influential source of the spread of invasive species. Limiting dispersal of exotics should begin by reducing the human vector for dispersal.

Summary

Long-term studies assist in the detection of new invaders by serving as the contextual baseline that is often absent in studies of invasive species. This temporal context allows examination of invader success, duration of dominance, ecological effects of the invader, and mechanisms of changes in ecosystems. In the 1980s and 1990s, rusty crayfish and rainbow smelt invaded three LTER lakes in northern Wisconsin. These species significantly altered the ecosystems by extirpating native species through predation and competition, changing patterns of biomass distribution and, in the case of rusty crayfish, altering physical habitat. As these invasions unfolded, the context provided by long-term research provided the information that revealed mechanisms of interaction. Predation by smelt on young-of-year cisco led to an extirpation of cisco augmented by the spatial overlap between adult rainbow smelt and larval cisco. Competition of young smelt with young-of-year yellow perch was the mechanism that led to precipitous declines in perch. Such findings, a result of having an emerging time series that includes key events and disturbances, is lacking in most cases of invasions and extinctions. Both rainbow smelt and rusty crayfish continue to invade regional lakes as well as systems well removed from the Northern Highlands. Interannual dynamics observed in the LTER lakes provide a strong predictive framework for understanding the ecological impacts, extirpation mechanisms, and regional dispersal of exotic invaders.

9

The Experimental Acidification of Little Rock Lake

Thomas M. Frost*

Janet M. Fischer

Patrick L. Brezonik

Mária J. González

Timothy K. Kratz

Carl J. Watras

Katherine E. Webster

Don't be too timid & squeamish about your actions. All life is an experiment. The more experiments you make, the better.
—Ralph Waldo Emerson (1842), in Porte (1982, p. 294)

L arge-scale manipulations have proven to be highly effective in developing a basic understanding of the functioning of ecosystems and in evaluating how they are affected by human-generated stresses (Schindler 1990, Carpenter and Kitchell 1993, Frost and Blood 1996). Consequently, large-scale manipulations are highly relevant to environmental policy and management (Carpenter et al. 1995), and they avoid the problems of extrapolating results from small-scale experiments that do not incorporate all components of the ecosystem (Schindler 1998). Although replication of the stress treatment is often impossible owing to costs or politics, the importance of unmanipulated reference ecosystems is recognized widely (Carpenter et al. 1995).

We added such a whole-ecosystem experiment to the repertoire of the North Temperate Lakes Long-Term Ecological Research (LTER) program with the experimental acidification of Little Rock Lake near the Center for Limnology's Trout Lake Station in northern Wisconsin (Fig. 1.5, top). The experiment allowed us to assess the effects of acidification on aquatic ecosystems by speeding up a process that normally occurs slowly from acid rain. Our interpretations of the

*Thomas M. Frost (see Memorial on page vii) died while this chapter was in progress. Janet Fischer completed, updated, and expanded the chapter and dedicates the chapter to Tom.

responses to acidification were strengthened by the use of a reference system and by comparison with other LTER lakes. Our results were relevant to policy makers and environmental managers and yielded new insights into more general issues such as biogeochemical cycles and ecosystem responses to stress. Here, we outline the basic design and rationale of the Little Rock Lake experimental acidification and highlight some of the insights that it generated.

Rationale and Genesis

During the late 1970s and early 1980s, acid deposition attracted considerable attention as an environmental stress that affects aquatic ecosystems in North America (Charles and Christie1991). Acid deposition was a particular concern because it affected ecosystems far from the pollutant source, including across national boundaries. Previous whole-lake experiments at the Experimental Lakes Area in western Ontario provided powerful insights about how acidification affects aquatic ecosystems (Schindler 1990); however, the generality of these findings for predicting responses of other lakes in other landscape types to acid deposition needed further exploration. In an effort to seek scientific guidance for policy making, the U.S. Environmental Protection Agency under the National Acid Precipitation Assessment Program supported the acidification experiment on Little Rock Lake, a two-basin lake located in close proximity to the LTER primary study lakes. John Eaton initiated the effort (Fig. 9.1). The project involved a multidisciplinary effort to characterize most aspects of the lake ecosystem and included researchers from the University of Minnesota, the University of Wisconsin-Superior, the Wisconsin Department of Natural Resources, the U.S. Geological Survey, the U.S. Environmental Protection Agency, and the Center for Limnology's LTER program at the University of Wisconsin-Madison (Fig. 9.2).

Little Rock Lake was chosen primarily because its bilobed shape made it amenable to division into two basins of similar morphology. In addition, the lake was typical of acid-sensitive lakes located within the Northern Highlands landscape. The hydrology of these lakes is dominated by direct inputs of precipitation, and their near-neutral pH and very low acid neutralizing capacity (ANC) represent lakes worldwide that are most susceptible to the effects of acid deposition (Watras and Frost 1989). Despite a dilute chemistry, however, there were no indications that the lake had been acidified by atmospheric deposition. Little Rock Lake differed in its fish community and hydrology from lakes used in previous acid manipulations. Lakes 223 and 302-South, sites of acidification experiments at Experimental Lakes Area in Ontario, are drainage lakes that contain coldwater fishes (Schindler et al. 1991). In contrast, Little Rock Lake is a seepage lake, with inputs dominated by precipitation and small inputs of groundwater during normal to wet periods and outputs of groundwater and evaporation (Rose 1993). The fish community is dominated by warmwater and coolwater species, primarily largemouth bass and yellow perch (Swenson et al. 1989). Thus, information on responses to acidification and during the recovery phase had the potential to add to the fundamental understanding of the impact of the external driver, acid rain, on lakes in the landscape.

Figure 9.1. The Little Rock Lake acidification project was initiated by John G. Eaton of the U.S. Environmental Protection Agency, Duluth Laboratory (mid-1980s, at Little Rock Lake). (J. Magnuson)

Basic Design

The general design for the Little Rock Lake experiment followed previous whole-ecosystem manipulations fairly closely (Brezonik et al. 1986). The two lake basins were separated with a vinyl curtain in August 1984, producing two basins with approximately equal surface areas (Fig. 9.3, 9.4). Following a year of baseline data collection, the treatment basin, to the north, was acidified with sulfuric acid, the dominant acid form in deposition throughout much of the world (Galloway et al. 1984, Schindler 1988). The treatment basin pH was decreased sequentially to three target pH levels, 5.6, 5.1, and 4.7, each maintained for two years. Acid additions began immediately after ice-out in spring 1985, using a fiberglass boat powered by an outboard motor and nicknamed the Acid Queen (Fig. 9.4, bottom). Acid conditions were measured throughout the treatment basin the day after the first additions, and more acid was added as needed to maintain target levels. This sequence was repeated at least twice weekly until the lake froze over in November. No acid was added to the lake during periods of ice cover. The overall acidification process was followed at each target level through 1990. During this time, the southern lake basin was not manipulated and served as a reference.

Although the power of whole-ecosystem experiments to reveal fundamental system phenomena is clear, standard statistical techniques are not appropriate for analyzing data from whole-ecosystem experiments. The Little Rock acidification project was central to the development of two specialized statistical methods for

Figure 9.2. Gathering of researchers from the Little Rock Lake acidification experiment in spring 1985 at the Trout Lake Station. From left to right: Carl J. Watras (WDNR, project manager), Neo D. Martinez (UW-Madison, graduate student), Patrick L. Brezonik (U. Minnesota, investigator), Michael E. Sierszen (UW-Madison, graduate student), Naomi E. Detenbeck (U. Minnesota, graduate student), J. Howard McCormack (USEPA, investigator), Burt Shephard (graduate student), Mhora Newsom (U. Minnesota, graduate student), William A. Swenson (UW-Superior, investigator), Kathy M. Jensen (USEPA), James G. Weiner (USFWS, investigator), James A. Perry (U. Minnesota, investigator), Paul J. Garrison (WDNR, investigator), Kesenia Rudensky (U. Minnesota, graduate student), Tom M. Frost (UW-Madison, investigator), Tim K. Kratz (UW-Madison, investigator), Todd Perry (U. Minnesota, graduate student), John G. Eaton (USEPA, project officer), William A. Rose (USGS, investigator), John Wachler (U. Minnesota, graduate student), Larry A. Baker (U. Minnesota, postdoctoral), Mike Kruse (UW-Superior), Katherine E. Webster (WDNR, investigator), Joan P. Baker (consultant). (J. Magnuson)

Figure 9.3. Aerial photograph of Little Rock Lake taken on August 28, 1990, at 1400 hours looking WNW showing the treatment (upper) and reference (lower) basins. The reference basin is visibly darker than the treatment basin, which had 50 percent less dissolved organic carbon than the reference basin. (C. Watras, Wisconsin Department of Natural Resources)

Figure 9.4. Top: Tom Frost at the barrier curtain sampling zooplankton from the treatment basin of Little Rock Lake in 1991. (J. Miller, University of Wisconsin-Madison University Communications) Bottom: Sulfuric acid being released into the boat wake via a plastic tube to the treatment basin in 1987. (J. Magnuson)

determining the likelihood that the differences observed between treatment and reference basins could have occurred by chance alone. The first, Randomized Intervention Analysis, uses a repeated re-sorting procedure with difference data (treatment minus reference) to test how likely an actual average value is compared to values calculated when data are reordered in time (Carpenter et al. 1989). The second, Time-Series Intervention Analyses, uses time series techniques to test whether a certain average difference value could have occurred relative to the overall time course of the data (Rasmussen et al. 1993). Stow et al. (1998) has similarly combined long-term and experimental perspectives to examine techniques for detecting ecosystem responses to stress.

During the acidification experiment and into the recovery phase, a wide range of physical, chemical, and biological response variables were measured concomitantly in the treatment and reference basins. Many responses to the acidification have been summarized by Eaton et al. (1992), Webster et al. (1992), Brezonik et al. (1993), Frost et al. (1999b), and others. Here, we highlight findings of the whole lake experiment and the lake's recovery through 2000.

Background Variability in Acid Conditions

Even before the intentional addition of acid began, interesting patterns in pH were observed (Fig. 9.5). The pH of surface water in the two basins tracked each other closely prior to any manipulation. Substantial intraannual variation in pH occurred, notably a dramatic depression in pH to values as low as 5.3 at the end of winter ice cover (Kratz et al. 1987b). Analyses of other chemical constituents revealed that

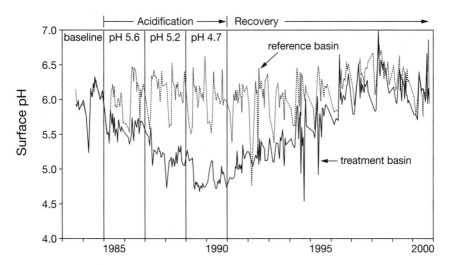

Figure 9.5. Surface pH values in Little Rock Lake's treatment and reference basins from 1983 to 2000. Target pH values for the three manipulation periods are listed at the top of the figure. Updated from Frost et al. (1998a) and Frost et al. (1999b).

these acid conditions were generated by a buildup of dissolved carbon dioxide to supersaturated concentrations under the ice, rather than by mineral acids like sulfuric acid. Furthermore, comparisons with the other LTER lakes in northern Wisconsin revealed that such high carbon dioxide levels were common during winter (Kratz et al. 1987b). These observations helped stimulate an extensive survey that revealed widespread supersaturation of carbon dioxide in lakes around the world (Cole et al. 1994).

Physical, Chemical, and Microbial Responses

Acid additions to the treatment basin generated an array of secondary stresses for the aquatic biota through a variety of physical and chemical mechanisms. These secondary stresses illustrate the importance of considering the interactions of multiple factors, even in highly controlled experiments that assess the effects of a single factor (Frost et al. 1999b).

Surface-water pH comparisons indicate that acid levels in the treatment basin were successfully maintained close to the target values throughout the experiment, except for a brief period of time in late summer of 1990 (Fig. 9.5). Interestingly, we had to add more acid than predicted from a straight titration because internal lake processes such as sulfate reduction generated alkalinity (Brezonik et al. 1993). As pH declined, the dissolved concentrations of several major cations increased (Fig. 9.6) (Brezonik et al. 1993, Sampson et al. 1995, Frost et al. 1999b). Calcium ions increased dramatically (Fig. 9.6, left top) and rose steadily during the entire treatment period to reach a concentration of 80 eq/L (1.6 mg/L) during the pH 4.7 period. The increase in calcium ions in this dilute lake may have ameliorated effects of low pH on some aquatic organisms (Brown 1982). In addition, concentrations of both magnesium and potassium ions increased by about 30% in the treatment basin during the acidification period (Fig. 9.6, left). The source of these added cations was the lake's sediments, and the mechanism was an ion-exchange process induced by the higher concentration of hydrogen ions in the water. The net effect of this process was to consume some of the acid, but it also resulted in an equivalent acidification of the sediments (Sampson 1999).

Concentrations of several minor and trace metals increased, including the potentially toxic aluminum, manganese, and iron (Fig. 9.6, right) (Frost et al. 1999b, Brezonik et al. 2003). In some cases, these metals reached concentrations that had been associated previously with stress in acidified systems (Schindler 1987). The heightened levels recorded for aluminum were shown to have adverse direct effects on the fishes (Eaton et al. 1992). In contrast, acidification to pH 4.7 had no effect on dissolved copper concentrations, and only small effects on lead and cadmium concentrations (Brezonik et al. 2003).

Other chemical changes were mediated by microbiological processes. The addition of sulfuric acid increased the concentration of sulfate from ~50 eq/L during the baseline period to a maximum of ~150 eq/L at pH 4.7. This increase in sulfide led directly to an increase in the rate of sulfate reduction (Urban et al. 1994, 2001) with a consequent generation of ANC (Baker and Brezonik 1988, Brezonik

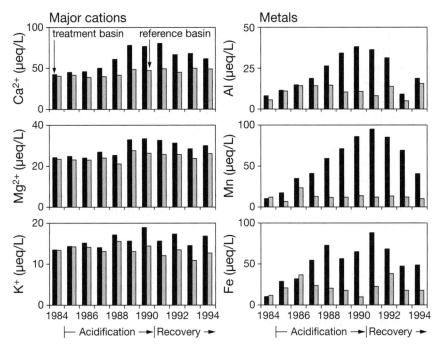

Figure 9.6. Average chemical concentrations of major cations and metals in the surface water of Little Rock Lake's treatment and reference basins from 1984 to 1994. Modified and expanded from Brezonik et al. (2003).

et al. 1993, Sampson and Brezonik 2003a). This process was a major source of resistance to acidification. In the treatment basin, in-lake processes removed about 46% of the total input of sulfate, including acid addition during the treatment phase. In addition, 36% of the sulfate inputs were lost by groundwater recharge, and 18% (representing 25% of the added acid) remained in the water column. In-lake sulfate loss processes contributed about 61% of the ANC generated in the treatment basin during the acidification phase, while calcium ion production from the sediments produced about 18% of the generated ANC. Increases in other cations accounted for another 18% of the ANC production. Accumulation of sulfur in reduced form led to dense populations of phototrophic sulfur bacteria in the hypolimnion of the acidified treatment basin (Hurley and Watras 1991, Watras and Bloom 1994).

Most dramatic among the microbially mediated chemical changes was an increase in the concentration of methyl mercury in the water column and biota of the treatment basin (Wiener et al. 1990, Bloom et al. 1991, Watras and Bloom 1992, Watras et al. 1994). Methyl mercury is the chemical species that biomagnifies in aquatic food chains and poses a health risk to piscivorous vertebrates. In the anoxic hypolimnion of the treatment basin, high concentrations of methyl mercury were observed in the water column in the zone where sulfate was microbially reduced and sulfide began to accumulate. Because sulfate-reducing bacteria methylate mercury as a metabolic by-product, and total mercury concentrations did not differ significantly between the reference and treatment basins, enhanced methyl mercury production and bioaccumulation was attributed to increased activity of sulfate-reducing bacteria following the acidification. Increases in methyl mercury propagated up the food web and were manifested as increased mercury levels in phytoplankton and zooplankton, as well as in the tissue of young-of-year yellow perch (Frost et al. 1999b). During recovery, the concentrations of aqueous sulfate and hypolimnetic methyl mercury and mercury in fish declined (Hrabik and Watras 2002).

Shifting chemical conditions ultimately affected physical properties related to water clarity. The treatment basin became more transparent and was obviously clearer to the naked eye (Fig. 9.3). This change occurred as the concentration of dissolved organic carbon decreased and was particularly pronounced by the third stage of the experiment at a target pH of 4.7 (Brezonik et al. 1993, Sampson and Brezonik 2003b). As a result, transmittance of UV radiation in the water column of the treatment basin became substantially higher (Williamson et al. 1996). By 1989, the water depth receiving 1% of UV-B radiation exceeded 2 m in the treatment basin compared with less than 0.8 m in the reference basin. However, short-term mesocosm experiments conducted in the lake in 1995 did not reveal significant effects of UV manipulations on the crustacean zooplankton (Frost et al. 1999b).

Biological Responses

The acidification of the treatment basin caused complex shifts in environmental conditions beyond a simple change in pH. Aquatic organisms were subjected to multiple chemical and physical stresses generated by the acid additions (Frost et al. 1999b). To understand biological responses to the acidification, we must comprehend the complexities of these physical-chemical shifts, as well as the intricacies of food web interactions.

As the acidification progressed, we observed a wide range of biological responses in the treatment basin. Most of these responses, however, did not appear to be directly linked with more acidic conditions. Rather, they seem to involve indirect responses that were driven largely by mechanisms related to changes in the food web (Webster et al. 1992). Here we present examples to illustrate these indirect processes.

One rotifer species, *Keratella taurocephala*, increased substantially with acidification (Fig. 9.7, top). This increase could not be tied with any direct response by *K. taurocephala* to lower pH conditions (González and Frost 1994) but instead appeared to be driven by a decrease in certain invertebrate predators. This linkage was graphically illustrated by a change in *Keratella* morphology. As implied by the literal translation of *taurocephala*, bull head, this species is characterized by large anterior spines resembling horns (Fig. 9.8, left). The increase in *K. taurocephala* abundance in the treatment basin was accompanied by an equally conspicuous decline in the average length of these anterior spines (Fig. 9.8, right), and both changes were associated with marked reductions in the abundance of the predators

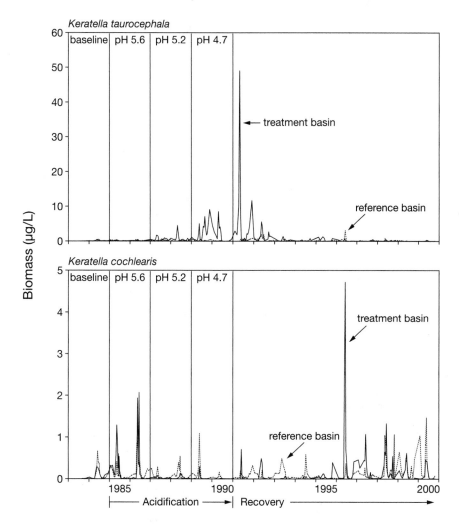

Figure 9.7. Rotifer biomass in Little Rock Lake's treatment and reference basins from 1983 to 2000. Top: *Keratella taurocephala* and Bottom: *Keratella cochlearis*. Target pH values for the three manipulation periods are listed at the top of each figure. Modified and updated from Frost et al. (1998a).

Asplanchna priodonta and *Mesocyclops edax*. Previous studies have reported induction of spines in other *Keratella* species by invertebrate predators, for example, Conde-Porcuma and Declerck (1998). Furthermore, in laboratory predation experiments, *Asplanchna priodonta* prefers short-spined morphs of *K. taurocephala* to the long-spined forms that predominated prior to acidification (González 1992). In summary, the *K. taurocephala* response in the treatment basin was complex and generated by changes in the food web.

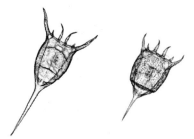

Figure 9.8. Variation in length of spines of *Keratella taurocephala* from Little Rock Lake in the summer of 1989. The long-spine form (left) was collected from the reference basin and the short-spine form (right) was collected from the treatment basin on the same date. From Gonzalez (1992) with permission.

Similarly, the marked increase in population abundance of the phantom midge larva *Chaoborus punctipennis* during the later stages of acidification could not be tied directly to shifting acidity (Sierszen and Frost 1993, Fischer and Frost 1997). Usually, *Chaoborus* increases are associated with decreases in fish predation (Eriksson et al. 1980, Yan et al. 1982, 1985). However, here it was linked to the decreases in an invertebrate predator, the cyclopoid copepod *Mesocyclops edax*, that preyed largely on first and second instars of *C. punctipennis* (Fischer and Frost 1997). Thus, a food web mechanism explains the increase in *Chaoborus*, rather than any direct response to pH.

Larvae of the hydrophilid caddisfly, *Oxyethira*, reached high densities in the treatment basin as the acidification progressed (Webster et al. 1992). This increase was associated closely with the most visually dramatic response of the acidification, that is, the development of extensive mats of filamentous green algae in the order Zygnemetales, primarily *Oedogonium* and *Mougeotia*. First observed in the treatment basin at pH 5.6, dense clouds of algae covered most of the littoral zone and significant areas out to 6 m deep in later years (Webster et al. 1992). Here, acidification benefited the caddisfly, not through a direct response to acidity, but rather through the increased availability of filamentous algae, an important food source for this benthic insect.

Unexpected Responses

A goal of the Little Rock Lake acidification experiment was to investigate the extent to which standard laboratory bioassays and other small-scale experiments could be used to predict biological responses in the lake. Major changes in the abundance of two rotifer species (*Keratella taurocephala* and *Keratella cochlearis*) were used to address this important issue (González and Frost 1994). A dramatic increase in *K. taurocephala* in the treatment basin was accompanied by a decrease in a previously more abundant congener, *K. cochlearis*, when pH dropped to 5.1 (Fig. 9.7). In laboratory assays, reproduction of each species was unaffected by pH at high food conditions. When food was limited, *K. cochlearis* exhibited reduced survivorship and reproduction at pH 5.1 or lower. At the same time, *K. taurocephala* survivorship was unaffected by pH or was greater at high pH, and its reproduction tended to be higher at intermediate pH (5.6). Thus, the decline of *K. cochlearis* might have been predictable, but only in assays that did not use ad libitum feeding conditions, a common laboratory technique. The direction of the *K. taurocephala* response would have been assigned incorrectly from laboratory tests. Laboratory bioassays successfully predicted responses of some zooplankton species to acidification. For example, *Daphnia galeata mendotae* is absent from most Ontario lakes with pH <5.5, and laboratory bioassays clearly indicate reduced survival at these pH levels (Keller et al. 1990). However, bioassays had limited ability to predict zooplankton community responses to acidification in Little Rock Lake.

Comparisons of laboratory bioassays in lake enclosures and treatment basin responses of largemouth bass provide an interesting twist to this issue. On the basis of earlier ecotoxicological studies, we assumed that the earliest life stages of eggs and fry would be the most sensitive life stage. Furthermore, different scales of experimentation all indicated that pH 4.7 had a detrimental effect on larval survival (Eaton et al. 1992). These tests, however, failed to identify the key interaction controlling the survival of largemouth bass, namely that the over-winter survival of young-of-year largemouth bass is highly sensitive to pH and aluminum. Few or no yearling bass were caught in springs following years when the treatment basin had been acidified to pH 5.6 or lower (Eaton et al. 1992). This complicated interaction would not have been recognized from first principles; thus, management actions based solely on bioassay data would have been inadequate to protect largemouth bass populations. Further, if the acidification extended longer than the six years of our experiment, major losses in the adult stages of this dominant fish predator would have occurred in the treatment basin. Delayed responses by long-lived fishes also occurred in the acidification experiment at the Experimental Lakes Area in Ontario (Schindler et al. 1985). Overall, these examples underscore the importance of whole-lake experiments for predicting ecological responses to environmental perturbations such as acidification.

Compensatory Dynamics among Species

Effects of environmental change on ecosystems can sometimes be tempered by compensatory dynamics, wherein loss of a sensitive species is offset by population increases among another species that performs similar functions (Schindler 1990, Howarth 1991, Frost et al. 1995, Tilman 1996). The Little Rock Lake experiment provided an opportunity to evaluate the role of compensatory dynamics in zooplankton community responses to acidification.

The vast majority of zooplankton species in the treatment basin exhibited substantial declines in abundance by the most acid stage of the manipulation, with a target pH of 4.7 (Brezonik et al. 1993). At the same time, a few species appeared to be favored by acidification. Three species, a rotifer (*Keratella taurocephala*), a cladoceran (*Daphnia catawba*), and a copepod (*Tropocyclops extensus*), increased so dramatically that the total biomass of these taxonomic groups appeared relatively insensitive to acid effects (Frost et al. 1995). Increases by these single species compensated for the losses of other species so that total zooplankton biomass declined in the treatment basin only during the later stages of the experiment (Fig. 9.9, top).

In a subsequent analysis, Fischer et al. (2001a) examined the role of compensatory dynamics in the responses of zooplankton functional groups, that is, groups of species that used similar resources and were vulnerable to the same predators. The responses of functional groups to acidification were highly variable, including examples of compensatory, independent, and synchronous dynamics among species within a functional group. Herbivorous copepods and medium-size herbivorous cladocerans exhibited compensatory dynamics, whereas other functional groups exhibited independent or synchronous dynamics. First-order autoregressive models indicated that groups that exhibited compensatory dynamics contained a combination of acid-sensitive and acid-tolerant species that competed with one another (Fig. 9.10). In contrast, groups that contained only acid-sensitive or acid-tolerant species exhibited more independent or synchronous dynamics. Overall, these findings pointed to the combined role of acid sensitivity and species interactions in determining responses of zooplankton functional groups to acidification.

Differences in the sensitivity of population versus community-level responses have important implications for early detection of stressors in ecosystems. In the treatment basin, compensatory dynamics among zooplankton species were substantial enough that, at the gross scale of community resolution, total zooplankton biomass response to acidification was small or absent until a pH of 4.7 was reached. In contrast, the species composition of that zooplankton community diverged substantially from that in the reference basin even during early stages of acidification (Fig. 9.9, bottom) (Frost et al. 1995). Similar resistance to stress has been documented in several previous investigations (Chapin et al. 1997). Surveys that fail to measure species responses to a stress will be unable to detect the early stages of effects of that stress.

Recovery following Acidification

A major unknown in acid deposition research relates to the capacity for acidified aquatic ecosystems to return to the pretreatment state as acid stress is reduced. Will the original communities re-establish, or will an entirely new community develop? We addressed these questions during the recovery phase of the Little Rock Lake experiment.

Acid additions were completed in late fall 1990, the second year of the pH 4.7 phase of the manipulation. After that year, we continued to monitor a wide range of variables to evaluate the treatment basin's recovery following acidification. In general, we focused our efforts on those features that were most sensitive to acidification.

Following the completion of acid additions in fall 1990, pH in the treatment basin exhibited a fairly rapid rate of recovery. By the end of 1996, pH values in the treatment basin had returned largely to the levels that occurred prior to manipulation (Fig. 9.5). Interestingly, the rate of pH recovery roughly mirrored the rate of acidi-

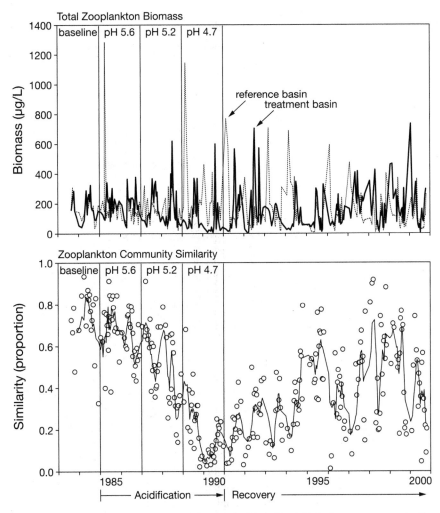

Figure 9.9. Top: Total zooplankton biomass in Little Rock Lake's treatment and reference basins from 1983 to 2000. Bottom: Similarity in zooplankton species composition between the treatment and reference basins calculated from 1983 to 2000. Open circles are the calculated similarity index values. The line is a five-point moving average. The community similarity index was calculated for crustacean and rotifer biomass data with the index of Inouye and Tilman (1988). Target pH values for the three manipulation periods are listed at the top of each panel.

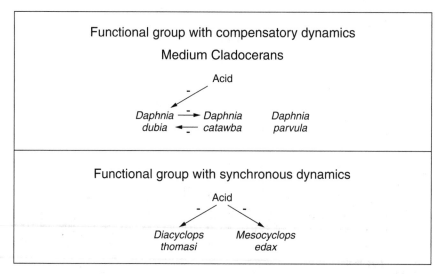

Figure 9.10. Interaction webs illustrating patterns of direct acid effects and competitive interactions in the Little Rock Lake treatment basin. Top: Medium-size cladocerans that exhibited compensatory dynamics. Bottom: Predatory copepods that declined synchronously. The interaction webs were fit using first order autoregressive models. Modified from Fischer (2001a).

fication, even though pH of the treatment basin was not manipulated by us during the recovery phase. Other chemical features, including major ions (Sampson 1999), minor metals (Brezonik et al. 2003), and nutrients (Sampson and Brezonik 2003b), showed similar patterns.

The relatively rapid recovery of chemical conditions provides an illuminating context for viewing delays in zooplankton community recovery. Total zooplankton biomass declined during acidification and recovered relatively quickly (Fig. 9.9, top). However, indices of zooplankton community similarity between treatment and reference basins, with some variation, persisted at low levels through at least 2000 (Fig. 9.9, bottom). By 1996, when chemical conditions had recovered, zooplankton community similarity remained low, indicating that the basins remained substantially different in community composition. From 1997 to 2000, similarity appeared to fluctuate below preacidification levels, suggesting that recovery of zooplankton community composition in the treatment basin was incomplete. A remaining question for recovery in the zooplankton community is whether the return to preacidified conditions is simply delayed or whether the community has been driven to an alternate stable state.

The timing of recovery of individual zooplankton species differed for species that had proliferated during acidification, the acid champs, and species that were adversely affected by acidification, the acid chumps. We had hypothesized originally that the three main acid champs (*Daphnia catawba*, *Tropocyclops extensus*, and *Keratella taurocephala*) would maintain their high numbers and resist the return of the preacidification community (Frost et al. 1998a). Instead the acid champs did return to substantially lower abundances by 1996, and we rejected our original hypothesis. At the same time, the recovery of some of the acid chumps was delayed. For example, while *Holopedium gibberum* recovered by 1993, the recovery of the copepod *Diaptomus minutus* was delayed until 1997. Factors that contributed to differences in the rate of recovery remain unclear.

Changes in Acid Tolerance

The Little Rock Lake experiment offered a unique opportunity to address the effect of acidification history on zooplankton community sensitivity to subsequent acidification. In 1995, when pH conditions in the treatment basin largely had recovered, we conducted acidification experiments simultaneously in the reference and the treatment basin in mesocosms that were 2000 L, 0.8 m diameter, and 4 m deep (Fischer et al. 2001b). In each basin, we compared zooplankton responses in acidified (pH 4.7) and control treatments. Zooplankton community responses to acidification were much stronger in the reference basin than in the treatment basin. Specifically, three out of six zooplankton species in the reference basin were affected adversely by acidification, whereas none of the zooplankton in the treatment basin were affected detrimentally by acidification (one species actually increased). Subsequent short-term reciprocal transplant experiments suggested that the differences in sensitivity of zooplankton to acidification could be attributed to rapid evolution of acid tolerance by zooplankton during the whole-lake experiment. Rapid evolutionary changes in acid sensitivity may make zooplankton communities less sensitive to subsequent pH stress; however, it is unclear how long these effects may last.

Comparisons with Other Lakes

The logistics of ecosystem level manipulations limit opportunities for replication. We compared our findings with results from other experiments and from systems that have been acidified by human causes. Several generalities emerge, giving confidence that we are not observing phenomena restricted to a single lake.

Effects of acidification on chemical processes and patterns that we observed in the treatment basin were similar to those that occurred in other lakes that had been acidified. Higher rates of sulfate reduction, increased concentrations of aluminum, reduced concentrations of dissolved organic carbon, and increased water clarity constitute findings that are common to many acidified ecosystems (Schindler et al. 1991, Schindler et al. 1996a, Williamson et al. 1996, Yan et al. 1996). One interesting contrast was the small increase in aluminum levels in the treatment basin compared to the higher values reported for other acidified lakes, for example, Schafran and Driscoll (1987). This difference may result from the hydrologic setting of Little Rock Lake, which permitted little contact between the lake's water and the underlying geological materials (Brezonik et al. 1993).

We also observed biological responses to acidification that paralleled those in other lakes that have been acidified, either experimentally or by acid deposition (Schindler et al. 1991). Despite differences in species composition, the experimental acidification of Lake 223 in the Experimental Lakes Area in western Ontario generated the same classes of effects on fish that we observed. Only early life stages were susceptible, while adult populations of the dominant fishes remained relatively unaffected during the first years in both experiments. The Lake 223 experiment was carried out over a long enough period that a decline of adult fishes was detected as individuals reached the end of their life spans (Schindler et al. 1985). For the dominant fish species in Little Rock Lake, the 6-year duration of the experiment was too short to detect declines in adult fish abundance, despite the diminished reproduction (Eaton et al. 1992). However, the relatively rare rock bass, with fewer than 600 individuals per basin, declined dramatically owing to complete recruitment failure after the first stage of acidification.

Remarkably consistent responses to acidification have been observed for a group of zooplankton species in a wide range of acidified lakes. One of the strongest positive responders in our experiment, *Daphnia catawba*, also became the dominant cladoceran species in Lake 223 at the same acid level (Schindler et al. 1991). *Keratella taurocephala* has been reported as the most obviously acidophilic rotifer species in several lake acidification studies (Frost et al. 1998a,b). However, responses of other Little Rock zooplankton species differed from those observed in other lakes. *Diaptomus minutus* had been reported as acid tolerant in enough previous studies that we had predicted that it would dominate after acidification (Brezonik et al. 1986). Instead, this species declined at the midpoint of the acidification (Brezonik et al. 1993) and showed little indication of returning during the first five years of recovery (Frost et al. 1998a,b). *Holopedium gibberum*, a cladoceran that occurs commonly in acid lakes, had been fairly abundant in the treatment basin during the early stages of the experiment (Frost and Montz 1988). We anticipated that *Holopedium* would persist throughout the acidification but surprisingly, it declined with acidification (Brezonik et al. 1993).

Perspectives

Ecosystem experiments like the whole-lake acidification in Little Rock Lake are predicated on a basic assumption that the dynamics of the reference system represent what would have happened in the treated system had it not been manipulated (Frost et al. 1988). In our project, and in many other whole-lake experiments, these inferences have been based upon what is observed in a parallel reference basin. In our case, the existence of similar data from seven nearby LTER lakes provided opportunities to test how effectively the reference and treatment basins reflected the range of conditions in the region (Carpenter et al. 1991). Similar comparison allowed us to evaluate how widely the pH depressions from carbon dioxide occurred in lakes throughout the region (Kratz et al. 1987b).

Developing and carrying out the Little Rock Lake acidification project involved substantial synergistic interactions with the North Temperate Lakes LTER program.

Much of the field work was conducted by the LTER field crew, and work could not have been initiated so quickly or carried out so smoothly without these ties. In return, the whole-lake acidification project provided a major experimental component in the North Temperate Lakes LTER program. The acidification project is a prime example of the types of associated research programs that are often generated at LTER sites. Approximately 170 scientists are listed as authors or coauthors of publications on Little Rock Lake from 1983 to 2003. Similarly, the U.S. Geological Survey's Water, Energy, and Biogeochemical Budgets (WEBB) program was founded, in part, because of potential interactions with the LTER group and its extensive data (Elder et al. 1992). Likewise, an Electric Power Research Institute investigation of mercury cycling in lakes was facilitated by the North Temperate Lakes LTER and the Little Rock Lake experimental acidification.

Additional benefits of the project were related directly to the acidification experiment. First, because the lake was closed to recreational angling during the experiment and recovery, demographic changes in the largemouth bass population caused by elimination of angling pressure could be assessed. Although the population abundance did not increase, average age increased from 2 years to 5 to 7 years following closure of the fishery (Swenson 2002). Largemouth bass biomass increased substantially, as well. Second, because mercury responses to acidification were of interest, a long-term measurement program was started in the reference basin. This long-term record is, to our knowledge, the longest continuous record of mercury concentrations in the water column of a lake anywhere in the world. These records reveal a declining trend in mercury concentration, perhaps in response to legislative decisions regulating air emissions of fossil fuel electric generating facilities in Wisconsin (Watras et al. 2000).

Finally, the Little Rock Lake project has involved limnologists in northern Wisconsin associated with limnology at the University of Wisconsin-Madison. Researchers conducting whole-ecosystem experiments regularly credit their inspiration to the groundbreaking work conducted by A. D. Hasler in the Northern Highlands Lake District in the 1950s and 1960s (Beckel 1987, Frost and Blood 1996, Magnuson 2002b). Hasler and his students performed a variety of whole-lake experiments, including the first on a divided basin lake, namely Peter and Paul Lake in the Upper Peninsula of Michigan (Hasler 1966). The Little Rock Lake experiment was designed with the knowledge and context of this history.

Summary

Ecosystem experiments are a powerful approach for evaluating how lakes are affected by human-generated stresses. To assess the effects of acidification on aquatic ecosystems, we conducted a whole-lake acidification experiment in Little Rock Lake with a treatment and a reference basin divided by a curtain. Acid additions to the treatment basin generated an array of secondary stresses to the aquatic biota through a variety of physical and chemical mechanisms. For example, water clarity increased dramatically owing to losses of dissolved organic carbon. As the acidification progressed, we observed a wide range of biological responses in the acidified basin.

Most of these responses did not appear to be direct responses to acidification. Instead, we found evidence for a multitude of indirect responses to acidification that were driven largely by food web dynamics. For example, strong interactions among zooplankton species led to compensatory dynamics, wherein losses of sensitive species were offset by population increases of species that perform similar functions. At the gross scale of community resolution provided by total zooplankton biomass, response to acidification appeared small until the most acidic stage of the experiment. Compared to the relatively fast recovery in lake chemistry, biological recovery of the acidified basin was delayed. A remaining question for recovery in the zooplankton community is whether the return to preacidified conditions is simply delayed or whether the community has been driven to an alternate stable state. Overall, the experiment underscores the value of a long-term, whole-lake manipulative experiment in promoting our understanding of ecological responses to and recovery from human-generated stresses.

10

Jumping in: Within-Lake Processes and Dynamics

David E. Armstrong
George H. Lauster
Beth L. Sanderson
David B. Lewis
Thomas M. Frost*

furuike ya
Kawazo tobikomu
Mizu no oto

an old pond . . .
a frog leaps in,
the sound of water
—Matsuo Bashō, in Shirane (1998, p. 13), with permission of Stanford University
Press

L akes are characterized typically by a variety of attributes, including
the amounts and temporal-spatial distributions of the lake's biological and chemical constituents. Complex interrelationships among external drivers,
basin characteristics, and internal processes determine the set of features that characterize the condition of a lake, referred to as the lake state or status (Fig. 10.1).
Thus, the influence of an external driver such as nutrient inputs on a lake attribute
such as plankton biomass is modified through internal processes such as primary
production, biogeochemical cycles, and food web interactions that are often lake-
specific. Accordingly, lakes influenced by similar external drivers may adopt widely
different states. A major component of our research at the North Temperate Lakes
Long-Term Ecological Research (LTER) program has been investigations into these
internal processes that influence lake status and dynamics.

Ultimately, we seek to develop a general framework for the importance of processes or adaptations across space and time, fundamental rules that must be rooted

*Thomas M. Frost (see Memorial on page vii) died while this chapter was in progress. The au-
thors completed, updated, and expanded the chapter and dedicate the chapter to Tom.

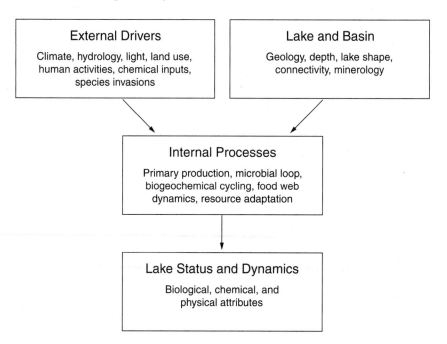

Figure 10.1. Lake status and dynamics are regulated through interactions among external drivers, lake and basin characteristics, and internal processes. In this chapter we focus on internal processes and their interaction with external drivers and basin characteristics.

in explanations of individual situations. Thus, observations recorded in our long-term data have played a key role in our understanding of lake status and dynamics. We attempt to answer questions such as "What generates or changes the chemical conditions observed at different times of the year?" and "What features allow organisms to persist or dominate in any one lake?" The approaches described in this chapter involve use of long-term, comparative, and mechanistic investigations, rather than theoretical models. These investigations focus on just one or a few lakes rather than across large spatial scales. This chapter provides several examples of the insights gained from this research. Examples are grouped into five themes: (1) biogeochemical processes, (2) chemical constituents that influence the distribution and abundance of organisms, (3) biotic interactions with the chemical and physical environment, (4) species and food web interactions, and (5) adaptations by organisms in gathering resources.

Biogeochemical Processes

Biogeochemical processes play a major role in determining the influences of external drivers on lake status and dynamics. Among the many important biogeochemical processes in lakes, our research has focused on the biogeochemical cycling of phosphorus and carbon, in keeping with our emphasis on advancing our understanding of lake metabolism. Specifically, we have explored the importance of internal recycling of the limiting nutrient phosphorus (P) on lake status through regulation of primary production and algal abundance. Our research on carbon (C) has centered on the interactions between internal processes and external drivers in regulating carbon dioxide (CO_2) and on whether lakes are a source or sink for atmospheric carbon.

Phosphorus Cycling

Our investigations of nutrient cycling have focused on phosphorus because, similar to many lakes (Schindler 1977), LTER lakes both in northern and in southern Wisconsin tend to exhibit phosphorus limitation. While inorganic nitrogen (N) concentrations are also relatively low, phosphorus limitation in these lakes is evidenced by several factors: (1) inorganic phosphate concentrations are especially low, often near the analytical detection limit during periods of active primary production; (2) algae and bacteria exhibit elevated alkaline phosphatase activities; (3) aqueous carbon dioxide levels are often above saturation (Cole et al. 1994), showing that inorganic carbon is not growth limiting; (4) ratios of carbon to phosphorus (C:P) and nitrogen to phosphorus (N:P) are higher than typical Redfield ratios (Redfield 1958) and indicative of phosphorus limitation (Hurley 1984, Poister et al. 1994). For these reasons, the response of algal production to P loadings is the focus of management efforts on phosphorus for Lake Mendota (Chapter 12).

Phosphorus loading is a key external driver of primary production and associated lake characteristics. Yet, the relations between phosphorus loading and lake characteristics are complex, influenced by numerous interactions among external drivers, internal recycling, and lake morphometry. We have had a special interest in how lakes receiving low inputs of nutrients sustain relatively high levels of productivity. The answer is tied, in part, to recycled production, that is, production through in-lake recycling of growth-limiting nutrients (Fig. 10.2). Because primary production and internal recycling are characterized by both seasonal and interannual variability, our analysis benefited from long-term (multiyear) data sets on primary production and nutrient sedimentation.

Recycled Production

Poister et al. (1994) investigated the importance and regulators of recycled production using a mass balance on phosphorus fluxes based, in part, on measurements of primary production, calculated nutrient uptake, and nutrient removal by sedimentation from the water column (Fig. 10.2). Three of our LTER lakes in northern Wisconsin, Trout, Crystal, and Sparkling Lakes, are typical of groundwater-fed lakes that receive low inputs of nutrients yet exhibit relatively high primary productivity (47 to 60 mmoles C/m^2.day; Adams et al. 1990). In these three lakes, recycled production was the main source of phosphorus driving annual primary production, representing about 86% of annual production. Within the range of interannual variations in production and sedimentation, differences among lakes were not apparent,

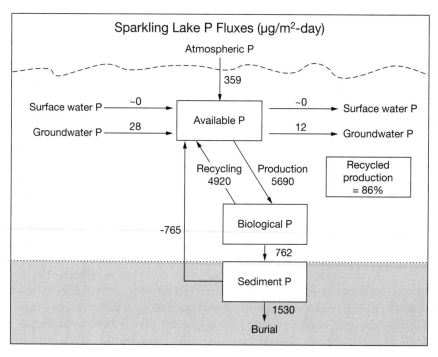

Figure 10.2. Calculated contribution of recycled production in Sparkling Lake in northern Wisconsin determined from phosphorus fluxes. Modified from Poister (1992).

even though Crystal Lake receives water and external inputs mainly as direct precipitation, while groundwater is the main source for Sparkling and Trout Lakes.

Recycled production is important in sustaining primary production. Algal abundance (size of the standing crop) is dependent on phosphorus concentration, because each algal cell requires a minimum amount of phosphorus. Thus, rapid recycling or turnover serves as a phosphorus source to sustain the standing crop of algae and to maintain primary production. Estimates of phosphorus turnover demonstrate the importance of recycled production to sustaining algal growth. Turnover times of about 20 days, calculated as the ratio of quantity of total phosphorus (moles) in the water column to the rate of phosphorus use in recycled production (moles P/day), indicate that phosphorus is re-used approximately 10 times per year. This analysis is conservative, because total phosphorus is assumed to equal bioavailable phosphorus. In comparison, the residence time of phosphorus (calculated as lake total P/annual input) in these lakes is about 1 year. Thus, if phosphorus were used only one time and then lost by burial in bottom sediments or outflow, annual primary production would be approximately one-tenth of the actual realized production. We conclude that rapid recycling of phosphorus is a key aspect of phosphorus supply and primary production in these P-limited lakes. The quantity of phosphorus in the photic zone limits the size of the standing crop, but turnover sustains the standing crop over the ice-free season.

While recycled production is a dominant source of phosphorus for primary production, we have limited information on the relative contributions of the processes that govern recycling, such as grazing by zooplankton, excretion by fishes, autolysis, sediment recycling, and bacterial decomposition. Particulate pheophorbide-chlorophyll relationships in the water column (Hurley and Armstrong 1991, Poister 1995) show that zooplankton grazing is an important contributor. However, owing to relatively high N:P ratios, phytoplankton is a P-deficient food for zooplankton. This phosphorus deficiency lessens the recycling of phosphorus relative to nitrogen by zooplankton grazing and contributes to high N:P ratios in the recycled nutrient pool (Elser and Hassett 1994). Lake ecosystems may tend to optimize recycled production when external nutrient sources diminish. However, recycling is also important in lakes influenced by high external P loadings, such as eutrophic Lake Mendota, Wisconsin. Data from Fallon and Brock (1980), Brock (1985), Adams et al. (1990), Poister et al. (1994), and Soranno et al. (1997) indicate that external P load is about 2 to 20 times higher and daily production is about 4 times higher in Mendota than in Sparkling, while recycling of phosphorus in the water column drives about 65% of the primary production in eutrophic Lake Mendota, compared to about 86% in mesotrophic Sparkling Lake. Even in Lake Mendota, blue-green algal blooms deplete epilimnetic phosphorus in summer. The resulting P-limited condition stimulates algal and bacterial phosphatase enzymes that enhance recycling of phosphorus, thereby contributing to algal production during the summer months.

Recycling from Sediments

Phosphorus removed from the water column by gross sedimentation is recycled partially after reaching the surficial sediments and can contribute to recycled production. Attempts to calculate phosphorus recycling from sediments in Crystal, Sparkling, and Trout Lakes by mass balance have been unsuccessful because sediment focusing and postdepositional migration of phosphorus in sediments lead to large uncertainties in calculating phosphorus burial rates from measured sediment accumulation rates (Poister 1992). Although our analysis of Sparkling Lake indicates that burial exceeds sedimentation, we know that substantial recycling occurs, that is, phosphorus sedimentation actually exceeds phosphorus burial. Evidence from other lakes shows that as much as 50% of the phosphorus deposited in surficial sediments is recycled (Fallon and Brock 1980, Jensen and Andersen 1992, Jensen et al. 1992, Shafer and Armstrong 1994).

Recycling of phosphorus from surficial sediments is related to both the extent of mineralization of the organic phosphorus deposited to surficial sediments and retention of the inorganic phosphorus formed by mineralization. The extent of recycling is less for phosphorus than for carbon and nitrogen because sorption and precipitation reactions preferentially retain inorganic phosphorus in the sediments (Hurley 1984). Release of inorganic phosphorus from anoxic sediments is tied to the reduction of sediment Fe under anoxic conditions (Mortimer 1941/1942). Basin morphometry can play a major role in determining interbasin differences in the development of anoxia (Robertson and Ragotzkie 1990b) because the approach to anoxia is related to the oxygen demand per unit volume of lake water (Asplund

1993). In aerobic sediments, phosphorus released from underlying anaerobic sediments is immobilized partially in an aerobic surficial sediment layer by sorption to oxidized iron phases (Caraco et al. 1992, Roden and Edmonds 1997).

LTER lakes in Wisconsin tend to divide into two groups, the northern soft-water lakes and the southern hard-water lakes, that differ in processes controlling the internal recycling of phosphorus from lake sediments. Contrasts between the two groups are linked to differences in regional geology and sediment chemistry. In the northern soft-water lakes, anaerobic sediments release inorganic phosphorus and reduced iron at high Fe:P ratios. In these lakes, phosphorus released into the water column is scavenged by oxidized iron at aerobic/anaerobic interfaces, or during fall and spring mixing. Even though phosphorus release into anaerobic waters is high, inorganic phosphorus levels in the water column are typically low at the onset of spring production, and the surface waters soon become phosphorus limited. Sparkling is an example of soft-water lakes. In southern hard-water lakes, inorganic phosphorus and iron are released from anoxic sediments at low Fe:P ratios. In these lakes, iron and inorganic phosphorus are uncoupled during transport into the water column by preferential iron retention in the sediments by sulfide and/or carbonate. Owing to low iron levels in the lake water, scavenging of inorganic phosphorus from lake water by iron is low, and relatively high inorganic phosphorus levels are maintained in the water column from fall mixing until the onset of spring production. Primary production is high in the spring and not limited by phosphorus until later in the year. Lake Mendota in southern Wisconsin is an example of a hard-water lake. Sparkling and Mendota may be viewed as end-members of soft-water and hard-water lakes. Amounts of phosphorus released and Fe:P ratios in anoxic waters vary across lakes with changes in lake characteristics and sediment chemistry, such as levels of sediment carbonate and sulfide.

Carbon Dioxide Supersaturation

Curiosity about the consistent annual cycle of pH values detected in the LTER lakes in northern Wisconsin has led to a reconsideration of the connection between lakes and drivers of lake chemistry. Our LTER researchers used long-term complementary data sets of pH, alkalinity, dissolved organic carbon (DOC), cations, and anions to demonstrate that another not directly measured chemical species, aqueous carbon dioxide, accumulates in lakes under the ice and is vented quickly into the atmosphere after spring melt (Fig. 10.3) (Kratz et al.1987b). These surprising initial findings of carbon dioxide accumulation led to a program of direct measurement of carbon dioxide and collaboration with others (Toolik Lake LTER, Hubbard Brook LTER, and other scientists) on the dynamics of carbon dioxide in lakes from a variety of regions around the world. The collaboration, along with data from other lakes, led to the finding that lake water carbon dioxide is not in equilibrium with the atmosphere and that, despite utilization of carbon dioxide by primary producers, lakes are often sources of carbon dioxide to the atmosphere (Cole et al. 1994). Cole et al. (1994) also estimated the flux of carbon from lakes at about 0.14×10^{15} gC/year. This carbon flux is a small fraction of the global cycle. However, at the

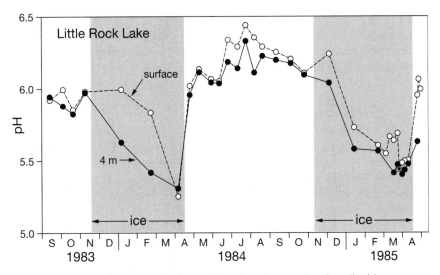

Figure 10.3. Declines in pH under ice resulting from increased carbon dioxide concentration provided evidence for carbon dioxide supersaturation in lakes. This time series from Little Rock Lake demonstrates the decline in pH at the surface and 4-m depth when ice prevented carbon dioxide release to the atmosphere. Redrawn from Kratz et al. (1987b) with permission of Her Majesty the Queen in Right of Canada.

local landscape scale, lakes may be important conduits of carbon from the terrestrial sources to the atmosphere (Kling et al. 1991).

This collaboration on carbon dioxide distribution led researchers to reconsider commonly held assumptions of in-lake processes and their interactions with external drivers from the land and air. A study of 27 lakes in the Northern Highlands Lake District examined whether above-saturation levels of carbon dioxide partial pressure (pCO_2) during summer were related to several watershed and lake variables (Hope et al. 1996). Dissolved organic carbon concentration in the surface waters was a good predictor of pCO_2. Additionally, DOC and pCO_2 both increased with increasing ratio of watershed area to lake area and with increasing percentage of wetland area in the watershed, suggesting that external carbon inputs drive the carbon dioxide concentration. The importance of riparian wetlands in determining lake DOC concentrations was indicated in a larger survey of lakes in northern Wisconsin (Gergel et al. 1999). These two studies suggested two complementary mechanisms of carbon dioxide accumulation in lakes: external inputs and internal production by respiration of DOC received from the watershed. Later research on two low-DOC lakes and two high-DOC bog lakes indicated that autochthonous carbon dioxide production was insufficient, such that allochthonous carbon dioxide inputs are necessary to account for the high carbon dioxide evasion rates (Riera et al. 1999). For bog lakes, the surrounding wetland may be the primary source. In clear, low-DOC lakes, within-lake respiration and primary production are more important, along with inputs of CO_2-rich groundwater when present (Kratz et al.

1997a). Similarly, in Allequash Creek, a lotic system flowing into Trout Lake, inputs of carbon dioxide from adjacent wetlands and respiration of organic carbon buried in the streambed were the primary sources of carbon dioxide to surface waters (Schindler and Krabbenhoft 1998, Elder et al. 2000).

Chemical Influences on the Distribution and Abundance of Organisms

Limnologists have long recognized the role of chemical constituents in structuring aquatic communities. Building on the classic paper by Redfield (1958), researchers have shown that the amounts and ratios of essential nutrients (C, N, and P) in lakes are known to limit primary production (Schindler 1977). More recent research involving our site has shown differences in C:N:P stoichiometry and recycling among trophic levels and between marine and freshwater ecosystems (Elser and Hassett 1994, Elser et al. 1996). The influence of chemical constituents on aquatic organisms depends ultimately on physiology that translates external inputs into organism abundance and distribution. For example, silica can limit the production of diatoms and other siliceous organisms, and, because groundwater is a dominant silica source, silica can limit diatom production in lakes where atmospheric precipitation is the dominant water source (Hurley et al. 1985).

Despite the wealth of information on limiting nutrients for algae, element supply also can play a critical role for other aquatic organisms. For example, silica influences the distributions of sponges, and the concentration of calcium affects the distribution of gastropods. For both groups of organisms, the research described later in this chapter illustrates how the availability of a nutrient can exert a critical influence on an organism while being above concentrations that directly limit their distributions. Low nutrient inputs above limiting levels change the risk of predation for snails and sponges and interact with predation to determine bacterial community composition. Besides the direct effect of a limiting nutrient on growth and distribution, the research highlights how external chemical inputs are, in turn, impacted by food web relationships to determine organism abundance and occurrence.

Bacterial Communities and Nutrients

Bacteria play a central role in lake metabolism and recycling of materials to inorganic forms, yet little is known about factors that influence the composition of bacterial communities, owing to their indistinct morphology. New methods borrowed from molecular biology and genetics are being adapted to ecological questions. Tracking changes in the bacterial community composition using molecular indicators, Fisher et al. (2000) found a change in community composition in response to combined phosphorus-nitrogen treatments but not to carbon. Community production was, however, responsive to added carbon, suggesting a complex relationship between community composition and activity. In treatments where bacterial grazers were removed, the bacterial community became less diverse but had higher overall productivity. Both resource availability and grazing pressure appear to affect bacterial structure and function. Results from this simple experiment indicate the complex nature of the microbial community ecology in lakes that is now being uncovered.

Sponges and Silica

Freshwater sponges occur commonly in lakes across northern Wisconsin. In historical surveys, Minna Jewell (1935) described the occurrence of sponges throughout the area and evaluated their distribution relative to lake chemistry. She documented that the common species *Spongilla lacustris* occurs in many lakes in the Northern Highlands Lake District, including most of the LTER primary lakes, and that lakes with this sponge exhibit a wide range of silica concentrations. Kratz et al. (1991a) extended her analyses to quantify the amount and distribution of silica in the sponges themselves. All freshwater sponges depend upon silica for their basic skeleton, which is made up of siliceous structures termed spicules (Frost 1991). Quantitative features of the structure of *S. lacustris* spicules were measured across habitats with different silica concentrations, and spicule structure changed systematically with silica availability. Spicule length was constant across habitat types, but their width decreased with lower concentrations of silica (Kratz et al. 1991a). However, spicules not only exhibit a simple geometrical response to silica availability but also increase in number in lower silica habitats as their width declines. Such a pattern is not consistent with a passive response to changing lake-water chemistry. Rather, the response may be an adaptation to maintain a high surface area of spicules to maximize the stiffening capacity of these structural elements. Sponge adaptations to low silica concentrations are not perfect, however. *S. lacustris* typically grows in fingers, and these are obviously less stiff in low-silica habitats. The thinner spicules appear to provide less resistance to predation on sponges, at least by snails. This contrasts with the typical situation for freshwater sponges, where predation appears rare (Frost 1991).

The systematic response by *S. lacustris* to different silica conditions also provides a paleolimnological tool for assessing historic chemical concentrations in lakes (Kratz et al. 1991a). Its use in the Northern Highlands has revealed a previously (Fig. 10.4) unreported systematic decline in lake-water silica concentrations that occurred over the past 10,000 to 12,000 years in many lakes in the region.

Recent surveys have compared present-day sponge distributions to those reported by Jewell (Colby et al. 1999). The presence or absence of *Spongilla lacustris* was fairly similar but not completely consistent between the surveys, done 60 years apart. Sponge populations appeared to have disappeared in some lakes and established themselves in others. More modern analyses of Jewell's chemical data and sponge distribution data indicated that chemical conditions in northern Wisconsin lakes did not exert a major influence on the overall occurrence of sponges. Sponges certainly shifted their body forms as chemical concentrations changed, but these changes did not affect the occurrence of the sponges (Colby et al. 1999).

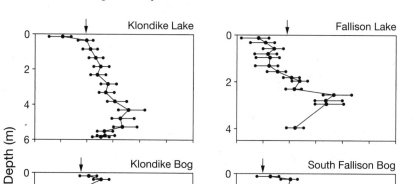

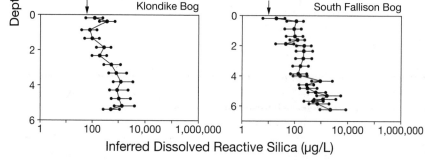

Figure 10.4. The response of sponges to silica concentration provides a basis for historical assessment of dissolved reactive silica availability from the paleolimnological record in two clear-water lakes (Klondike and Fallison Lakes) and two bog lakes (Klondike Bog and South Fallison Bog). Lake concentrations in 1980s are indicated by the arrows. Error bars indicate 95% confidence intervals. Redrawn from Kratz et al. (1991a) with permission of the American Society of Limnology and Oceanography.

Snails and Calcium

Like sponges, freshwater snails may experience ion concentrations that are low enough to exert some critical influence without being so low as to preclude the persistence of snails entirely. The ecological significance of the ion concentrations in which snails grow has been debated (Lodge et al. 1987). The traditional view has been the overriding importance of calcium for snail distributions and physiological processes. Some studies have confirmed that snail abundance and growth may be related positively to calcium levels in the water relative to the concentrations of other elements (Nduku and Harrison 1980, McKillop 1985). Snail density and shell calcium levels are not always related positively to environmental calcium concentrations (Hunter and Lull 1977, Rooke and Mackie 1984).

Snails, then, are capable of coping with environments with low ion concentrations. Though generally absent from acidic bogs (pH <6), snails are ubiquitous among circumneutral north temperate lakes. Surveys in LTER and other lakes in northern Wisconsin have revealed gastropods in waters with calcium concentrations below 2 mg/L (Mittelbach et al. 1992, Lewis and Magnuson 2000). As with sponges, the consequences of being in an extremely low, but not limiting, ion concentration environment constitutes an intriguing question.

Calcium may exert a critical influence on snails by affecting shell characteristics. Calcium, an important component of mollusc shells, binds with bicarbonate to produce crystalline layers of calcium carbonate (Wilbur 1976). The shell of freshwater snails is a conspicuous defensive structure. Environmental conditions, such as calcium concentration that affects shell strength, may influence snail interactions with their predators. LTER researchers have investigated whether the strength, or crushing resistance, of snails is greater in lakes with more calcium (Lewis and Magnuson 1999). Two pulmonates species (*Helisoma anceps* and *Physa skinneri*) from high-calcium lakes had stronger shells than counterparts from low-calcium lakes. A species with a vastly different evolutionary history (*Amnicola limosa*) typically had strong shells across all lakes, even in those with low calcium concentrations. Thus, for some, but not all, snail species, shell strength correlates with lake calcium. Determining whether differences in calcium availability are the actual mechanisms that underlie differences in shell strength requires more investigation. Nevertheless, living in an environment that causes weak shells has consequences, such as increased vulnerability to shell-damaging predators like crayfish and pumpkinseed (Osenberg and Mittelbach 1989, Lewis and Magnuson 1999).

Biotic Interactions with the Chemical and Physical Environment

Organisms vary substantially in their distributions and abundances in lakes. These differences occur from lake to lake, across time within a lake, and from location to location within a lake. Our research has described the occurrence patterns of many species in the lakes investigated and frequently has identified the factors controlling these patterns. We have explored whether these differences are controlled by direct influences of external drivers, the inherent geological setting and morphometry of the lake, or internal interactions with other organisms. The results indicate that these different influences often combine to determine organism distributions, and that the importance of an influence may change depending upon spatial scales used. When two influences are important, such as temperature and prey abundance or morphometry and algal production, then predicting the response of organisms to disturbances such as climate change or species invasion becomes more complex and may be increasingly lake-specific.

Dinoflagellates and Lake Morphometry

Dinoflagellates are conspicuous components of the phytoplankton communities of many northern Wisconsin lakes. One species, *Peridinium limbatum*, is a major element of the plankton of two lakes, Crystal Bog and Trout Bog, but the substantial between-habitat differences were puzzling (Sanderson and Frost 1996). Despite similar environmental conditions in the two bogs, *P. limbatum* abundance was nearly 1000 times higher in Crystal Bog than in Trout Bog. Experimental tests in mesocosms in the two habitats failed to demonstrate any differences in the extent of herbivory or the degree of nutrient limitation that might have explained the population differences. Comparisons of bog morphometries and oxygen conditions, however, explained the observed differences. *P. limbatum*, like many dinoflagellates, generates resting cysts as one portion of its annual life cycle. These cysts must arise successfully from a lake's sediments to maintain a population over more than one year. In Trout Bog, most sediments deeper than 7 m are overlain by anoxic waters throughout nearly the entire year. Crystal Bog, in contrast, with a maximum depth of less than 3 m, is mixed and oxygenated throughout most of the year. Most Trout Bog sediments are therefore a trap for dinoflagellate cysts, and relatively few return to the surface water to re-establish planktonic populations. Basin differences between the bogs, a glacial legacy, combined with anoxic conditions, explain observed differences in dinoflagellate populations (Sanderson and Frost 1996). We expect that the controls identified in Crystal Bog are applicable to other shallow bogs where bottom waters do not become anoxic.

Fish and Winter Oxygen Depletion

Anoxic conditions provide part of the explanation for the distribution of fish communities across northern Wisconsin lakes (Chapter 4). Many northern lakes are subjected to winter anoxia under the ice that eliminates most fish species in a process referred to as winterkill. Anoxia can be attributed to a combination of factors and partially explains the distribution of fish communities in northern Wisconsin. The extent of winter oxygen depletion varies among years and lakes. The multilake and multiyear data set of the LTER program provides an excellent opportunity to determine the important processes driving the variation in winter oxygen depletion rates. Depletion rates varied as much as 50% among years over 10 years (Asplund 1993). Temporal coherence in depletion rates among lakes with similar morphometry suggests that deep and shallow lakes differ in the mechanisms that influence depletion rates. The ratio of sediment area to lake volume best explained the variability in mean depletion rates among lakes. Summer chlorophyll was useful in predicting differences in depletion rates among lakes but did not explain variability among years.

An external driver, light, played an important role in oxygen depletion; oxygen depletion rates in several lakes were related strongly to extinction of light by snow and ice during the winter (Asplund 1993). The importance of light indicated that the oxygen production and consumption under the ice were important processes determining interannual variation in fish winterkill. LTER researchers have detected a climate change effect on ice duration (Chapter 7). Climate change can therefore influence the occurrence of winterkill by changing snow cover, ice thickness, and ice duration. The importance of snow and ice variation and lake productivity to winter oxygen depletion suggests that this depletion process, and the occurrence of winterkills, is sensitive to global warming and regional climate change.

Fish and Temperature

Temperature and oxygen levels have the potential to influence fish metabolic rates and survival through mechanisms other than winterkill. Rudstam and Magnuson (1985) incorporated the effects of these factors into models of the vertical distributions of two zooplanktivorous fishes, yellow perch (*Perca flavescens*) and cisco (*Coregonus artedii*) in five Wisconsin lakes. Their analysis revealed that, although temperature and oxygen explained a significant portion of the variation in fish distribution in some lakes, fish distribution differed substantially from model predictions in most lakes. The availability of prey may have been a major factor influencing the distribution of fishes. In oligotrophic Sparkling and Trout Lakes, cisco occurred at cooler temperatures than expected, exhibiting lower metabolic rates that may be more advantageous when food is less available. Yellow perch distributions were more variable than cisco and exhibited substantial day-night differences. For example, in oligotrophic Crystal Lake, yellow perch occupied deep, cool regions of the lake during the day and shallower, warmer depths at night. This pattern likely was influenced by the presence of benthic invertebrates in the hypolimnion of the lake. Perch appear to feed in the cooler waters during daylight hours and to move into preferred thermal conditions at night because feeding in the hypolimnion is less profitable when light is unavailable. Fish distribution appears to be influenced not only by temperature and oxygen but also by food availability and light-mediated changes in behavior. These results provide another example of a strong externally influenced driver, temperature, interacting with an internal biological driver, food availability, to influence lake status as reflected in fish distribution.

The ability to assess fish behavior has been limited by the ability to measure fish distribution and abundance in lakes. Estimates of the distribution and abundance of open-water fishes were improved dramatically through the development of remote sampling and measurement technologies during research on LTER lakes. Rudstam et al. (1987) and Jacobson (1990) used data collected with a single-beam, 70–Hz sonar system to develop a method for filtering and analyzing echo peak probability density functions to estimate fish size and abundance. This method was used subsequently to investigate the dynamics of zooplankton and fish interactions in Lake Mendota (Luecke et al. 1990) and perch dynamics in Crystal Lake (Sanderson et al. 1999).

Benthic Animals and Sediment Habitats

Benthic animals are especially sensitive to the physical and chemical conditions provided by their sediment environment. Processes such as competition and predation and factors such as physical structure and oxygen availability exert influences that vary with animal taxa. The strength and spatial scales of these processes may cause aggregation or exclusion of selected animals at certain spatial scales. Two investigations in Trout Lake explored the spatial scales of benthic animal aggregation and the relations of their distributions to sediment type and the presence of roots.

Spatial aggregation was investigated by measuring distributions of benthic animals and specific taxa at four levels of increasing area among three locations around the lake (Sagova and Adams 1993a). Aggregation patterns differed with both taxonomic groups and levels of spatial scale. Oligochaetes were distributed randomly at smaller scales but aggregated among sites. Chironomids, particularly red chironomids, were aggregated at smaller scales but distributed randomly among sites. Mod-

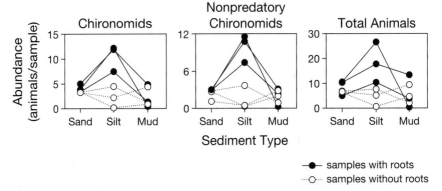

Figure 10.5. Comparison of benthic invertebrate abundances among sediment types and presence or absence of macrophyte roots at three littoral sites in Trout Lake, Wisconsin. Modified from Sagova and Adams (1993b).

eling of distributions of taxonomic composition, abundance, and animal size indicated that chironomids were influenced primarily by interspecific competition at small scales but by environment at larger scales. Predatory chironomids were aggregated owing to intraspecific competition, while controls on red chironomid distributions shifted from intraspecific processes at small scales to the environment at larger scales. Oligochaetes always were influenced by the environmental conditions.

The second analysis focused on the importance of spatial-scale differences in sediment type and plant distributions on benthic animals (Sagova and Adams 1993b). Plants can modify the sediment environment through oxygen release by their roots and serve as a refuge from predation, a source of sediment stability, and a source of food for the benthic food web. Differences in sediment composition or texture can modify the effect of macrophytes owing to differences in sediment structure, food concentration, and diffusion of oxygen. The results indicated that the additional oxygen provided by macrophyte roots is a key structuring factor for benthic animal communities (Fig. 10.5).

These investigations provided insight into the complex relationship of benthic communities to their environment and into the dependence on the scale of the investigation. Continuing research is providing information on the impact of invasion of LTER lakes in northern Wisconsin by rusty crayfish (*Orconectes rusticus*) on the long-term dynamics of the benthos (Chapter 8).

Plankton and Water Clarity

Phytoplankton plays a major role in controlling water clarity in lakes. Seasonal and interannual variations in water clarity are thus linked to cycling of phytoplankton populations. Crystal Lake was used as a test site to explore the causes of plankton-related cycles in water clarity in an oligotrophic lake (Sanderson 1998). Secchi disk transparency in Crystal Lake exhibited cyclic changes over the course of several five-to-six-year periods. As the name implies, water clarity is usually high in Crys-

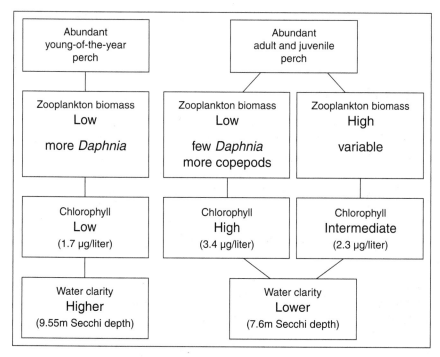

Figure 10.6. Conceptual diagram illustrating the patterns leading to interannual changes in Crystal Lake water clarity. Values in parentheses indicate average summer values. Redrawn from Sanderson (1998) with permission of B. Sanderson.

tal Lake, averaging 11 m during the summer in some years. In other years, however, average values decreased to as low as 6 m and raised a substantial question about the causes of these interannual variations. We evaluated the potential for variations in external nutrient loading, climate, predation by fish, and nutrient cycling by fish and zooplankton to mediate the observed fluctuations in algal biomass and consequently water clarity (Sanderson 1998). Overall, external factors did not explain these patterns. Rather, long-term dynamics were related to internal trophic processes that link dynamics in the fish with those in the phytoplankton and zooplankton (Fig. 10.6). Our results suggest that phytoplankton were sensitive both to nutrient regeneration by zooplankton and to shifts in grazer community composition to domination by *Daphnia*. Other researchers have noted the importance of *Daphnia*, rather than overall zooplankton biomass, to increased water clarity (Carpenter and Kitchell 1993). Furthermore, dynamics in zooplankton and phytoplankton communities were linked with cyclic population dynamics in the dominant fish species, yellow perch (Sanderson 1998). Lower zooplankton biomass and lower chlorophyll corresponded with years of abundant young-of-year yellow perch. This apparent contradiction with low abundances of both zooplankton and phytoplankton was explained by the higher *Daphnia* abundance in years with young-of-year perch. Ultimately, our results link long-term dynamics of a lake to internally generated processes in the food web, a phenomenon often obscured by variability in external drivers.

Species and Food Web Interactions

Aquatic ecologists have long sought to understand the factors that influence the diversity and abundance of organisms that occur in lakes. This section focuses on the basic biological interactions that affect organism abundance: the relationships between organisms and their food, their predators, and their competitors. Species interactions are often a key internal process determining a lake's biological attributes. The results of these investigations have provided explanations of organism variation among lakes, among years, and among seasons. Time was central to the conclusions of two studies described, where the duration of the system, that is, temporary ponds, and the duration of the data collection changed our understanding of species interactions. In the examples presented later in this chapter, an internal process, food web relationships, strongly influenced organism occurrence in both time and space. In the case of snails, these species interactions mediated responses to an external disturbance: the invasion by rusty crayfish (*Orconectes rusticus*) (Chapter 8).

Snails, Predators, and Crayfish Invasions

Resource acquisition by predators varies through time and among lakes. This variation among lakes derives from the unique way that internal processes respond to external drivers in each lake. The principal external drivers in the examples below are landscape features that either inhibit or promote the invasion of lakes by crayfish (Chapter 8). Invasion by crayfish, in turn, influences resource acquisition by resident species, namely snail herbivory on periphyton and fish predation on snails. First, *Amnicola limosa*, the most common snail species and arguably the most important benthic grazer in the northern Wisconsin LTER lakes, demonstrates a tradeoff between growth and safety (Lewis 2001). This tradeoff influences both the consumption of algae by snails and the consumption of snails by crayfish (*Orconectes* species). When risk from crayfish is low, that is, in lakes with few or no crayfish, snails prefer sand and cobble sediments rich with algal food resources (Fig. 10.7, top). Also, snails use bottom sediments in spring when crayfish are inactive owing to cool water temperatures (Fig. 10.7, bottom). Conversely, when crayfish are abundant and active, chemical cues released by crayfish induce snails to crawl up structurally complex habitat, such as macrophytes. In this habitat, predation from crayfish is reduced greatly, but algae are about 10-fold less abundant than on the risky sediments. Consequently, snails grow slowly and develop weak shells (Lewis and Magnuson 1999, Lewis 2001). Second, theoretical studies suggest that these changes in snail behavior and morphology, induced by crayfish, may influence the acquisition of snails by a fish predator, the pumpkinseed (*Lepomis gibbosus*) (Fig. 10.8). Owing to the influence of crayfish on snail behavior and morphology, pumpkinseed more easily locate and ingest snails (Lewis 2000). Thus,

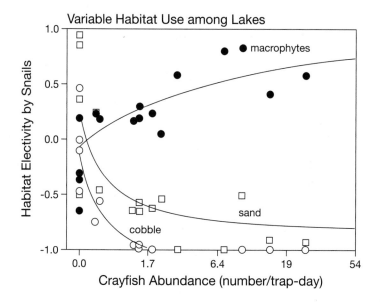

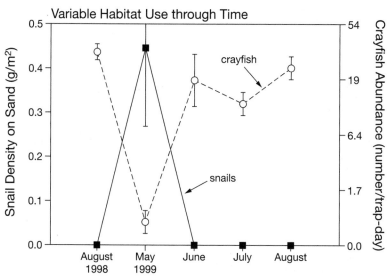

Figure 10.7. Variable habitat use by snails reflects a tradeoff between acquiring food and minimizing predation risk from crayfish. Top: Snail preference for macrophytes, coble, and sand habitats in lakes with different densities of crayfish in northern Wisconsin. Preference is expressed as electivity ranging from 1 (most preferred) to –1 (least preferred). Bottom: Changes in snail densities with season related to the crayfish activity indicated by crayfish caught per trap in Big Lake, Wisconsin. Redrawn from Lewis (2001) with permission of the Ecological Society of America.

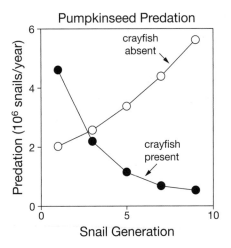

Figure 10.8. Predation of snails by pumpkinseed over time is influenced by the presence of crayfish. Results are from a discrete-time difference equation model (births minus deaths each generation) of snail population dynamics. Shown here are the dynamics of the loss term for predation to 25-g pumpkinseed with and without crayfish. Initial snail and pumpkinseed population sizes are based on estimates for Allequash Lake. Redrawn from Lewis (2000) with permission.

immediately following a crayfish invasion, lake-wide predation of snails by pumpkinseed increases. This enhanced consumption of snails by pumpkinseed, however, exceeds the estimated growth rate of the snail population, which crashes as a consequence. Thus, the invasion of a lake by crayfish may cause a short-term, one- to three-year, enhancement followed by a longer-term depression of resources for a competing predator.

Rotifer Reproduction and Food

Although purely heterotrophic, zooplankton usually have access to a wide diversity of food resources and exhibit a variety of mechanisms of choosing from their various resources. Herbivorous zooplankton routinely process a variety of sestonic food particles that vary in size, shape, and digestibility. A wide range of selective behaviors is exhibited by different suspension-feeding species (Porter 1977, Reynolds 1997), but the choices made by herbivores that confront diverse algal resources, and the consequences of those choices, are not understood fully. LTER researchers have conducted several projects focused on evaluating this interface. Sierszen and Frost (1992) developed models to quantify the energetic advantages and disadvantages of selective feeding behavior by suspension-feeding zooplankters. They concluded that selectivity became more advantageous as variation in particle food quality increased. Gonzalez and Frost (1992) demonstrated that the seasonal dynamics of rotifers in Little Rock Lake are driven by a fundamental shift in the availability of food resources. A substantial decrease in the reproductive output of rotifers in early summer was tied directly to a lack of algal food, and experiments demonstrated that this decline was not tied to life histories of the rotifers themselves. Reproduction could be returned to high levels by providing increased algal food to rotifers in microcosms in the lake.

Zooplankton Competition

In an assessment of interactions between large- and small-bodied zooplankton species, Schneider (1990) determined the relative importance of different forms of competition between *Daphnia* and rotifers. Large daphnids frequently have been observed to reduce rotifer populations in natural and laboratory situations. *Daphnia* affect rotifers by two mechanisms. *Daphnia* feeding can reduce the availability of algal food for rotifers in exploitative competition (Neill and Peacock 1980). *Daphnia* can suppress rotifers directly by capturing them in their feeding currents, causing direct mortality in a form of interference competition (Gilbert and Stemberger 1985). The relative importance of these two mechanisms was largely unresolved, and experiments with animals from northern Wisconsin lakes were conducted to test the relative importance of the different mechanisms (Schneider 1990). This work revealed that most of the effects of *Daphnia* on rotifers could be attributed to interference competition, rather than to exploitation competition, that is, *Daphnia* food consumption (Fig. 10.9). This result has been confirmed in other evaluations of rotifer-*Daphnia* interactions (MacIsaac and Gilbert 1990).

Time Scales of Species Interactions

One fundamental issue in evaluating the nature of among-species interactions in ecology involves understanding the factors that control the extent of biotic interactions. Time for interactions to develop is one obvious factor that could influence overall community processes, and the widespread occurrence of temporary ponds in the forested areas in the Northern Highlands provided an opportunity to compare the influence of time among ponds that varied systematically in their length of duration. Schneider and Frost (1996) used a set of these ponds to test the hypothesis that the extent of biotic interactions in the invertebrate community, among-species predation and competition, should increase as the time that a habitat exists increases. The amount of time that a pond remained wet is analogous to the amount of time over which succession could occur in other habitats. Across a series of ponds that persisted from less than two weeks to more than two months, the number of interacting species and the extent of biotic interactions increased systematically and dramatically as pond duration increased (Fig. 10.10). In temporary pond habitats, the extent of biotic interactions increased as the time available for the active development of a community increased.

Interactions among Fishes

The dynamics of fish populations are notoriously variable across time and space and are influenced strongly by internal processes such as predation and by external

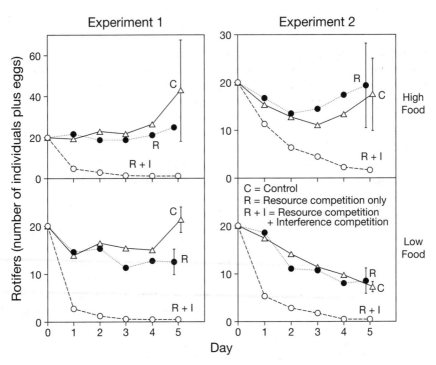

Figure 10.9. Influence of food resource abundance (*Cryptomonas*), exploitation competition, and exploitation and interference competition together between the rotifer (*Keratella cochlearis*) and *Daphnia* in experimental mesocosms. Controls had no *Daphnia*. Experiments were conducted twice at both high- and low-food concentrations. Redrawn from Schneider (1990) with permission of the American Society of Limnology and Oceanography.

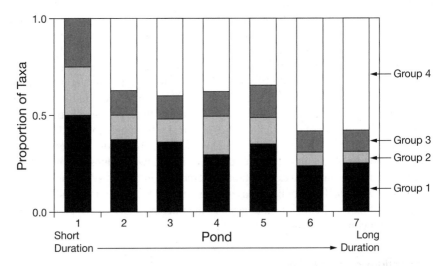

Figure 10.10. Time available for interactions influences community structure. Proportion of four taxa groups in ephemeral and permanent ponds ordered by increasing length of pond durations from 3 to more than 100 days. Groups 1 to 3 are overwintering taxa differing in season of recruitment. Group 4 taxa are nonwintering spring migrants. Redrawn from Schneider and Frost (1996) with permission of the North American Benthological Society.

drivers such as climate. Our understanding of the importance of these processes often is limited by the temporal extent of the data available. In Sparkling Lake, predation by walleye on yellow perch, darters, and minnows varied among years in response to among-year differences in growth rates and abundance of young-of-year yellow perch (Lyons 1987, Lyons and Magnuson 1987). In 1981, the young of year were abundant but grew too fast to be eaten by small walleyes, and in 1982, the young of year were not abundant, and walleye preyed heavily on minnows and darters. In contrast, in 1983, young perch were abundant but slow growing and were the dominant prey item for walleye. This work, early in the LTER program, hinted at the complexity and large among-year variability in species interactions.

Our interpretation of both the patterns and the drivers of species interactions has evolved as the length of our time series has grown. This is exemplified by the dynamics in Crystal Lake, where we have followed fish community structure since 1981. Initially, we viewed the formation of strong year classes of perch as a random event (Magnuson et al. 1990). With time, however, we began to suspect that the once "variable" recruitment success of perch instead belonged to a longer-term, cyclic process.

Theoretical models of populations confirm that cycles can occur in populations, yet observations of such cycles in a natural population are rare. This may result from the variability introduced by environmental influences and the insufficient length of time needed to see cyclic patterns emerge from year-to-year variation. Since 1981, we have observed three cases of cohort dominance spaced at consistent five-year intervals (Sanderson et al. 1999). Young-of-year perch were abundant in 1981 and 1982, 1986 and 1987, and 1990 and 1991; these cohorts can be seen growing across years (Fig. 10.11). Few to no young of year were found during the in-between years. Mechanisms that might be responsible for the temporal pattern include competition between young-of-year and juvenile perch, cannibalism on young of year by either juvenile or adult perch, and potential for reproduction by adult perch. Our analyses show that the presence of young-of-year perch is related negatively to juvenile abundance and positively to adult abundance. The presence of young-of-year fish in a given year logically depends on whether reproductively mature perch are present. Adults are not reproductively mature in Crystal Lake until they reach approximately 135 mm in total length. In years when few to no reproductively mature adults are present, no young of year are present (e.g., 1984 and 1985). Juveniles usually are abundant in years with no young of year. Any young-of-year perch that may have been produced in these years likely would have been affected negatively by competitive or cannibalistic interactions with the juvenile perch.

Combined, these mechanisms can explain the periodicity we have observed in Crystal Lake's yellow perch population. The cycle is initiated when perch become reproductively mature. As new cohorts of young perch grow, they prevent the success of young-of-year perch spawned from the reproductive adults in the following years through competition and cannibalism. This block in reproductive success continues until they themselves become reproductively mature, and the cycle then repeats. Our data record does not demonstrate the uniform oscillations of theoretically modeled populations but is long enough to document three complete cycles.

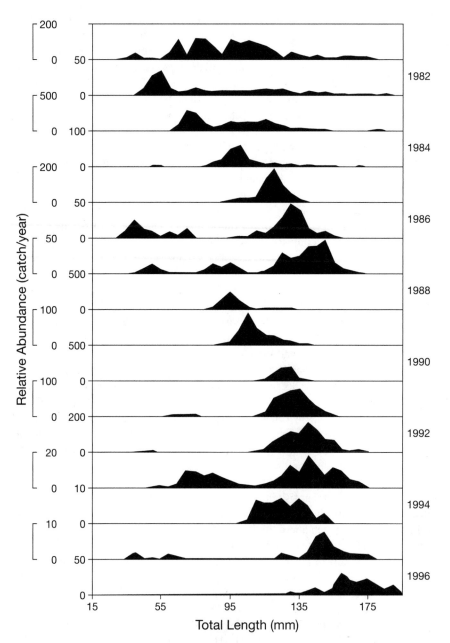

Figure 10.11. Length frequencies of yellow perch each year from 1981 (top) to 1996 (bottom). A strong-year class of perch dominates the catch at approximately 5-year intervals. Redrawn from Sanderson et al. (1999) with permission of the Natural Research Council of Canada.

From these cycles we can begin to construct the dynamics that lead to this periodic behavior in the perch population of Crystal Lake and make predictions regarding how these dynamics may change with the invasion of the exotic rainbow smelt (Hrabik et al. 1998).

Organism Adaptations for Gathering Resources

For the northern Wisconsin LTER lakes, the effects of glaciation are still felt thousands of years later. The glaciers left behind geologic deposits containing slowly weathering minerals, resulting in a region of solute-poor lakes. We have mentioned previously the phosphorus-limited character of the northern lakes. The northern lakes are also low in other elements owing to the predominance of precipitation in the water budgets of some lakes and the solute-poor groundwater inputs in other lakes (Chapter 2). LTER researchers have examined how organisms make choices in these systems where nutrients, and consequentially food resources, are low. Mixotrophy and arid plant adaptations have been the answers for some plants, while feeding flexibility has been important for some animals. These examples demonstrate how resource adaptations allow organisms to thrive in many systems where unfavorable basin characteristics, and hence low resources, normally would limit their growth.

Metabolism in Macrophytes

Research on the carbon uptake mechanisms of aquatic macrophytes illustrates one such surprising adaptation. Crassulacean Acid Metabolism (CAM) is associated frequently with plants that occur in arid environments such as deserts and some grasslands; it has been documented only recently in some aquatic plants (Keeley 1983). Boston and Adams (1983) documented that crassulacean acid metabolism occurred in several species of submersed isoetid plants that occur throughout many northern Wisconsin lakes and evaluated its ecological significance. This type of metabolism was comparable to C_3 daytime carbon gain, contributing 45% to 55% of the annual carbon gain for studied populations (Boston and Adams 1986). Isoetids were also capable of gathering carbon from lake sediments via uptake through roots (Boston et al. 1987). Adaptations to gather carbon explain the puzzle presented by the occurrence of an adaptation for arid environments in aquatic habitats. Isoetid plants are particularly successful in oligotrophic lakes, where the concentrations of dissolved inorganic carbon are usually low. Regardless of the habitat, whether arid terrestrial or oligotrophic aquatic system, crassulacean acid metabolism is an adaptation for obtaining carbon that is important in arid situations because carbon can be obtained with a minimal loss of water, and in oligotrophic lakes where inorganic carbon availability, but not water loss, is an issue (Boston and Adams 1985).

Omnivory in Carnivorous Zooplankton

Cyclopoid copepods, a major component of most lake plankton communities, are omnivorous raptorial feeders (Williamson 1991). They usually are thought to be carnivorous in their adult stages (Wetzel 1983). Adrian and Frost (1992) evaluated the diet of a common North American species *Tropocyclops extensus* (reported as *prasinus mexicanus*). Algal materials made up the majority of its diet. Further comparisons with two other cyclopoid species that, along with *T. extensus*, represented a substantial size gradient suggested that algal food might be greatest in the diets of small-bodied copepods (Adrian and Frost 1993). Utilization of algal food may be an adaptation to poor predation success owing to this copepod's small body size. This predominance of algal food in the diets of small cyclopoid copepods represents another surprise confirmed by a detailed LTER organismal study.

Macrophyte Mixotrophy

Other surprises derive from carnivorous plants and from freshwater sponges. Most aquatic organisms use a single basic mechanism for gathering energy. Plants are usually primary producers that obtain energy through photosynthesis and gather their nutrients from mineral sources. Animals are consumers that gather both energy and nutrients from organic sources. Some organisms, however, mix their mechanisms for gathering resources. A focus of one LTER project was the mixotrophy of carnivorous plants and freshwater sponges. The relative contributions of the different resource-gathering mechanisms for these organisms have been assessed, and the factors that influence the occurrence of one mechanism relative to the other have been evaluated.

The only aquatic carnivorous plants are bladderworts, in the genus *Utricularia*. These plants supplement the nutrients that they gather through normal plant uptake mechanisms by capturing animals. Several *Utricularia* species are abundant in the Northern Highlands Lake District. They are conspicuous members of the macrophyte communities of Crystal Bog, Trout Bog, and Little Rock Lake and occur occasionally in all of the LTER primary lakes in northern Wisconsin. Knight and Frost (1991) quantified the proportion of plant biomass that was dedicated to prey-capture bladders in *Utricularia macrorhiza* across a series of northern Wisconsin lakes and tested the hypothesis that the plants would devote more biomass to capturing prey as the potential nutrients available for normal plant uptake decreased. Specifically, they expected that the proportion of tissue dedicated to bladders would increase as the total dissolved minerals in a lake, as indicated by specific conductance, declined. In a survey of six lakes, they observed a significant correlation between the number and biomass of prey-capture bladders and the specific conductance of the water. The nature of this relationship was, however, the opposite of the predicted pattern. The biomass devoted to capturing prey by the bladderwort increased as the specific conductance of a lake increased (Fig. 10.12) (Knight and Frost 1991). This result contradicted expectations and led to assessments of the cost and benefits of carnivory for this species.

Bladderwort tissue devoted to capturing prey does not photosynthesize as efficiently as does other plant tissue; the cost of carnivory is substantial in many situations (Knight 1992). In the most costly cases, models suggested that growth of plants with prey-capture bladders would be only one-fifth the growth that would occur if all plant tissue were devoted to normal photosynthesis. In other cases,

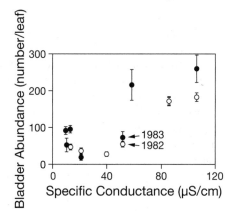

Figure 10.12. Mixotrophy in bladderworts (*Utricularia* species) varies among lakes in northern Wisconsin. Biomass dedicated to prey capture bladders increased with specific conductance that is related to concentration of nutrients and other minerals. Redrawn from Knight and Frost (1991) with permission of the Ecological Society of America.

carnivory was less costly, and plants with prey-capture tissue would have reached more than 80% of the growth achieved by plants with all photosynthetic tissue. Despite its costs, carnivory provides substantial benefits to the bladderwort. As much as 80% of plant-tissue nitrogen could be derived through the capture of animal prey (Knight 1988). A simple tradeoff between carnivory and photosynthetic capacity apparently does not exist for *Utricularia macrorhiza*. Rather, a positive feedback seems to exist between primary production and prey capture such that plants increase the proportion of tissue devoted to prey capture as both the resources available for capture and a plant's overall photosynthetic capacity increase (Knight and Frost 1991).

Sponge Symbiosis

Another situation in which prey capture and photosynthesis are combined occurs in symbiotic associations between algae and invertebrates. These symbioses occur widely in aquatic habitats worldwide and frequently make up important components of freshwater and marine ecosystems. Most conspicuous among these are associations between corals and dinoflagellates that are essential for the development of tropical reefs throughout the world (Reisser 1992). Symbiotic algae occur commonly in several sponge species in many different lakes and streams in northern Wisconsin, and we have used these organisms to explore fundamental aspects of the ecology of algal-invertebrate symbioses.

One basic question regarding algal-invertebrate symbioses involves the controls on the relative contribution of autotrophy and heterotrophy in their overall energy budget. This symbiosis is compelling particularly for sponges because as much as 75% of the growth of *Spongilla lacustris* in one habitat was derived from algal production (Frost and Williamson 1980), but sponges in other habitats have few or no endosymbiotic algae. A survey across northern Wisconsin lakes revealed that the amount of algal biomass in *S. lacustris* measured as chlorophyll varied significantly among habitats, differing by more than a factor of three (Frost, personal communication to T. Kratz). Observed differences were not related to a habitat's water clarity, and explanations for the amount of algal biomass present in these mutualistic associations were not obvious. Similarly, direct comparisons of the contributions of autotrophy and heterotrophy for three sponge species in a single lake, measured as the rates of primary production and filter feeding (clearance rate), revealed substantial among-species differences in both processes (Fig. 10.13) (Frost and Elias 1990). Organisms that co-occur under conditions with the same light and food availability exhibit fundamentally different energetics. Thus, explanations for the proportion of sponge and algae in these mutualistic associations are not straightforward. These studies further illustrate the functional importance of symbiotic relationships. In addition, these efforts discovered that one sponge species common in northern Wisconsin and elsewhere in eastern North America, *Corvomeyenia everetti*, hosted a previously unreported form of symbiont, a yellow-green alga. The occurrence of this association demonstrates two separate but convergent developments of mutualistic sponge-algal symbioses (Frost et al. 1997).

Parallels between the carnivorous plants and sponges with algal symbionts are interesting. Despite their obviously independent evolutions, both appear to have converged on a common dependence on photosynthesis as the primary source of energy. Green sponges placed in the dark usually grow to a much smaller size than sponges in the light, and carnivorous plants cannot survive in the dark. Prey capture in both cases apparently is focused largely on obtaining nutrients rather than gathering energy. Nutrient acquisition adaptations allow these organisms to thrive in dilute, nutrient-poor lakes.

Summary

The status of a lake, as reflected in biological and chemical attributes, is linked to both external drivers and internal processes. For instance, phosphorus cycling through the food web is influenced by biogeochemical cycles of oxygen, iron, and sulfur, as well as by phosphorus loading, and varies among lakes differing in geochemistry, basin morphometry, and food web structure. Biological communities vary with features such as lake chemistry, morphometry, resource availability, and species interactions. The ability of external chemical inputs to influence organism abundance can be increased or decreased by internal processes. Physiological adaptations such as mixotrophy or crassulacean acid metabolism allow organisms to mitigate the effect of limiting nutrients and increase their range, while food web relationships may exacerbate the limiting conditions of calcium for snails and sponges. Since the beginning of the North Temperate Lakes LTER program, short-term research efforts have characterized the biological communities and behavior of organisms in a variety of lakes. More recently, long-term observations in this suite of lakes have linked lake characteristics to a variety of internal processes,

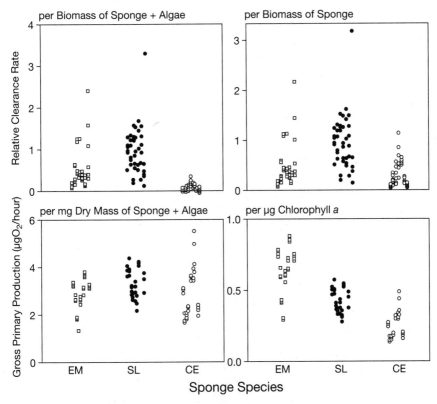

Figure 10.13. Symbiotic associations between algae and sponges at the same light and food availability differ among sponge species, indicating their differences in the contribution of autotrophy versus heterotrophy to growth in the same lake. EM = *Ephydatia muelleri*, SL = *Spongilla lacustris*, and CE = *Corvomeyenia everetti*. Redrawn from Frost and Elias (1990) with permission of the Smithsonian Institution Press.

thereby forming the basis for further research at a variety of spatial and temporal scales. Observed patterns often are explained on the basis of a variety of internal factors such as resource availability and acquisition strategies, biogeochemical cycles, and species interactions. Our understanding of the causes of changes in lake attributes strengthens our ability to predict the influences of disturbances such as human activity, exotic invasions, and climate change. The complexity of interactions among organisms and chemical constituents underscores the general need for caution in predicting the influences on lakes of disturbances that arise from changes in external drivers.

11

People in a Forested Lake District

Bradley S. Jorgensen
David Nowacek
Richard C. Stedman
Kathy Brasier

We cannot solve the problems that we have created with the same thinking that created them.
—Ascribed to Albert Einstein

Human-lake interactions in the largely forest landscapes of northern Wisconsin and the urban agricultural landscapes of southern Wisconsin (Chapter 12) are important components of the North Temperate Lakes Long-Term Ecological Research program. Some of the research questions in the north concern attitudes and behaviors toward lake-centered recreation and lakeshore dwellings, impacts on lake ecology, and feedbacks to human activities.

Bringing together the social and natural sciences is recognized as a necessity in ecosystem research where both the human and the biophysical systems interact in significant ways (Lee et al. 1990, Force and Machlis 1997, Machlis et al. 1997). Such an enterprise can be difficult (Heberlein 1988), but it provides opportunities for changing the way environmental questions are conceived and studied in each of the sciences.

The following discussion deals with social science perspectives on lakes and how these approaches may link and interact with the natural sciences to provide an understanding of the regional dynamics of lakes at various spatial and temporal scales. We provide a short history of land and water use in the Northern Highlands Lake District, and Vilas County in particular, where LTER study lakes are located. Human habitation of the area is described in terms of the spatial arrangement of land ownership that underpins property tax payments. Also discussed are differences in the experiences of individuals and groups that give rise to different perceptions of lakes and varying human conceptions of a sense of place. The formation of lake associations in Wisconsin is discussed as a means to highlight expressions of group behavior as an important unit of analysis in LTER research. Finally, we illustrate selected areas of developing research between the social and natural sciences.

Historical Legacies

During the early stages of long-term ecological research, members of our research team clarified how long-term studies were necessary to uncover what they called the invisible present (Chapter 1) (Magnuson et al. 1983, Magnuson 1990). This concept pointed to processes that extend over time scales sufficiently long to make them unobservable to our senses. Like the hour hand on a clock, some changes occur too slowly to be perceived without systematic monitoring. In social science, the present can be invisible in another sense, namely when the past is unrecognized in the present. This invisibility occurs primarily when past human actions persist into the present, frequently in the form of unintended consequences of the past that shape present actions. But, because we receive these consequences as a given part of our situation, we fail even to inquire into their origins. In this sense, the past that persists into the present is invisible because historical legacies appear to be a natural part of the present. The term "legacy" is most commonly used in social science today to denote such phenomena, especially regarding earlier public policy (Weir et al. 1988, Pierson 1994).

The following two sections on Ojibway life and economic and natural resource policy are intended to lay out specific legacies of past human actions, both in the deep and in the more recent past, that form the general social framework within which our study lakes are embedded. These legacies include the original surveying of our study area in 1864, a treaty with the Ojibway in 1842, public acquisition of large tracts of land resulting from land policy in the 1800s, and zoning laws put in place in the 1930s. In addition to pointing to such enduring elements of this framework, the section on economic and natural resource policy concludes with a brief discussion of the contemporary development of lake classification policy.

The Ojibway

Long before European settlement, the Ojibway lived in the area around Trout Lake. The General Land Office of the United States first surveyed the area around our study lakes in September of 1864 (General Land Office Surveys 1864). Surveyors walked through the woods and around the lakes, marking square mile sections. They carried a standardized chain that measured a mile when laid end to end 80 times. The surveyors kept notebooks where they recorded their measured steps through the forest and sundry observations about the surrounding landscape. The notebook for the survey just to the west of Trout Lake noted that:

> There are numerous lakes. Some are of large size. Trout River, which runs through the center of the township, flows out of Trout Lake and is a wide shallow stream in many places widening into small lakes. It has but little current and is unfit for logging purposes. There is but little pine timber in the township, it being principally birch, aspen and small scrubby pine. A few indians live at the outlet of Trout Lake

and have about one acre in cultivation. The land is of poor quality and not adapted to farming purposes. (General Land Office Survey 1864: town 41, range 6-east)

These observations are characteristic of the view the surveyors were taking toward the land, namely its suitability for agriculture and timber harvest. Recognition of the unsuitability for farming was ironic because 60 years passed before boosters, politicians, University of Wisconsin officials, and others finally ceased the promotion of agriculture in the northwoods (Hurst 1964).

The official survey was a consequential step in the modern history of our study area because the geographical boundaries necessary for a market in land were set. Although much additional surveying has occurred subsequently, many land transactions still follow original survey lines. The survey was far from the first consequential act of the past that continues to affect the present. In the Treaty with the Chippewa of 1842, the Ojibway or Chippewa ceded territories to the United States that encompassed our study lakes. Wisconsin was not yet a state and would not be until 1848.

Even though the Ojibway were ceding this land, they reserved usufructuary rights that they continue to exercise to this day. One main way that these rights are still exercised is through spear fishing, a traditional harvesting technique that only the Ojibway and not other Wisconsin citizens are permitted to practice. Of our seven core lakes, three have been speared regularly: Sparkling, Big Muskellunge, and Trout. From 1985 to 1998, the total number of fish harvested by spearing has been 186 from Sparkling Lake, 3101 from Big Muskellunge Lake, and 4256 from Trout Lake (Great Lakes Indian Fish and Wildlife Commission 2004). Thus, the international Treaty with the Chippewa of 1842 continues to affect the ecology of the LTER study lakes.

Economic and Natural Resource Policy

Natural resources have played a significant role in the economic development of Wisconsin and of the United States in general. The period of rapid U.S. economic expansion during the late 1800s and early 1900s was unusually resource intensive. This resource intensive thrust of U.S. development derived from the rapid depletion of U.S. natural resources, rather than from a unique endowment of natural resources (Wright 1990).

Wisconsin's northern forests were among the resources quickly depleted during this period. Their depletion was driven by an innovative policy by the state to compensate for the limited availability of money capital. The state used natural resources to attract scarce capital by opening and securing investment opportunities through low-priced sales, grants, and franchises. Timber was the focal resource in northern Wisconsin, particularly from 1875 to 1895. During this period, the development of transportation capacity by water and rail was driven by the goal of bringing Wisconsin timber to the expanding markets of Chicago and the Great Plains (Hurst 1964).

By 1900, the timber boom in Wisconsin had run its short course. The heavily forested northern parts of the state had been so denuded that they became known as the cutover. Lands that had been stripped of trees were left idle, partly because attempts to reforest frequently were incinerated by the pandemic of forest fires that fed on discarded slash left after the timber cuts. Many investments in reforestation simply were not made because risk from fires was high and the period of investment maturity was long. While fire control was an apparent solution, the scale of organization was beyond the capacity of any individual or company. Large-scale organization was necessary because neglect of such control on lands adjacent to one's own vitiated whatever precautions one might take. Beginning in 1895, the state of Wisconsin first organized fire suppression on a sufficiently comprehensive scale (Hurst 1964). By 1910, the State Board of Forestry had established the Trout Lake headquarters both for fire suppression and reforestation (State Board of Forestry 1910).

A promise held out for the cutover lands was their conversion to agricultural production. In a number of northern Wisconsin counties, and Vilas County in particular, the popular expectation that the plow would follow the axe largely went unfulfilled. Although the demand for food during World War I created a momentary boom in agriculture in the area, the general economic downturn that followed the war was particularly severe for agriculture. The weakness of the economy continued throughout the 1920s and only grew worse in the depression of the 1930s. Even had the economic conditions for agricultural been more advantageous, the geological history of Vilas County had left some of the poorest soils for agriculture in Wisconsin. Agriculture was not to provide economic redemption for Vilas County (Hurst 1964).

With the exhaustion of the timber economy came a decline of the local economy and a fiscal crisis for local government (Hurst 1964). Revenues declined as the value of land diminished and many owners fell tax delinquent. These fiscal stresses were exacerbated by the haphazard patterns of settlement. Settlers were scattered across the countryside, forcing a duplication of services that could have been provided more cheaply to a population more concentrated in towns (Hurst 1964). Local authorities, bound to provide an array of services, especially education, confronted higher per capita costs for these services than were the case in other parts of the state.

The state, recognizing the collective irrationality of individual settlement decisions, ushered in a new era of resource management in 1929 by delegating to the counties the authority to zone land uses. Vilas County was among the first counties to exercise this new authority, passing its first zoning ordinance in 1933. Most of the lands surrounding LTER lakes were zoned for forestry and recreation; building of year-round family dwellings was permitted only on Trout and Sparkling Lakes. Farming was not permitted anywhere within several miles of our study lakes (Vilas County News-Review 1934). Although the state's intent was to empower the counties to rationalize settlement patterns and thereby cut the costs of services, the authority granted to the counties was quite general and thus open to novel future uses. In succeeding years, this general authority has been adapted by local officials to a series of new challenges that have arisen as the economic base of the region has turned toward tourism and recreation. Part of the development of zoning policy over time has been the proliferation of zoning categories. The zoning law of 1933 recognized only three categories of land use (Vilas County News-Review 1934). Today there are 10. Zoning is no longer a means of land use control exercised exclusively by government officials. Lake associations in particular have turned to zoning to control land use surrounding their lakes. In 1973, the Trout Lake Property Owners Association petitioned the county to shift its privately owned shorelines into a more restrictive zoning category. Although this initial effort failed, the association was successful in a second bid for rezoning in 1981. Lake features have come to play a central role in determining development patterns (Schnaiberg et al. 2002). Later we discuss briefly a novel form of zoning specifically for lakes currently under development in Vilas County.

In addition to zoning, the state of Wisconsin established another mode of coping with idle lands, taking direct title to them either by assumption of title by tax default or by direct purchase. By these means, the state began constructing a Forest Reserve under the direction of a State Forest Commission, established in 1903 (Hurst 1964). State acquisition of cutover lands was a politically sensitive affair. Many boosters of the northern economy persisted in the assumption that lands cleared of trees would be turned to agriculture and that agriculture would create a sound base for sustained local economic development (Hurst 1964). Directors of the Forest Reserve were aware of this view and the political potency of its adherents. They were careful to acquire lands that soil surveys indicated were particularly ill suited for agricultural development, including the lands around Trout Lake. Most of the lands that today make up the state forests surrounding our LTER lakes were purchased between 1908 and 1912 from four logging companies: Yawkey-Bissell, H. W. Wright, Alexander Stewart, and Land, Log, and Lumber companies (State Board of Forestry 1910).

When such state acquisitions were ruled unconstitutional by the Wisconsin Supreme Court, in 1915, the state was not forced to sell off its already acquired reserve lands. However, the acquisition of more lands ceased, and acquisition of remaining idle lands by the counties and the federal government was encouraged. For this reason, lands just to the east of our study lakes shifted from state ownership to county ownership. The state Supreme Court ruling was reversed by constitutional amendments from 1924 to1927, when the combination of poor soils and slack agricultural prices had weakened opposition to public acquisition of lands unsuitable for agricultural settlement (Hurst 1964).

These two responses to the cutover, the creation of zoning authority at the local level and the large-scale acquisition of idle lands by county, state, and federal authorities, are among the most important features of the social situation surrounding our study lakes. These authorities are the dominant modalities for managing lake resources. While these patterns of public ownership appear stable into the foreseeable future, zoning is presently being altered in ways that may influence our primary study lakes. Zoning previously has been used only to categorize and thereby regulate the uses of land. Currently, this pattern is changing through the development of a lake classification system so that policy might differentiate among types of lakes. The development of the classificatory scheme and its potential effects on our study lakes and on other lakes in Vilas County are among our current areas of research.

Figure 11.1. Housing density in Vilas County, Wisconsin, at 10-year intervals from 1940 to 1990 (Hammer et al. 2004). Purchased from the Applied Population Laboratory, Department of Rural Sociology, University of Wisconsin-Madison.

Distribution of People

Today, Vilas County is experiencing significant population growth, and land values are much higher than at any other time in its history. An increasing trend toward lakeshore development in Vilas County has accompanied a growing focus on tourism and recreation. A study of recreational homeowners and regional development identified recreational housing as the greatest growth sector in Vilas and other Wisconsin counties (Preissing et al. 1996).

Vilas County has experienced a rapid increase in housing density from 1940 to 1990 (Fig. 11.1). In 1970, more than half of Vilas County had housing unit densities of less than 10 units per square mile, and more than 10% of the land surface had fewer than 5 housing units per square mile. By 1990, the most recent year with data, no block groups had less than 5 housing units per square mile.

The pace of change has increased since 1990, and a review of the Vilas County zoning records revealed 2,960 new housing units constructed between 1985 and 1995.

Between 1985 and 1989, an average of 184 new homes per year were constructed in Vilas County. Between 1990 and 1995, this average nearly doubled to 342 new constructions per year (Anderson 1996). Much of this development is concentrated along lakeshores. Approximately 74% of all structures visible from orthophotos are covered by shoreland zoning, being within 100 m of lakes or 300 m of streams (Rodgers 1998).

High densities of housing units were adjacent to lakeshores (Fig. 11.2). Most of the remaining undeveloped shoreline is publicly owned. The Northern Wisconsin's Lakes Study (Wisconsin Department of Natural Resources 1996) compared levels of shoreline development in the 1960s with that of 1996 and found an overall increase of 60% and an average of 10.4 dwellings per mile of shoreline frontage.

Lakes and Property Ownership

Evidence of the significance of lakes to people is apparent in the distribution of the human population with respect to the location of lakes (Voss and Fuguitt 1979). Residential properties on LTER lakes occur around the western and southern boundaries of Trout Lake, the northwestern corner of Sparkling Lake, and two relatively small areas on Big Muskellunge Lake. Some property owners have their primary

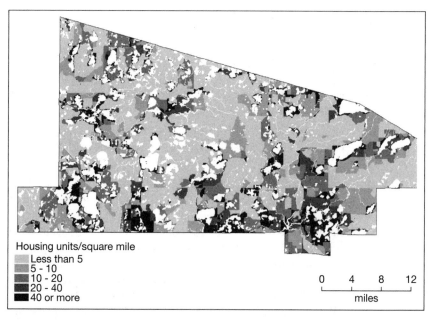

Figure 11.2. Housing density in Vilas County, Wisconsin, for 1990 (Hammer et al. 2004). Purchased from the Applied Population Laboratory, Department of Rural Sociology, University of Wisconsin-Madison.

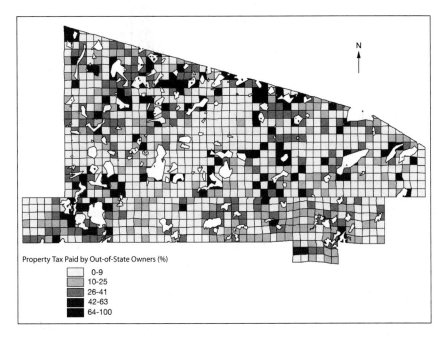

Figure 11.3. Distribution of property taxes paid by out-of-state owners in Vilas County, Wisconsin in 1997 (Vilas County Tax Data).

residences outside Wisconsin. In 1997, 27% of those taxes were billed to owners whose primary residence was outside Wisconsin (Illinois 18%, other states 9%), while the greatest proportion of county property taxes (73%) was paid by owners who were Wisconsin residents (Vilas County 44%, other counties 29%). The largest number of nonresident land owners live in Illinois, and the Chicago area in particular. A second group of out-of-state residents have their primary residences in Florida, California, Arizona, and Texas. This group lives in the warmer states during winter and migrates seasonally to the Northern Highlands lake region. These out-of-state owners are especially attracted to lakes, purchasing property more frequently near lakes than state residents.

According to our analysis of county property tax records, out-of-state residents pay a greater proportion of property taxes on or near lakes than farther away (Fig. 11.3). The darkest areas indicate square-mile sections where out-of-state residents pay more than their countywide average of 26% of property taxes. The lighter-shaded sections indicate areas where out-of-state owners pay a percentage of taxes equal to or below their countywide average. These sections tend to be located at greater distances from lakes. Thus, while Wisconsin residents may be responsible for paying a greater proportion of county property taxes in total, they pay a lower proportion in areas near lakes.

Sense of Place

Landscapes are objects of human beliefs, feelings, and actions. Research on sense of place addresses these concepts (Jorgensen and Stedman 2001). Sense of place is the meaning and importance attached to a given spatial setting by an individual or group (Tuan 1977), including cognitions and beliefs and feelings formed through experience with the setting (Ryden 1993). Our experiences over time with a setting lead not only to beliefs about a place but also to feelings of attachment to the place (Williams et al. 1992) and a responsibility to care for a place (Syme et al. 2002). The physical attributes of a setting contribute to individual and group perceptions of the place, but they do not determine them.

Envisaging how the development of a sense of place may occur in the LTER lakes region of northern Wisconsin is easy. As a region with abundant lakes and publicly owned forests, many regarded the area as a pleasant vacation place with abundant natural amenities. However, these meanings are neither stable over time nor homogeneous among individuals who interact with the spatial setting. The dominant meanings of the region have changed over time in ways consistent with historical changes in the demography and human activities in the region (Gough 1997). For example, from a European perspective, the area around the LTER lakes historically changed from a wilderness populated by savages to a place of vast timber resources. Following the cutover period, the area became a place to build a human community around agriculture and then a ruined place when agriculture failed. Yet, much of the contemporary meaning to be found in the setting is a function of this history. For example, the large amounts of public recreation lands are a direct function of the failure of agriculture and the reversion of these lands to the public domain.

Stedman (2000) articulated a model of sense of place as a collection of cognitions, attitudes, and identity based on experience with a physical setting, that is, a particular lake and characteristics of that lake. He sampled 1,000 owners of property in Vilas County, Wisconsin, and discovered four primary belief domains. People viewed lakes as (1) up north; (2) a community of neighbors; (3) places of pristine, unspoiled nature; and (4) places that had been impacted by human use.

These four cognitions serve as the foundation for place attachment, conceptualized as one's level of personal identification with the setting, and place satisfaction, or attitude toward the setting. Experience with one's lake drives both variables: satisfaction and attachment are fostered by a number of experience variables. Property ownership, as opposed to visiting one's lake, is associated with higher levels of attachment. Interestingly, however, year-round residence, as opposed to seasonal residence, is associated with lower levels of attachment. This finding counters much of the conventional wisdom in place theory (Tuan 1977) but makes sense when the role of cognitions as a mediating variable is considered. Year-round residents are less likely to view their lake as up north, a cognition that is associated with increased attachment. Characteristics of the lakes themselves are important factors, and these are explored later in the chapter.

Lake environments can be important as objects of identification, emotional attachment, and as prime settings for undertaking valued activities (Jorgensen and

Stedman 2001). But, the relationship between an individual's place identity, attachment, and behavioral dependence with other variables relevant to lake management can be complex. In our LTER lakes, these place variables were predicted by property owners' attitudes toward housing development on their lakes, attitudes toward natural vegetation on the properties, and the importance of their lakes in their perceptions of their properties (Jorgensen 2002). However, place variables were not associated with the amount of property development (e.g., docks, developed beaches, boathouses), the year the property was acquired, or the age of the property owner. Stronger place identity, attachment, and dependence were associated with negative evaluations of shoreline housing, more positive attitudes toward preserving and maintaining the natural vegetation on the property, and greater importance attributed to the lakes in owners' appreciation of the their properties. However, lake importance was a significantly larger predictor of identity than of attachment to the property or its exclusivity as a place to conduct important activities. Given the central place of lakes in property owners' identity, the range of potential lake and property management options may be thoroughly scrutinized by owners, and the scope for attitudinal and behavioral change tightly circumscribed.

Finally, place variables demonstrate interesting relationships to place-protective behavior. Respondents with high levels of attachment were more likely to seek to protect their lakes against unwanted environmental change. The opposite relationship was obtained for satisfaction; respondents who were highly satisfied with their lakes were less likely to engage in the protective behaviors described.

In an example specific to the LTER lakes, the Trout Lake Property Owner's Association petitioned the Vilas County Board of Supervisors to rezone the privately owned shoreline on Trout Lake from General Business to Single Family. The latter land-use classification is a more restrictive one that prohibits commercial development. The language used in the petition emphasized sense-of-place attributes through the use of descriptive phrases like "one of the most outstanding lakes in northern Wisconsin" (Druse and Neufeld 1981). Such evaluative statements were further elaborated in the petition with references to specific attributes such as water clarity, shoreline aesthetics, public ownership of the shore, and long-term family heritage.

Many examples link sense of place with actions taken to preserve that place in the face of external threats such as condominium development, water level regulation, or poor lake water quality. The following section highlights the role of collective action that groups of individuals operate to maintain and perpetuate a shared sense of place.

Lakes and Collective Behavior

Unlike the previous section, which focused on the individual's attachment to a place, the focus in this section is on the groups formed around lakes: how and why they are formed, their operation, and their effects. Individual property owners are limited in the resources they have available and the types of activities that they can engage. People tend to form groups, such as lake management organizations, because groups can gather resources, implement plans, and regulate problems more efficiently and effectively than individuals. The Trout Lake Property Owner's Association mentioned earlier is an example of a lake management organization, one of a number of such organizations in Wisconsin. Because these organizations can be effective in making changes, we need to understand their formation, operation, and effects both in the Northern Highlands Lake District and throughout the rest of Wisconsin.

Wisconsin's Lake Management Organizations

More than 600 lake management organizations exist in Wisconsin. The first was formed in 1898 around Lake Geneva, in the southeastern part of the state; the majority formed after 1970 (Klessig 1973, Schrameyer 1997). In 1996, 58 of the 72 Wisconsin counties (81%) had at least one. Vilas County has 561 (9.3%) of the 6,040 named lakes in Wisconsin and is home to 61 (10.9%) lake management organizations (55 lake associations and 6 lake districts).

Lake management organizations in Vilas County, and elsewhere in Wisconsin, take a variety of forms. These include (1) mandatory and voluntary lake associations; (2) lake districts that are special-purpose units of government with statutory authority to tax themselves for lake management purposes (legislative authority given in 1973); (3) sanitary districts that are special-purpose units of government with the authority to collect resources for implementing public sewer systems; and (4) committees or commissions of local governments (counties, towns, villages, and cities) (Klessig 1973, Wisconsin Department of Natural Resources 1995a, 1995b, 1995c).

Activities of these organizations include management of aquatic plants, lake aeration, boating safety patrol, and fish stocking; regulation of recreational boating; education of property owners; and the creation and implementation of management plans in conjunction with state agencies and private firms. A lake management organization may lobby for specific state policies that affect its lake. Many of them are members of the statewide organization, Wisconsin Association of Lakes, that assists its members with lobbying efforts and gathering information and resources. Groups of individuals interested in creating a lake management organization may seek assistance from several state agencies, including the Wisconsin Department of Natural Resources and the University of Wisconsin Extension (Wisconsin Department of Natural Resources 1995a, 1995b, 1995c).

Only 4% of the lakes in Wisconsin have a lake management organization (600 associations and 15,000 lakes) (Wisconsin Department of Natural Resources 2004). This prompts one to ask why such organizations develop on some lakes but not on others. On the basis of preliminary research using secondary data from the Wisconsin Department of Natural Resources (2004) and key informant interviews, we are able to identify some plausible explanations. First, organizations may form in response to the discovery of an environmental problem or to lakeshore owners' perceptions of an environmental problem with their lake. Lake management organizations may provide, or may have access to, the resources necessary to investigate and remedy such problems.

Second, threats to the status quo of a lake may lead to the formation of a lake management organization. For example, a conflict over land use and development may lead to conflict between lakeshore residents and the developer. The lakeshore residents may create a lake association to raise the resources necessary to fight the developer. As discussed earlier, such collective behavior may stem from a perceived threat to a shared sense of place.

Third, several lakeshore owners may decide that a specific project should be undertaken, such as lake aeration or boating regulation that requires a collective rather than an individual effort. These types of endeavors may be responses not to specific environmental issues but to land owners' desires for a certain type of social experience, such as catching walleye rather than panfish or canoeing rather than motor boating.

Fourth, lakeshore owners may hear about other associations forming in response to one of these factors, or they may see that several nearby lakes are threatened. To protect their own lake, they may form an association as a preemptive measure.

Finally, some lake associations exist primarily so that members may have social gatherings, such as picnics and parties. For example, the Alma-Moon Lake District formed out of an existing association and maintained the association as a separate organization for the explicit purpose of socializing. Frequently, lake associations tend to have life cycles; that is, they form in response to a specific issue, and, once the issue is solved or addressed, the association continues to exist at a low level of activity. These maintenance organizations keep people in contact with each other, often through social gatherings, and allow for quicker mobilization in response to future problems (Tarrow 1994).

While these reasons may create the desire to act collectively, several other factors increase the likelihood an organization will be formed. First, a critical mass of lakeshore owners must perceive the problem and agree that an organization is the means to address it. Second, the lakeshore owners must feel that they have the time and resources to devote to the organization. Finally, the likelihood of formation will increase if some of the lakeshore owners have the knowledge of how civic organizations are created and the relationships with other organizations that may assist them. Together, the perception of a problem, the desire to address it, and the means to organize affect the likelihood of an organization's formation. Those lakes that have all of these factors present are more likely to create a lake management organization.

Another question we are pursuing concerns lake management organizations' relationship to the state and the state policies that affect their creation and function. The formation of new lake districts has declined since the 1970s, and the number of new lake associations has been increasing (Fig. 11.4). These trends may be a response to changes in state policy, specifically, the abolition of state funding for lake districts in the early 1980s. In the late 1980s, funding was restored, but the funding was made available to any municipality, qualified lake association, county, or lake district. This change in state funding meant that the incentives for forming a lake district, such as exclusive access to state funds, were removed. However, groups of people who desired public funding could still form less bureaucratic, more flexible organizations such as lake associations.

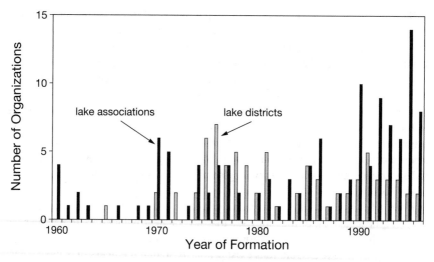

Figure 11.4. Number of lake districts and lake associations in Wisconsin by year of formation from 1960 to 1996 (Vilas County Tax Data).

The involvement of lake associations in the lake classification process is another important area we are studying. Lake classification is a triage process, in which lakes are categorized according to a number of ecological and sociological criteria, such as water clarity, physical dimensions, and extent of development. Local governments then use these classifications to target policies and land use regulations. Lake associations and local governments have been instrumental actors in gathering state resources for these efforts. The state has infused substantial amounts of money into the process of lake classification, perhaps as a response to the weaknesses of the previous policies governing lake districts. Lake districts, while they have the authority to impose taxes, do not have legislative authority over land use.

The study of the role of ecological factors in the formation of lake associations, as well as ecological responses to the behavior of these groups, is one area of research we are undertaking. Some additional research projects are discussed in the following section.

Linking Natural and Social Sciences

The social sciences have been involved in the North Temperate Lakes LTER program since 1997. In our early stages of thought and integration to combine social and natural science in the LTER, we started to move forward on simple, pragmatic grounds through recognition of overlapping areas of theory and method. While significant conceptual and methodological obstacles may slow achievements during this initial stage, our statement of intent in itself was a first step to realize or develop theories and methods of an emerging interdisciplinary science.

The importance afforded time and space, that is, or history and setting, is at the core of natural and social science integration within the LTER program. Both domains of science are traditionally ill equipped, theoretically and methodologically, to achieve an overarching view of human-environment interactions. We recognized our limitations and past failures in efforts to understand the interplay between social and natural phenomena. Integration was framed from the outset as a process of cooperative learning, rather than one to obtain the simple sum of our separate endeavors.

We later discuss three examples that provide insights that could not be achieved if we were working solely from either the social or the natural sciences. These examples have been identified by the natural sciences as important domains of human-environment interaction. Each area of research implicates the role of human perception in structuring the biophysical world in light of broader sociohistoric dynamics that operate in specific geographic contexts. Through the process of human perception at various levels of analysis by individuals and groups, biophysical contexts are linked to social structural patterns such as social strata, group membership, and socioeconomic relations. The following examples highlight different facets of human perception arising from the interaction between the social and biophysical domains.

Coarse Woody Habitat

The various themes that give meaning to places have implications for limnological properties of lakes like the distribution and abundance of coarse woody habitat and riparian forest cover. Coarse woody habitat refers to trees and tree fragments that have fallen into the lake from the riparian zone. Woody habitat in streams has been well researched in the natural sciences, and a reasonable body of knowledge has developed about some of the factors that contribute to its abundance in aquatic systems (Harmon et al. 1986, Gurnell et al. 1995, Christensen et al. 1996).

Coarse woody habitat can be an important component of lake ecosystems, and one that is influenced by human activity (Christensen et al. 1996). Thus, characteristics of this habitat can be conceived of as indicators of human and ecological interaction. The types of things that homeowners do on their property frontages, such as clearing and building, indirectly influence the presence of woody habitat through effects on riparian forest density. People who remove the material from the lake water directly affect coarse woody habitat abundance.

To better understand the human dimensions of coarse woody habitat in lakes, shoreline property owners on eight lakes in Vilas County were surveyed in 1998. The lakes included in the study were Big Muskellunge, Diamond, High, Plum, Razorback, Sparkling, Trout, and Witches Lakes. These lakes were chosen because they have a mix of privately and publicly owned shorelines and reflect variation on several characteristics of ecological and social significance (Table 11.1). The response rate to the mail questionnaire was 66% and produced a sample of 319 shoreline property owners. These individuals were predominantly male (80%), had gross annual incomes in excess of $35,000 (78%) and a median age of 59 years, and were mostly residents of either Wisconsin (59%) or Illinois (24%). Nearly one-quarter of the shoreline owners listed their property as their permanent residence. On aver-

Table 11.1. Characteristics of eight lakes in Vilas County, Wisconsin, included in the 1998 coarse woody debris (habitat) survey by the North Temperate Lakes LTER.

Lake	Area (km²)	Shoreline Length (km)	Lake Order (approximate)	Lake Association
Big Muskellunge	396.2[a]	16.1[b]	−1	No
Diamond	46.5[b]	3.5[b]	−1	Yes
High	299.9[b]	13.6[b]	1	No
Plum	427.9[b]	21.5[b]	2	Yes
Razorback	154.4[b]	9.7[b]	−1	No
Sparkling	64.0[a]	4.3[a]	−3	No
Trout	1607.9[a]	25.9[a]	2	Yes
Witches	24.4[b]	3.6[b]	−3	Yes

[a]http://lter.limnology.wisc.edu.
[b]Perennial lakes and ponds in Vilas County, Wisconsin (2001).

age, the remaining 75% visited their properties for 75 days out of the year (42 days in summer, 18 in fall, 5 in winter, and 10 in spring). Seventeen percent of respondents were members of an environmental group, and 30% were members of a lake association.

Analysis of the questionnaire responses indicated that the significance of woody debris (the survey had used debris rather than habitat) to shoreline owners was reasonably low. When asked to identify the most important lake issues, woody debris was not cited widely compared with issues such as water quality, habitat preservation, and shoreline development (Table 11.2). Moreover, just over half of the sample (55%) had either rarely thought about logs in their lakes or not thought about them at all, prior to confronting the issue in the questionnaire.

Nevertheless, field observations suggest that woody habitat is less common or even scarce where private lakeshore property and structures such as cabins and docks

Table 11.2. Important lake issues identified by shoreline property owners on eight lakes (Table 11.1) in the 1998 coarse woody debris (habitat) survey.

Lake Issue	Percent of Sample
Water Quality	65
Habitat Preservation	49
Fish Stocking	46
Lakeshore Development	45
Boating Ordinances	35
Recreational-User Conflicts	32
Crowding	28
Nonnative Plants and Animals	17
Property Prices	15
Tree Logs in the Lake	8

Table 11.3. Intentions of shoreline property owners to remove logs from the lake water on their property on eight lakes (Table 11.1) in the 1998 coarse woody debris (habitat) survey.

Log Characteristic	Percent Never or Unlikely to Remove	Percent Unsure	Percent Definitely or Probably Remove
Obstruction to Boating	20	10	71
Danger to Swimming	21	9	69
Above Water Level	54	14	32
Near Lakeshore	59	10	30
In Shallow Water	60	11	29
Near Property	55	17	28
One of Many Logs	56	20	24
Looks Ugly	66	15	19
Looks Big	69	16	15

are located. This is consistent with our survey data showing that approximately 25% of shoreline owners reported having removed at least one log from the water, and 64% of these respondents had done so as recently as 1996. Despite the low saliency of woody debris abundance as a lake issue, a moderate proportion of property owners participated in removing the material from lakes. Moreover, 41% of the sample reported that they still would not remove woody debris perceived to be dangerous to either boats or to humans. Logs that represented swimming or boating hazards were more likely to be removed than woody debris that was close to owners' properties or that was simply large, unaesthetic, or visible (Table 11.3).

Responses to the survey suggested that property owners participated in a range of behaviors on or around their lakes. The most frequent activities of respondents in the previous year tended to be passive, low-cost recreational pursuits (wildlife observation and sightseeing), motor boating, and open-water fishing (Table 11.4). Furthermore, property owners who had larger, more expensive houses on their land, as well as those who participated in motorized, water-based recreation such as motor boating, water skiing, and jet skiing, held the strongest removal intentions for woody debris.

Lakeshore owners were surveyed about the type of development presently existing on their properties as well as the characteristics they would ideally like to have. Most owners reported having a dock, trees, and natural vegetation on the shoreline of their properties (Table 11.5). Relatively smaller proportions of owners reported the presence of landscaping and retaining walls. Owners' ideal expectations for their properties tended toward development options, but the trend was not strong.

Coarse woody habitat density and riparian forest density can be conceived as characteristics of properties on a lake, a whole lake, or an entire lake district. Lakes and the human communities situated around them differ in terms of economic and social variables (property values, recreation patterns) that have analogous levels of aggregation at the individual and group level. Investigating relations among variables over different units of analysis is often of interest. The social sciences have

Table 11.4. Activities by shoreline property owners on eight lakes (Table 11.1) in the 1998 coarse woody debris (habitat) survey.

Activity	Percent of Sample
Wildlife Observation	79
Open-Water Fishing	75
Motor Boating	74
Sightseeing	61
Photography	60
Hiking	56
Canoeing or Kayaking	51
Picnicking	38
Water Skiing	38
Ice Fishing	29
Sailing	28
Snowmobiling	22
Snow Skiing	21
Hunting	20
Camping	11
Jet Skiing	8
All-Terrain Driving	5

developed methods for this purpose that might be employed by the natural sciences. For example, researchers may want to know whether the relationship between woody habitat density and riparian forest density is different at the level of the individual lakeshore property and that of the whole lake. Riparian forest density can be regarded as an indicator of shoreline development in forested lake settings. Measurements of coarse woody habitat abundance and forest densities might be taken at randomly selected shoreline properties on a number of different lakes. Values for the variables at the lake level might be garnered by aggregating the property-level

Table 11.5. Actual and ideal property characteristics of shoreline property owners on eight lakes (Table 11.1) in the 1998 coarse woody debris (habitat) survey.

Property Characteristic	Percent with Existing Property Characteristic	Percent Expressing Ideal Property Characteristic
Trees on the Shoreline	91	4
Private Dock or Pier	90	6
Natural Shoreline Vegetation	89	5
Clear View of the Lake	69	6
Winterized House	65	13
Detached Garage	50	9
Grassed Area on Shoreline	47	6
Seasonal Cabin	44	2
Developed Beach	29	11
Detached Boat House	22	11
Landscaped Shoreline	15	7
Retaining Wall on Shoreline	9	8

data for each lake to arrive at mean estimates of both woody habitat and riparian forest densities. With data for each unit of analysis, correlations could be calculated for the two variables within each level.

Although easily implemented, the preceding design and analysis confounds units of analysis. The bivariate correlations based on the aggregated and individual property data reflects both lake and property level relationships. The relation between coarse woody habitat abundance and forest densities at one level of analysis is not independent of the other level. Data analytic techniques of the social sciences can remove the lake effect from the property-level relationships and the property effect from the lake-level means (Kenny and La Voie 1985). Moreover, multilevel covariance structure analysis (Muthen 1994) might be used to examine effects at both the lake and the property levels within the same empirical model. With this technique, relationships at both levels of analysis can be estimated simultaneously by making use of the between and pooled-within sample covariance matrices.

Social analysis of perceptions of coarse woody habitat needs to be sensitive to the implications lakeshore property development and recreational patterns may have for maintaining a sense of place. People who use the lake edge for a range of passive pursuits and those who use it for active pursuits perceive the presence of large logs in the lake differently. These behaviors both express, and contribute to, a sense of place. Such an approach has the potential to provide the natural sciences with a fuller description of woody habitat abundance than might be suggested by ecological models alone.

Lake Position in the Landscape

Important or interesting constructs of one discipline may at first glance have little significance for another. Indeed, lake position in the landscape (Chapter 3) represents one such idea that provides limnological understanding of the heterogeneity among lakes in a lake district but on first thought seems to have little consequence in the human perceptual system. The relative elevation of most lakes in the Wisconsin landscape is perhaps below most people's perceptual threshold. Despite the name Northern Highlands Lake District, lakes may be only a few meters different in altitude along an underground hydrological flow path. However, to the extent that lake position coincides with other dimensions of human perception, such as a small, brown bog lake versus a large, clear-water lake, the idea has some utility in bridging discipline-bound conceptual models.

Lake position covaries with lake area such that larger lakes tend to exist at lower elevations in the landscape (Chapter 3) (Kratz et al. 1997b, Reed-Andersen et al. 2000c). Moreover, the amount and type of recreational facilities that differ among lakes are associated with lake area. Variation in boating activity on 99 lakes, including some of the primary LTER lakes, was related strongly to lake area, the presence of fishes, and accessibility to the lake by means of boat ramps, camping grounds, and quality of access roads (Reed-Andersen et al. 2000c). The number of houses and resorts per unit shoreline on a lake tended to increase from small lakes high in the landscape to large lakes low in the landscape (Chapter 3) (Riera et al. 2000).

Lake position had only a marginally significant relationship with the average number of boats on a lake when lake area and the social characteristics of lakes were taken into account. Position did, however, have a strong relation with average boat density on lakes, but this relation reflected the fact that no boats were observed on bog lakes typical of the highest positions in the landscape (Reed-Andersen et al. 2000c).

Importantly, relations with landscape position may be mediated through lake area. Likewise, a lake's size and associated properties are antecedent to the types of social variables that characterize natural lakes. In this sense, lake area can be cast in a classic mediator role as an intervening or process variable (Baron and Kenny 1986) between lake position and human-derived features such as recreational facilities and roads. Interestingly enough, lake area correlates with a number of other ecological and physical landscape characteristics and can conceivably represent a link between many variables relevant to investigations focused on the lake as the unit of analysis. Thus, although people may not directly perceive lake position, certain landscape features related to the spatial organization of lakes such as lake area may be salient aspects of perception.

Certain social characteristics of lakes may be endogenous to particular physical characteristics such as lake area, position in the landscape, and shape, but the presence and behavior of lake associations suggests more complex relationships. Lake associations may develop for a number of reasons, for example, resident perceptions of environmental degradation or other changes that threaten shared senses of place. Because lake associations may operate to organize social action to protect environmental values, water quality parameters and other indicators of environmental integrity presumably would show signs of improvement over time to the extent that lake organizations are effective in achieving their stated goals. The relationship between ecological variables, on the one hand, and lake association presence, focus, and extent of activity, on the other, introduces nonrecursive or reciprocal elements that may be discernable only over time. While the ecology of a lake may change as a result of a lake association's performance, this social group is likely to react to its shared perception of those changes in light of their collective sense of place (Lake Mendota example, Chapter 12). However, the reaction depends on the extent that the group's perceptions of environmental quality vary systematically with objective indices of the ecological variables. Human preferences for shoreline landscapes have been linked to perceptions of the physical features of shoreline vistas (Nasar 1987), but little work has been done to determine landscape correlates of sense of place (Jorgensen 2002). Nevertheless, the narrower focus of research like that conducted by Nasar (1987) can inform our understanding of human preferences for lakeshore environments. For example, individuals preferred lakeshores that were well kept and vegetated but not those with visible industry, streets, urban structures, and rocks and breakwalls.

Stedman (2000, 2002) examined the relationship between landscape attributes and sense-of-place components. Landscape attributes were related to place satisfaction or attitude. Higher satisfaction, that is, more favorable attitudes, was associated with lakes with lower density of shoreline housing and no public access. Limnological lake properties also affected attitudes: higher levels of satisfaction were associated with lakes that had lower levels of chlorophyll and low turbidity and that tended toward the blue-green end of the color spectrum. In contrast, place attachment was relatively unaffected by in-lake characteristics.

Furthermore, Stedman and Hammer (2005) have built models that make sense of human response to rapid social and environmental change by integrating perceptual data with lake attributes such as lakeshore development, chlorophyll, and turbidity. They discovered that lakeshore property owners perceived water quality reasonably accurately but that conditions of dense shoreline development were linked to perceptions of poor water quality, even when water quality data did not support these perceptions.

Perception, whether at the level of the group or the individual, needs to be incorporated into models that posit links between humans and physical and geographic characteristics of lakes. Indeed, perceptions of lake features may have relationships at each level of analysis that are not consistent (Kenny and La Voie 1985). For example, individual perceptions may be important in explaining why individuals choose to join lake associations, but shared group perceptions may be more important in explaining why lake associations adopt a particular stand on an issue.

Exotic Species Invasion

Some areas of study relevant to both the natural and the social sciences reveal limnological findings that are imbued with meanings to ecologists that are likely to differ from those held by others. One example is the characterization of the rainbow smelt as a species that disturbs the structure and functioning of a lake ecosystem (Chapter 8). Magnuson and Beckel (1985), for example, consider smelt as a form of biological pollution, introduced by humans into Sparkling Lake and responsible for the collapse of native cisco populations.

Despite acknowledged scientific knowledge, anglers may think that smelt are good for the lake and important to their experience of the place. They may be able to point to a range of benefits that smelt provide, and some of these may be claimed even as environmental benefits. Yet, from an ecologist's point of view, such perceptions and attitudes might be considered ill formed or misguided. Consequently, a researcher may advocate education programs to rectify the misinformation. Given conflicting interpretations of the species, rainbow smelt invasions into the lakes of northern Wisconsin may represent an example where limnological knowledge has no greater claim to ascendancy in the human community than that afforded any other viewpoint.

A number of preliminary on-site interviews have been conducted with smelt anglers. Smelting is a recreational activity and occurs in the late hours of the evening during the relatively short time of the year when the smelt are migrating into the shallows and streams to spawn. The smelting experience often involves groups of people getting together to enjoy the outdoors and the company of family and friends and eating the evening's catch. Smelting is an occasion for anglers to supply themselves with bait prior to the opening of the fishing season. Given these benefits, just over half of the smelters interviewed believed that there were not enough places in northern Wisconsin where smelt could be caught, particularly on inland lakes. Negative impacts that smelt may have on native fishes largely were unrecognized.

Some anglers felt that people had been smelting in northern Wisconsin for as long as they could remember and did not perceive smelt to be an unfamiliar aspect of the region or an introduced species. Because anglers were familiar with smelt and valued the benefits the fish provides, they may have been more inclined to believe that smelt should be managed in a sustainable manner in much the same way as other fish resources.

Understanding the social dimensions of smelt invasions requires ecological information. Knowledge of limnological processes helps us see links to the human system and to form hypotheses about interactions of human dimensions with characteristics of lakes. For example, to examine exotic species invasion as a function of angling behaviors and attitudes requires some knowledge of the time required for species to become established in new environments and detectable by scientific methods. Without such knowledge, concluding whether the appearance of an exotic resulted from angling practices, other processes, or chance is difficult.

Practices such as emptying unused bait into a lake or cleaning the catch at an uninfected lake can serve as pathways for smelt transmission (Chapter 8) (Hrabik and Magnuson 1999). Given the importance of smelt gametes to invasions, anglers who fish at one lake or stream but clean their catch at the lake near their residence are worthy of study. Understanding what anglers think of smelt, what they do with smelt, and where they do it may help identify lakes at risk of future smelt invasion. Investigating the importance of smelting to sense of place assists in developing ameliorating strategies that are likely to garner community support and, therefore, have the greatest effect on reducing the spread of exotics.

Summary

The Northern Highlands Lake District where our LTER lakes are located in northern Wisconsin has undergone a shift in the nature of human-environment relations from extractive uses of natural resources to dominance of in situ benefits. This change is apparent in the significance of lakes in focusing the spatial distribution of humans and in the dynamic of property ownership and development. These trends are likely to result in adverse environmental impacts such as invasion of exotics, loss of woody habitat, eutrophication from septic leakage or municipal sewage treatment, land clearing, and wetland drainage. These consequences of social change may occur in conjunction with climate changes (Chapter 7).

Human-environment relations can be understood by exploring the interpretive systems of values and beliefs that individuals and groups attribute meaning to natural objects such as lakes. Any explanation of the dynamics of lakes in the landscape requires linking natural and social sciences. More specifically, human behavior in natural settings and ecosystem responses to externally driven changes need to be viewed in the context of social values and the meaning of places to individuals within specific social groups. Without an understanding of the interplay between humans and the physical and biological worlds, future problems often may appear as surprises and their solutions involve reactions to perceived threats to maintain a sense of place.

Reciprocal interactions between people and lakes help determine the long-term dynamics of lakes. The management approaches required to protect the lakes and associated areas within the Northern Highlands Lake District primarily involve the management of human activities. The myriad experiences people have with various settings in the region serve to give the place meaning. The history of the area reveals how dominant meanings of the region have changed over time as a function of changing demography and human activities. Today, lake environments can feature strongly in how people think about the region and what they do there. In many cases, people come to identify with lakes and form emotional attachments to them. However, these meanings and behaviors are not always consistent with protection of lake ecosystems. Rather, when human activities and motives are factored into ecological understandings of specific phenomena such as coarse woody habitat distribution and abundance and exotic species transmission, a more complete picture can be achieved. Moreover, understanding these interactions can inform lake and property management strategies aimed at protecting valued attributes of the region.

12

The Ongoing Experiment: Restoration of Lake Mendota and Its Watershed

Stephen R. Carpenter
Richard C. Lathrop
Peter Nowak
Elena M. Bennett
Tara Reed
Patricia A. Soranno

I think they are the most beautiful bodies of water I ever saw.

The first one we came to [Lake Monona], was about ten miles in circumference, and the water as clear as crystal. The earth sloped back in a gradual rise; the bottom of the lake appeared to be entirely covered with white pebbles, and no appearance of its being the least swampy.

The second one that we came to [Lake Mendota] appeared to be much larger. It must have been 20 miles in circumference. The ground rose very high all around;— and the heaviest kind of timber grew close to the water's edge.

—Surgeon's Mate John Allen Wakefield, 21 July 1832 (Thayer 1983, p. 118)

When Wakefield's battalion pursued the Sauk warriors led by Chief Black Hawk through the uncharted wilds of southern Wisconsin, the lands surrounding the lakes supported wetlands, prairies, oak savannas, and forests (Curtis 1959). Wakefield's account, and paleolimnology, tells us that the lakes were clear with light-colored sediment (Hurley et al. 1992, Kitchell and Sanford 1992, Kitchell and Carpenter 1993). By the late 1840s, clearing of land for agriculture was well under way (Lathrop 1992a), as the city of Madison and the newly founded University of Wisconsin were developing on Lake Mendota's shore. By 1880, newspaper accounts regularly reported blooms of noxious algae and fish kills in the Madison lakes (Brock 1985). Subsequent studies of lake sediments show dramatic changes in the lake's plankton at about this time (Hurley et al. 1992, Kitchell and Sanford 1992, Kitchell and Carpenter 1993). Within four decades (about twice the current extent of the LTER program) in the 1800s, the landscape was transformed from native vegetation to farms and a prosperous town. At the same time, the lakes were converted from clear oligo-mesotrophic waters to enriched, green eutrophic waters with severe blooms of blue-green algae. By the time E. A. Birge arrived in the 1880s to establish a limnology program at the University of Wisconsin (Beckel 1987), both the landscape and the lakes were dramatically changed. The insights that Birge and other U.W. scientists gained from the study of the Madison lakes, regardless of their condition, shaped the early development of limnology in the United States. However, the linkages between the state of the land and the state of the lakes were not articulated for nearly a century after Birge conducted his pioneering work.

Although land use and land cover in the watershed are now completely different from the presettlement condition (Lathrop et al. 1992, Soranno et al. 1996), the geologic setting of the lakes has changed little from that witnessed by Wakefield. The watershed is calcareous, sediments are calcium-rich, and the lakes have hard water. "Whitings" (events when precipitation of calcium carbonate granules gives the water a whitish cast) are common in summer. Dams to control water levels, draining of wetlands, and falling groundwater levels have altered the hydrology of the region. The four largest Madison lakes are fed largely by surface flow. Waters of several streams gather in Lake Mendota, and the Yahara River flows on to feed the downstream lakes Monona, Waubesa, and Kegonsa before continuing to the Rock River, a tributary of the Mississippi. Two smaller lakes studied by the North Temperate Lakes Long-Term Ecological Research program (LTER) are fed by groundwater. Lake Wingra was originally a wetland draining into Lake Monona (Bauman et al. 1974). A dam converted this wetland into a shallow lake in the early 1900s. Originally the lake was spring-fed, but in recent decades urban runoff from storm sewers has become the dominant hydrologic input. Fish Lake, a seepage lake in an agricultural watershed, is similar to Lake Mendota with respect to watershed land cover and aquatic community structure, but different hydrologically.

Here we show that the history of the Madison lakes is linked intimately to the history of the lands around them. Consequently, current programs for restoring the lakes depend on altering land use in ways that reduce runoff of nutrients, especially phosphorus (P). As in the Northern Highlands Lake District, we have a remarkable opportunity to study lakes in a landscape context (Chapter 3). This perspective provides us with rich opportunities to link the social and natural sciences (Chapter 11) by studying the impacts of human behaviors and decisions on lake ecology and the human behaviors and policy initiatives evoked by changing lake ecosystems. Massive management interventions can be viewed as long-term experiments, creating unique opportunities to learn from the responses of lakes to landscape modification. LTER plays a key role in the design of the management interventions and the assessment of their effects on the lakes.

We focus on the water quality of Lake Mendota and the history of research and management of a core problem, eutrophication. Lake Mendota is the best studied, and since the mid-1950s the most intensively managed, of the lakes in the region. The Lake Mendota case is well suited to the theme of this book because long-term changes in land use and the lake are intertwined closely. The changes in the lake's ecology cannot be understood without knowledge of the social changes and the evolution of management policies in the surrounding landscape. The future of the lake depends heavily on society's attitudes toward it, as reflected in management policies and the behaviors of individual developers, farmers and anglers.

Historical Trends

The Lake, Its Catchment, and Environmental Problems

The condition of Lake Mendota and its management are highly visible to state government officials and the general public of the region. The quality of freshwater resources are important in a state with more than 15,000 lakes and a lake-centered recreational industry that is a major sector of the economy. Lake Mendota is bordered on the south and east by Wisconsin's capital city and leading university. Two books and hundreds of scientific papers have been written about the lake (Brock 1985, Kitchell 1992).

The lake (surface area of 40 km², maximum depth of 25.3 m) is located centrally in the Yahara River watershed, a rich agricultural region (Fig. 12.1). Agricultural runoff is the main cause of Lake Mendota's eutrophication (Lathrop 1992a, Soranno et al. 1996, Lathrop et al. 1998). Urban areas, especially construction sites, are an important and growing source of phosphorus-rich runoff to the lake. The rapid population growth in the watershed is a double-edged sword because the increase in runoff from construction sites and urban areas also may increase the political and economic pressures for lake restoration. Because outflow from Lake Mendota is the major source of phosphorus to downstream lakes, its restoration is viewed as a critical step toward the restoration of the entire Yahara chain of lakes (Lathrop 1990).

Management of Lake Mendota's biological problems such as species invasions and extirpations has been limited to harvesting of nuisance species such as exotic Eurasian water milfoil and carp. The principal strategy has been first to repair the physical-chemical template, primarily the eutrophication problem, and then to move on to the lake's biological problems. For example, reductions in runoff should decrease the frequency and intensity of blue-green algal blooms (Lathrop et al. 1998). However, ongoing issues such as overabundant carp, loss of littoral zone habitat, low densities of benthic invertebrates, losses of littoral zone fishes, and, in some years, overabundant aquatic macrophytes in shallow waters need immediate attention. Moreover, new invaders, for example, zebra mussels, may create management problems in the future. We briefly describe the major management issues for Lake Mendota eutrophication, habitat loss, and biotic change. We then provide a history of eutrophication management, leading to the current large-scale management experiment to restore the watershed.

Phosphorus Inputs and Eutrophication

Blooms of blue-green algae have plagued Lake Mendota since the 1880s (Brock 1985). During summer, water clarity is often poor, and surface scums of blue-green algae are common (Lathrop and Carpenter 1992a). The eutrophic condition of the lake is caused by overenrichment with phosphorus that originates from a variety of sources.

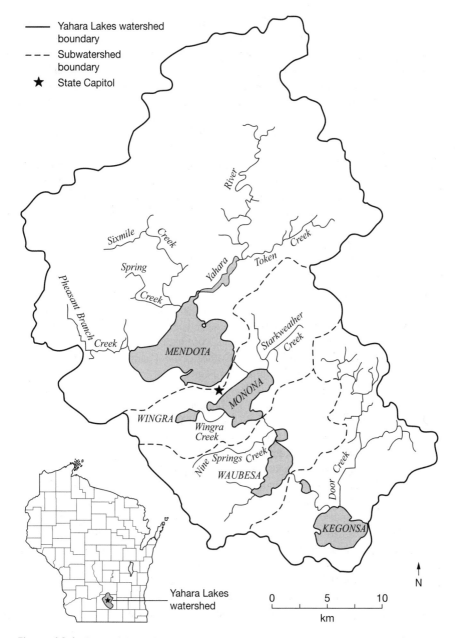

Figure 12.1. Map of the Yahara watershed showing the major lakes and their watersheds. Redrawn and updated from Lathrop et al. (1992).

The bulk of Madison's sewage effluents has been discharged downstream of Lake Mendota since the early 1900s (Lathrop 1992a). However, sewage from developing communities upstream of Lake Mendota began entering the lake in increasing quantities in the 1920s. Dissolved phosphorus concentrations increased substantially after the mid-1940s in streams that receive effluents from these growing communities (Lathrop 1992a). These effluents continued entering the lake until the sewage was diverted in late 1971, reducing inputs of dissolved phosphorus by about 30% (Lathrop 1990). Diversion of the sewage left runoff from the watershed as the overwhelmingly dominant source of phosphorus to Lake Mendota. Trends in fertilizer use and soil phosphorus content (Bennett et al. 1999) suggest that nonpoint inputs are likely to increase unless managers intervene. Imports of phosphorus into the watershed in the form of fertilizers and food for livestock exceed exports of phosphorus in the form of farm products. The excess phosphorus accumulates in soils; most are already rich in phosphorus. Thus, management of eutrophication now emphasizes reduction of phosphorus input from runoff.

The arable lands of Dane County, which includes most of Lake Mendota's catchment, were converted fully to agriculture by the 1870s (Fig. 12.2). Agricultural runoff has been a significant source of nutrients to the lake since that time (Lathrop 1992a). Soon after World War II, changes occurred that increased runoff of nutrients to the lake. Use of artificial fertilizers increased (Lathrop 1992a). More land was allocated to corn production (Fig. 12.2). Cultivation of corn exposes soil to erosive flows for longer periods of time than cultivation of other grains (Lathrop 1992a). Since the 1970s, mass of phosphorus in the agricultural soils of the watershed has been building up by about 575×10^3 kg per year (Bennett et al. 1999). This buildup is explainable by the imbalance of the watershed's P budget. Each year, about 1307×10^3 kg of P are imported in fertilizer, animal feed, and other forms, but only 732×10^3 kg of P are exported as produce, meat, dairy products, and runoff to the lake (Bennett et al. 1999), with all inputs to the lake averaging 34×10^3 kg of P (Lathrop et al. 1998).

At present, watershed managers focus on two primary sources of the phosphorus being delivered into Lake Mendota—agricultural land and urban/suburban construction. Both of these sources illustrate how social and biophysical factors interact to create complex policy problems. Soil P concentrations and their variance are substantially higher in agricultural soils, including former agricultural soils subject to development (Bennett 2002). "Hot spots" of high soil phosphorus may play a key role in the amount and timing of phosphorus runoff (Heckrath et al. 1995). Yet the pattern of soil P is a result of social-ecological dynamics.

Agricultural phosphorus comes from fertilizers and animal manures that can run off from barnyards or be applied to cropland. From a policy perspective, control of phosphorus coming from agriculture requires both technical fixes (e.g., buffer strips, terraces, barnyard drains) and behavioral changes (e.g., implementation of conservation practices such as incorporating field-applied manure within 72 hours and accounting for nutrients in manure). For watershed managers, the question has been how to find the appropriate mix of policy tools, such as education, financial incentives, regulation, and technical assistance, to induce the adoption of technical and behavioral fixes. Watershed managers traditionally have identified tracts of land, called critical areas, that are targeted because of their biophysical susceptibility to

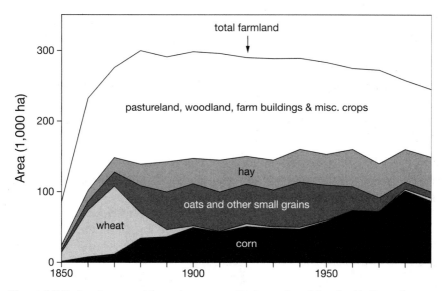

Figure 12.2. Land area used for major crop production and total farmland in Dane County, Wisconsin by decade since 1850. Redrawn from Lathrop (1992a) with permission of Springer-Verlag Publishers.

sediment and phosphorus loss. The owners of these tracts are offered cost-share assistance and technical assistance with the implementation of conservation practices. Past targeting of program resources has been guided by the assumption that farmer behaviors are identical across these critical areas.

Yet, variance in farmer behavior is at least as large as variance in biophysical settings they manage (Nowak and Madison 1998). Thus, identifying critical areas using biophysical attributes only may substantially over- or underestimate the potential runoff of phosphorus from a given site. Equivalent human behaviors in diverse biophysical settings do not have equivalent implications of phosphorus delivery to aquatic ecosystems. Moreover, certain human behaviors in noncritical biophysical settings can contribute significantly to eutrophication. We are investigating how farmer behaviors are distributed in heterogeneous spatial patterns across the landscape and how this patch structure can be used to inform conservation policy. What are the implications for P-loading of spatially heterogeneous phosphorus management behaviors across areas with varying soil phosphorus test levels or probabilities of phosphorus delivery to Lake Mendota?

The second primary source of phosphorus, transitional land that is being developed for housing and commercial use from agricultural production, also stresses the importance of studying the watershed as an interconnected system at multiple scales. While management practices may decrease the amount of soil being eroded from construction sites, the amount of phosphorus in that soil may have increased with additional use of phosphorus in agriculture (Bennett et al. 1999). Moreover, the probability of extreme events, whether anthropogenic or natural, is an important factor in phosphorus delivery to the lake, and an increasing frequency of these events (Chapter 7) may offset any gains made from phosphorus reduction policies focused on the average land owner. Thus, the history of land use, behaviors of farmers and developers, susceptibility of the terrain to runoff, and probabilities of extreme events all have substantial implications for nonpoint pollution policies. Policies that ignore any of these factors risk failure. In the Lake Mendota watershed, we are just beginning to understand the social-ecological interactions that drive water quality.

Habitat Loss

Losses of wetland habitat from Lake Mendota's catchment have been significant. In 1847, the water level of Lake Mendota was raised 1.5 m by a dam at the outlet (Lathrop et al. 1992). The higher water altered littoral habitat, flooded surrounding wetlands, and created new wet areas upstream in the watershed. Currently, about half of the original wetlands in the watershed remain; the rest were filled for urbanization or ditched and drained for agriculture (Lathrop et al. 1992). These remaining wetlands are now protected by law but are subject to change as urbanization continues to damage wetlands, primarily through changes in groundwater levels and erosion from construction sites.

Littoral zone habitats have been extensively altered. They now cover less surface area, and submerged aquatic plant beds contain fewer native species (Lathrop et al. 1992, Nichols et al. 1992, Nichols and Lathrop 1994). From the late 1800s to the 1950s, about 25% of the lake area was covered with rooted plants growing to depths greater than 5 m. Dominant species included wild celery (*Vallisneria americana*) and diverse pondweeds (*Potamogeton* spp.). By the early 1960s, these communities were drastically altered. Abundance of native species declined steeply, owing in part to displacement by the invasive, nonnative Eurasian milfoil (*Myriophyllum spicatum*). The maximum depth limit of plant growth declined to 3 m or less as water clarity decreased. In shallower water, the biomass of milfoil rose to levels that required more intensive weed harvesting efforts by the city and later the county (Lathrop et al. 1992).

Biotic Change

Landscape change caused by people is the root cause of the ecological changes in Lake Mendota (Lathrop 1992a, Soranno et al. 1996, Lathrop et al. 1998). Increasing human populations in the watershed have exacerbated the impacts of human activities. The ecological trends in the biota are linked to intensive fishing, introduction and stocking of fishes, and locally intensive management of symptoms of eutrophication, such as harvesting or herbicide applications to control macrophytes. Transport of boats from infested waters to Lake Mendota places the lake at risk of invasion by zebra mussels (*Dreissena polymorpha*) and other aquatic nuisances. Before addressing management of the landscape, in the next section of the chapter, we address in this section the biological changes caused by intensifying human use of the lake.

Species diversity of aquatic macrophytes declined after the 1950s. Many species of *Potamogeton* are no longer present (Nichols and Lathrop 1994). Declines in species favored by waterfowl, such as wild celery and sago pondweed (*Potamogeton pectinatus*), may have contributed to declines in migratory waterfowl. The invasive Eurasian milfoil began declining after the mid-1970s (Carpenter 1980, Nichols and Lathrop 1994) and is no longer growing in extensive monotypic stands (Nichols et al. 1992, Deppe and Lathrop 1993). We do not know whether species diversity will increase in Lake Mendota following the decline of Eurasian milfoil. Diversity has increased in nearby Lake Wingra in recent years, coincident with declining Eurasian milfoil (Trebitz et al. 1993, Nichols and Lathrop 1994).

Many species of small-bodied littoral zone fishes have been extirpated in Lake Mendota (Lyons 1989). Some of these species were lost in association with the loss of the diverse macrophyte beds that are their preferred habitat. Others may have been casualties of increased piscivore stocking in the 1960s (Magnuson and Lathrop 1992). Now a single species, brook silverside (*Labidesthes sicculus*), dominates the littoral fish community. Abundance of silversides is highly variable, and this variability may affect the forage base of piscivorous fishes (Magnuson and Lathrop 1992).

Several fish species have been introduced to Lake Mendota. The European common carp (*Cyprinus carpio*) has contributed to declining water clarity and loss of macrophytes (Lathrop et al. 1992, Nichols and Lathrop 1994). Other important fish invaders include the freshwater drum (*Aplodinotus grunniens*) and the yellow bass (*Morone mississippiensis*) (Lathrop et al. 1992, Magnuson and Lathrop 1992). Yellow bass were first recorded in Lake Mendota in 1957 but have been rare since a major dieoff in 1976. Drum were uncommon in Lake Mendota during the mid-1900s but have reached moderate densities since the 1970s.

A hundredfold decline in density of profundal benthic invertebrates (*Chironomus* and *Chaoborus* midge larvae) occurred between the mid-1950s and the mid-1960s in Lake Mendota (Lathrop 1992b,c). Loss of this invertebrate production may affect fish species such as yellow perch (*Perca flavescens*), a native species that supports a major sport fishery (Lathrop et al. 1992). During the same period, the fingernail clam *Pisidium*, once common throughout the lake's profundal sediments, was extirpated (Lathrop 1992b,c). More recently, declines of unionid clams throughout the littoral sediments have been noted (LTER, unpublished data).

Few major changes in the plankton have occurred in the past century (Lathrop and Carpenter 1992a,b). Recent sediments of the lake record the invasion of *Eubosmina coregoni*, a European exotic zooplankter (Kitchell and Sanford 1992). This invader exhibited seasonally moderate abundances during the mid-1980s but has since been rare (Lathrop and Carpenter 1992b). We can speculate that phytoplankton species composition has shifted to greater dominance by nuisance bluegreen algae as the lake has become more eutrophic. Unfortunately, the lake already was eutrophied by the time the first limnological studies were done in the 1890s (Brock 1985, Lathrop and Carpenter 1992a). Although no direct observations of phytoplankton are available for the period prior to European settlement, phytoplankton pigments in dated sediment cores suggest broad changes in the relative importance of the major divisions of algae (Hurley et al. 1992). Historical observations (opening quote) and the sediment record indicate major changes in the plankton around the time the watershed was settled (Hurley et al. 1992, Kitchell and Sanford 1992, Kitchell and Carpenter 1993).

Major Management Efforts for Eutrophication

The history of Lake Mendota's management is largely a story of evolving efforts to understand and control eutrophication. Here we present a summary of that history, culminating in the LTER program's role in a massive management experiment by state and local government agencies to reduce phosphorus runoff from the watershed.

Improving Understanding of the Phosphorus Problem

Water quality problems became a major issue for Lake Mendota by the 1940s (Hasler 1947). Around that time, a special study was commissioned to address the causes of algal blooms in the lake (Carpenter and Lathrop 1999). However, legal and political maneuvering focused attention on diverting Madison's sewage from the downstream lakes, where water quality problems were far worse.

By the mid-1960s, concern was again focused on Lake Mendota. The Lake Mendota Problems Committee was formed in the early 1960s, drawing on university and agency scientists and managers to develop a plan for improving the condition of the lake (Carpenter and Lathrop 1999). As a consequence, all sewage was diverted from Lake Mendota by 1971 (Sonzogni and Lee 1974, Lathrop 1992a). This committee also addressed the role of nonpoint pollution of Lake Mendota, but little action was taken (Carpenter and Lathrop 1999).

Beginning in 1975, the Dane County Regional Planning Commission began preparing a plan to control nonpoint pollution inputs to Lake Mendota. This plan was linked to the federal government's Water Pollution Control Act Amendments of 1972 and the Clean Water Act of 1977. Although a number of sources of nonpoint pollution were identified, corrective action was limited (Carpenter and Lathrop 1999).

To more effectively address the ongoing water quality problems in Lake Mendota and the other Yahara lakes, the Dane County Board created the Dane County Lakes and Watershed Commission in 1988. In 1990, the governor signed into law the commission's composition, duties, powers, and organization. The commission was formed in response to pending state legislation that would create a special watershed commission to address the fragmentation of watershed management activities related to the Yahara River chain of lakes. The Lakes and Watershed Commission was empowered to improve water quality and the scenic, economic, recreational, and environmental value of Dane County's water resources. The commission can recommend programs, plans, and projects to the County Board for approval. In addition, the commission can recommend that minimum standards be passed for water quality benefits and may suggest boating regulations. The commission has financing powers subject to County Board approval to achieve these objectives. To date, the Dane County Lakes and Watershed Commission's major accomplishments that have direct water quality benefits are the adoption of countywide standards for construction site erosion control and for wetland zoning.

In 1980, the western half of Lake Mendota's drainage basin (Sixmile-Pheasant Branch Creeks) was designated as a priority watershed project by the Wisconsin Department of Natural Resources. This designation allowed state funds to be used to share the cost of nonpoint pollution reductions with individual landowners and municipalities. Funding came from a major state bonding initiative as part of the department's Nonpoint Source Water Pollution Abatement Program, created in 1978 by the state legislature and more recently called the Runoff Management Program.

Because the Sixmile-Pheasant Branch project was one of the earliest to be conducted in the state and because corrective measures were voluntary, participation by landowners was minimal. In addition, many of the management practices that were tried turned out to be ineffective at reducing phosphorus inputs to the lake (Lathrop et al. 1998, Carpenter and Lathrop 1999). For example, the relationship between P-loading and runoff volume did not change substantially in the Pheasant Branch subwatershed between 1976–1980 and 1990–1994 (Lathrop 1998). Between these periods, watershed management practices were instituted in the subwatershed, and marginal lands were taken out of production. Reductions in phosphorus inputs, however, were not apparent. Perhaps they were masked by other factors such as higher phosphorus concentrations in soils (Bennett et al. 1999), increased construction activity, or degradation of riparian lands. Phosphorus runoff may be dominated by a few key hot spots that have high soil phosphorus and high runoff potential (Sharpley 1995, Pionke et al. 1996, Bennett 2002). Perhaps the voluntary nature of the Sixmile-Pheasant Branch project tended to attract farmers whose fields were already low P runoff areas and therefore did not lead to changes in areas that were producing the majority of the runoff.

The Biomanipulation Project

By the mid-1980s, managers, scientists, and the public were frustrated by slow progress in the restoration of Lake Mendota (Lathrop et al. 1998, Carpenter and Lathrop 1999). Sewage diversion, a necessary and beneficial step, had produced insufficient benefits to water quality. The Sixmile-Pheasant Branch project was limited in scope and involved untested management practices. Massive amounts of phosphorus were stored in the lake sediments, and internal recycling of phosphorus was sufficient to sustain massive algal blooms each summer (Stauffer and Lee 1973, Soranno et al. 1997). During summer, recycling of phosphorus from sediments occurs at higher rates than inputs from the watershed (Soranno et al. 1997). Unlike inputs from the watershed, which vary considerably from year to year (Lathrop et al. 1998), recycling rates from sediments are relatively constant from year to year (Soranno et al. 1997). The internally cycling phosphorus is not susceptible to management, except by decreasing phosphorus inputs to the lake over many years. Internal recycling of phosphorus can be controlled directly only by internal interventions that change nutrient cycles within the lake (Cooke et al. 1993).

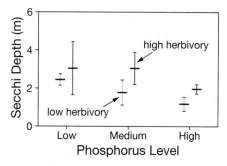

Figure 12.3. Mean Secchi disc transparency during summer in Lake Mendota for years of low, medium, and high phosphorus and with small-bodied grazers (low herbivory) or large-bodied grazers (high herbivory). Data for transparency, phosphorus levels, and zooplankton were obtained for 60 summer seasons between 1900 and 1993. Means and 95% confidence intervals are shown. Redrawn from Lathrop et al. (1996) with permission of the Canadian Journal of Fisheries and Aquatic Sciences.

The time was right for a new approach, and management and science interests converged on the idea of a biomanipulation experiment for Lake Mendota (Kitchell 1992, Lathrop et al. 2002). Biomanipulation is the reconfiguration of fish stocks by increasing biomass of top predators, thereby decreasing biomass of fishes that feed on zooplankton and increasing zooplankton grazing of phytoplankton (Shapiro et al. 1975). Biomanipulation appeared to be an alternative attack on Lake Mendota's eutrophication problem that potentially could sidestep the obstacles posed by nonpoint pollution and phosphorus recycling (Lathrop et al. 2002). Whole-lake experiments near Land O' Lakes, Wisconsin, had shown that trophic cascades from fish to phytoplankton could control biomass of algae (Carpenter et al. 1987). Historical records of water clarity from Lake Mendota suggested that changes in the food web could increase water transparency during summer by about a meter, as measured by Secchi disc (Lathrop et al. 1996) (Fig. 12.3). These findings suggested that Lake Mendota's water quality could improve with biomanipulation.

At about the same time, the U.S. Congress increased funding to state sport fish restoration activities through enhancements to the Federal Aid for Fisheries Restoration Program (Addis 1992). These new funds opened a window of opportunity for innovative management. Scientists, managers, and the public were willing to try something new. Local sport fishing clubs were enthusiastic about the idea of enhancing game fish populations to improve water quality.

The biomanipulation involved massive stocking of walleye (*Sander vitreum*) and northern pike (*Esox lucius*), combined with restrictive bag and size limits (Johnson et al. 1992, Johnson and Staggs 1992, Lathrop et al. 2002). A total of 2.7×10^6 walleye fingerlings and 1.7×10^5 northern pike fingerlings were stocked from 1987 to 1999 in Lake Mendota (Fig. 12.4). In addition, approximately 100×10^6 walleye and northern pike fry were stocked, but stocking was discontinued after three years because fry survived poorly. The heavy stocking rates of walleyes in the early years

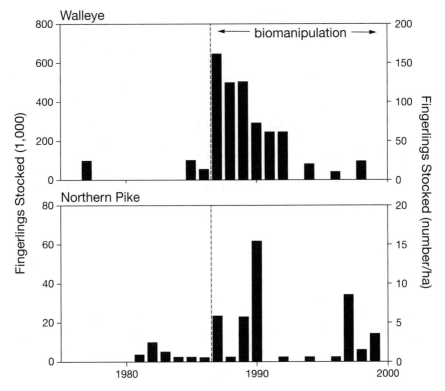

Figure 12.4. Annual stocking of walleye and northern pike in Lake Mendota, showing the increase with the biomanipulation experiment. Redrawn from Lathrop et al. (2002) with permission of Blackwell Publishing.

of the project represented a major share of the state's walleye hatchery production; this was a controversial commitment of resources diverted from popular walleye stocking programs in the northern regions of the state where many of the fish were raised (Johnson and Staggs 1992). Northern pike stocking rates were almost an order of magnitude lower because obtaining fingerlings from local hatcheries was difficult. Most of the walleyes were stocked from 1987 to 1992; northern pike stockings were heaviest in the early and later years during the 1987–1999 period of evaluation. Managers had hoped that the heavy stocking rates from 1987 to 1992 would build up the adult spawner population sufficiently to allow for natural reproduction to sustain the population at high densities.

As a result of the stocking, the biomass of both piscivore species substantially increased in the lake (Lathrop et al. 2002). Northern pike biomass was highest in 1990; walleye biomass reached a maximum in 1998. In general, the combined biomass of both species ranged about 4 to 6 kg/ha from 1989 through 1999. While the combined piscivore biomasses of stocked fish indicated a substantial increase occurred from pre-biomanipulation biomasses, the levels are lower than levels reported for other biomanipulation projects, for example, >20 kg/ha (Benndorf 1990, 1995).

The unprecedented fisheries manipulation and associated research created a high profile in the media that led to increasing angler effort and greater harvest of sport fish (Johnson and Staggs 1992, Lathrop et al. 2002). Fishing effort directed at walleyes in Lake Mendota increased more than sixfold from 1987 to 1989 and remained high at about 2 angler-hours/ha/month, throughout the 1990s (Johnson and Carpenter 1994, Lathrop et al. 2002). Walleye harvest during this same period increased severalfold and diminished the impact of the biomanipulation on the lake's food web. Catch rates of northern pike tracked increases in northern pike abundance, increasing rapidly from 1987 to 1990 in response to the stocking efforts (Lathrop et al. 2002). Catch rates decreased after 1990, as northern pike biomass and abundance declined when stocking was reduced.

During the biomanipulation period, water clarity was remarkably good in some years. This increase in water clarity was associated with dominance in the zooplankton of a highly effective grazer, *Daphnia pulicaria* (Fig. 12.5). A thorough statistical analysis using data from 1976 to 1996 suggested that episodes of high water clarity were explainable by a combination of low runoff, calm summer weather (low mixing of phosphorus to surface waters), and the impact of *Daphnia pulicaria* (Lathrop et al. 1999). Dominance of *D. pulicaria* since 1988 appears to have been triggered by a massive summerkill of cisco (*Coregonus artedii*) in 1987 (Vanni et al. 1992, Rudstam et al. 1993, Lathrop et al. 2002). The biomanipulation possibly has played a role in the suppression of cisco and maintenance of *D. pulicaria*. However, past dieoffs of cisco have been followed by many years of low cisco densities and high *D. pulicaria* densities (Lathrop et al. 1996). Thus, the clear-water years associated with the biomanipulation could be explainable by fortuitous circumstances unrelated to the biomanipulation itself (Lathrop et al. 1999).

A major lesson of the Lake Mendota biomanipulation is that food web structure cannot be altered by piscivory without severe restrictions on angling. People are the true top predator in Lake Mendota (Kitchell and Carpenter 1993). A corollary of this insight is that social sciences should be a part of the planning of any future management interventions. The capacity of biomanipulation to improve water clarity in this large eutrophic lake was never given a definitive test because angler harvest interfered with the manipulation's impact on food web structure. The best evidence of food web impact on phytoplankton in Lake Mendota remains the long-term record and informative but rare and serendipitous events (Vanni et al. 1992, Rudstam et al. 1993, Lathrop et al. 1996, Lathrop et al. 2000).

Nevertheless, the program had a number of benefits, including transfer of sonar and modeling technologies to managers in the Wisconsin Department of Natural Resources and direct evaluations of new survey and census techniques, stocking efficacy, and regulations (Rudstam and Johnson 1992, Staggs 1992). Among the most valuable outcomes was the experience in interagency and interdisciplinary collaboration that has created new links between the University and the state's natural resource management agency and has inspired some project alumni to develop collaborative management experiments in other states. Biomanipulation now

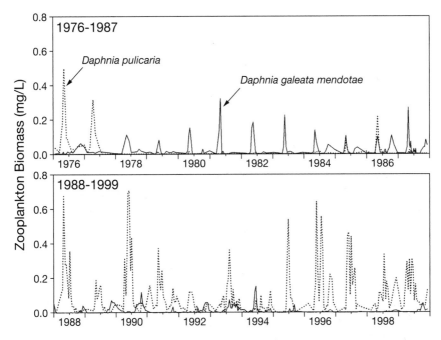

Figure 12.5. Biomass concentrations (mg dry mass/L) for *Daphnia galeata mendotae* and *D. pulicaria* from 1976 to 1999 in Lake Mendota. Redrawn from Lathrop et al. (2002) with permission of Blackwell Publishing.

is a tool used widely for water management (Hansson et al. 1998). The Lake Mendota experience contributed to the understanding and improvement of biomanipulation methods for large, deep lakes.

If a second biomanipulation is attempted in Lake Mendota, harvest should be controlled rigorously, and predators like northern pike, which grow well in the lake, should be augmented (Kitchell 1992). However, the lake presently has only a small population of cisco, and remaining planktivorous fishes, including yellow perch, appear unable to suppress *Daphnia pulicaria* (Johnson and Kitchell 1996). Since 1999, yellow perch have attained relatively high densities in the lake in spite of the augmented piscivore biomass (Lathrop et al. 2002), although angling pressure on the perch has been intense. Consequently, populations of this effective grazer remain high. Comparative studies suggest that the water-quality benefits of heavy grazing were near the maximum possible level in Lake Mendota during the 1990s (Carpenter et al. 1996). Thus, biomanipulation alone would not likely improve water quality beyond levels attained in the 1990s. However, while biomanipulation could stabilize the food web and prevent a resurgence of cisco or another planktivore, the reason that yellow perch recently increased after a decade of low abundances is not clear. A massive increase in planktivory would eliminate *Daphnia pulicaria* and substantially diminish water clarity (Lathrop et al. 1996, 1999). Thus, managers have reason to sustain low levels of planktivory in the lake.

The Priority Watershed Program

By the mid-1990s, the need for further decreases in phosphorus input to restore Lake Mendota was evident. Internal manipulations like the biomanipulation project were not likely to improve water clarity beyond levels attained in the 1990s. In fact, agricultural practices in the watershed were likely to increase eutrophication if the status quo was maintained (Reed-Anderson et al. 2000a). Controls of nonpoint pollution needed to be more comprehensive and more effective than those implemented during the Sixmile-Pheasant Branch project of the early 1980s. In order to succeed, a nonpoint pollution control program for Lake Mendota would have to be among the largest and most ambitious ever attempted. Opportunity to create such a program emerged from the efforts of Wisconsin Department of Natural Resources managers, the Land Conservation Department and the Lakes and Watershed Commission of Dane County, and others, leading to designation of the Lake Mendota Priority Watershed Project in late 1993. The designation as a priority watershed brought new regulatory and cost-sharing capabilities to the battle against nonpoint pollution. The program coincided with expanded scientific understanding of sources of phosphorus in the watershed (Soranno et al. 1996, Bennett et al. 1999), farm practices related to nonpoint pollution (Nowak and Madison 1998), and the economic value of aquatic resources in the region (Carpenter et al. 1998, Wilson and Carpenter 1999).

The long-term database for Lake Mendota was used to calculate the reductions in phosphorus loading needed to reduce the probabilities of bluegreen algal blooms in Lake Mendota (Lathrop et al. 1998). The long-term data showed that the probability of bloom conditions during summer was about 60% (Fig. 12.6). In other words, a visitor to the lake can expect to encounter a blue-green algal bloom on three days out of five, during summer. However, the data indicate that this probability can be cut to about 20% if phosphorus input rates can be cut in half. In other words, if phosphorus inputs are reduced by 50%, then a visitor to the lake could expect to encounter a blue-green algal bloom on only one day in five during summer. This result was important in planning nonpoint pollution reductions for Lake Mendota. Input reductions of about 50% were adopted as the goal of the priority watershed project (Betz 2000).

The watershed plan of the priority watershed project, approved in the summer of 1997, recommended that approximately $17.8 million be allocated to reduce nonpoint pollution inputs in order to meet water quality goals for the lake (Betz 2000). Of that money, $8.7 million was to come from state bonding monies and other general-purpose revenues. The remainder of the money was to be sought from local sources, including municipal governments and landowners. In the case of individual farmers, the cost of many approved practices would be shared, with the state providing 70% and the landowner providing 30%. Municipalities that implemented phosphorus controls in existing urban areas could expect to have half the costs covered by the state. For developing and planned urban areas, municipalities were expected to cover all costs of implementing management practices that were estimated at $5.6 million.

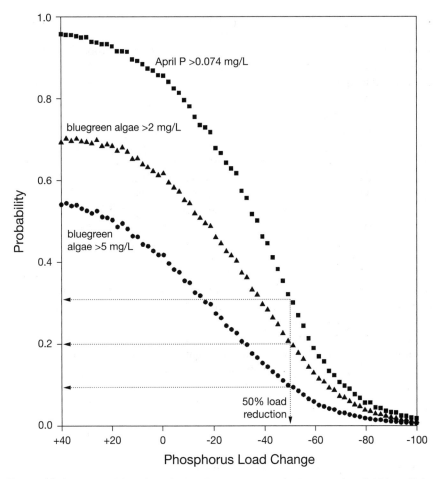

Figure 12.6. Probabilities of total phosphorus concentration greater than 0.074 mg/L in April, blue-green algal concentrations greater than 2 mg/L, and blue-green algal concentrations greater than 5 mg/L versus percentage change of the annual phosphorus input to Lake Mendota. Probabilities corresponding to a 50% reduction in inputs are marked by the dotted lines and arrows. Redrawn from Lathrop et al. (1998) with permission of the Canadian Journal of Fisheries and Aquatic Sciences.

Funding shortfalls have been a concern during the implementation phase of the priority watershed project. Because the priority watershed program was popular around Wisconsin, only a fraction of the original state monies may become available for Lake Mendota. Constrained budgets of municipalities may lead to underfunding of needed management practices. However, because most of the P-loading comes from rural sources (Lathrop et al. 1998, Betz 2000), allocating the available funding to reducing those sources will be a high priority. In addition, even though the sign-up for practices by farmers will be voluntary, mandatory implementation of best management practices will be required for all critical sites. Large animal feedlot operations, where hundreds of cattle are aggregated with too little land to dispose of their manure, and uplands where erosion problems are severe provide examples of rural sites that produce high phosphorus loadings. Finally, new agricultural nonpoint pollution regulations currently in state legislative review may give additional impetus for helping achieve goals of the priority watershed project for phosphorus loading reductions to the lake.

What level of phosphorus input is appropriate for Lake Mendota? Economic cost-benefit analysis may help answer this question (Wilson and Carpenter 1999). In such an analysis, the economic benefits of activities that cause pollution, such as agriculture and development, are compared to the benefits derived from uses of clean water, such as drinking, irrigation, or recreation. The latter benefits are difficult to measure, and social scientists disagree about the most appropriate methods (Wilson and Carpenter 1999). The policies recommended by economic cost-benefit analysis are extremely sensitive to the data used to estimate value of water quality (Carpenter et al. 1999). If the value attached to the Lake Mendota restoration is $30 million, then the economically optimal policies are similar to those adopted by managers in planning the Lake Mendota Priority Watershed Project (Fig. 12.7). If the value of the restoration is only $10 million, however, then cost-benefit analysis indicates that loadings should be allowed to rise until the lake is highly eutrophic. Economic valuations of ecosystem services can support arguments for either restoration or degradation of ecosystems. In a contingent valuation study, the value attached to improved water quality of Lake Mendota was roughly $50 million (Stumborg et al. 2001), supporting the extensive restoration effort planned by the lake's managers. Therefore, for Lake Mendota, support for large-scale restoration seems substantial.

Evaluating Lake Mendota's Responses to the Priority Watershed Project

The Lake Mendota Priority Watershed Program is expected to evoke change in both the ecosystems and the social systems surrounding the lake. Many of these changes will affect the program's ultimate success in abating eutrophication. LTER researchers are following key ecological and social variates through the course of the program. The LTER research turns the priority watershed project into an experiment by evaluating baseline conditions, monitoring reference lakes that are not affected by the project, evaluating the performance of the program's components, and detecting any surprising responses. LTER researchers have conducted many of the limnological and modeling studies needed to plan the project, and they have worked closely with state and county staff in establishing the phosphorus loading targets and project plan. The nonpoint pollution control practices are now being implemented. As the implementation phase of the project advances, we will measure limnological and social responses in Lake Mendota and three nearby LTER lakes as reference systems, one, Fish Lake, in an agricultural watershed, and two, Lake Wingra and Lake Monona, in urban watersheds. Response variates include changes

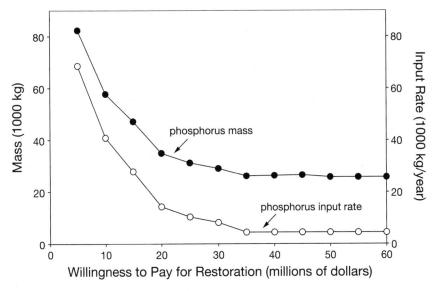

Figure 12.7. The amount that people are willing to pay for restoration of Lake Mendota determines the management targets for phosphorus mass in the lake and phosphorous input rates. Calculated from the model of Carpenter et al. (1999) using data from Lathrop et al. (1998).

in land cover, land use, and runoff management practices, changes in behavior that affect phosphorus inputs to the lake, changes in phosphorus inputs, and changes in nutrients, phytoplankton, zooplankton, benthos, and fishes. Responses could take decades, because time lags occur from installing the manipulations, the lake's long hydraulic residence time of 4 to 6 years, and the large amounts of phosphorus stored in soils and sediments. As the North Temperate Lakes LTER program makes consistent measurements through this period, we will obtain a comprehensive, interdisciplinary assessment of this large ecosystem management action.

Adaptive management begins with interactive modeling to clarify problems; increasing communication among scientists, managers, and the public; screening policies to eliminate alternatives that are not likely to do any good; and identifying knowledge gaps that could cause errors in the models (Holling 1978, Walters 1997). Management experiments are then designed and implemented to fill the knowledge gaps and reveal key impacts at the space and time scales that will be the actual setting of future management (Walters 1997). Walters (1986) distinguishes passive adaptive management, in which scientific monitoring is used to guide decision making, from active adaptive management, in which models are used to expose contrasting views of ecosystem dynamics and carefully chosen experiments are used to discriminate among the models and improve future management actions. The priority watershed project is an example of passive adaptive management. It was not designed to learn how to cope with future problems of Lake Mendota in an evolving way, as occurs in active adaptive management. Rather, the focus was on a passively adaptive reaction to a current (though tenacious) problem. At this point, the most important experiment is the Lake Mendota Priority Watershed Project itself; by reducing phosphorus input we will learn a great deal about the response potential of the lake and likely will improve water quality.

Over past decades, management of Lake Mendota has cycled through phases when control of sewage effluents was emphasized, to periods when treatment of symptoms such as the harvest of macrophytes and carp was given priority, to the use of biomanipulation, to the present focus on nonpoint phosphorus control (Carpenter et al. 1998, Carpenter and Lathrop 1999). Similar cyclic transitions are known from diverse environmental management programs (Gunderson et al. 1995). Management effort for the lake has exhibited considerable resilience, owing to the public's continuing interest, the persistence of scientists at the University of Wisconsin-Madison and at the Wisconsin Department of Natural Resources, and the actions of local governmental managers. Vigorous management intervention seems likely to continue in the future, and scientists likely will be attentive to emerging issues. Success in controlling eutrophication will depend on how we cope with the next set of surprises. These may come from species invasions, changing hydrology and climate, or a shift in social values and priorities.

Unsolved Problems and Potential Surprises

Surprises are, by definition, impossible to predict, but we envision a number of plausible events that could change greatly the context of management for Lake Mendota. A number of species have invaded the Great Lakes region, and many of these could reach Lake Mendota and thrive.

Zebra mussels for example, have invaded the Great Lakes, the Mississippi River, and some smaller lakes in southeastern Wisconsin. On the basis of its water chemistry and heavy boat-trailer traffic from infested waters, Lake Mendota is likely to be invaded by zebra mussels (Ramcharan et al. 1992, Padilla et al. 1996). One individual was found in Lake Monona in summer 2001; more may be present in the Madison lakes. By filtering phytoplankton, zebra mussels are likely to increase water clarity when the lake is well mixed, but surface blooms of nuisance bluegreen algae would not be eliminated (Reed-Andersen et al. 2000b). Other impacts to the lake could outweigh any benefits of improved water clarity. Potential impacts of zebra mussels on zooplankton, larval fish, and fish recruitment are unknown. Increased water clarity could increase benthic algae and submerged macrophytes. The interaction between water clarity and submerged macrophytes has caused macrophytes to vary in all the Yahara lakes in past years (Nichols and Lathrop 1994). On the other hand, macrophyte growth could decrease, because the young mussels settling on the plants could weigh them down. Ironically, increased water clarity could divert public attention from the nonpoint pollution of the lake, which is the ultimate cause of the algal blooms. Zebra mussels may appear to alleviate some symptoms but are not likely to be a permanent solution to the lake's water quality problems.

A warming global climate undoubtedly will impact Lake Mendota in many ways that are yet to be understood fully (Chapter 7). A trend toward earlier thawing of the lake in the spring is under way (Robertson et al. 1992, Magnuson et al. 2003). Global warming is expected to cause water temperatures to increase faster in the spring and to increase summer epilimnetic temperatures (Chapter 7) (Robertson and Ragotzkie 1990a, De Stasio et al. 1996). Because hypolimnetic temperatures may decrease in some lakes with a warming climate, water column stabilities should be greater in warmer summers. In these summers, internal mixing and hence internal phosphorus loading are less, resulting in reduced algal densities and greater water clarity (Lathrop et al. 1999). However, longer periods of summer stratification could produce higher phosphorus concentrations that are transferred to the entire water column at fall turnover and carried over for algal uptake the following spring and summer. Because the duration of summer stratification is likely to increase, bluegreen algal blooms could occur for longer periods of time as a result of global warming. Increased water temperatures negatively affect fish when habitat temperatures exceed a species' preferred temperatures. Thermal habitat will increase for warmwater fish, while oxygenated habitat will decline for coldwater and perhaps coolwater fish (De Stasio et al. 1996, Magnuson et al. 1997b). Further research on the cumulative impact of these changes and others on Lake Mendota continues. The complexity of potential effects of a warmer and wetter and a warmer and dryer climate on the expression of eutrophication on water clarity was diagrammed in Fig. 5.8 in Gitay et al. (2001).

Land use and land cover in the Lake Mendota watershed have changed substantially (Soranno et al. 1996). Wegener (2001) used air photos to reconstruct the land use and cover history of the watershed since the 1930s and compared the results with the long-term record of water-level fluctuations in the lake. The rise and fall of water levels has become significantly more rapid. This "flashiness" of the lake level is associated with increased area of impervious surfaces in the watershed. Lakeshore property owners see the effect through increased frequency of flooded lawns and basements. The Laboratory of Limnology on Lake Mendota experienced three record floods during the 1990s. Intensive rain events exacerbate the flooding (Chapter 7).

The social context of Lake Mendota is changing, as well. The number of farmers in the watershed is declining as the mean size of farms increases and urbanized areas expand (Department of Agriculture, Trade, and Consumer Protection 1995). If expanding farm size is associated with larger livestock operations and more intensive tillage of corn, nutrient runoff from agriculture may increase (National Research Council 1993). Developing urban areas tend to produce far more nutrient runoff than stable urban areas, because construction increases soil erosion (Novotny and Olem 1994). Further, this construction is taking place on former agricultural land with soils that contain substantially augmented levels of phosphorus (Bennett et al. 1999). Therefore, nonpoint nutrient management is chasing a moving target, and increases in input rates may offset benefits of watershed restorations to some extent. On the other hand, growth of the urban population may increase the electorate's interest in clean lakes as an important amenity, thereby creating more pressure for lake restoration.

C. S. Holling has quipped that most phenomena in ecology are predictable, except for the interesting and important ones (Gunderson 1998). The future of Lake

Mendota falls in the latter category. The North Temperate Lakes LTER program gives us the opportunity to observe, understand, and learn from events as they unfold and to regularly evaluate and improve our ability to predict. With LTER participation, the management program for Lake Mendota becomes an experiment, subject to rigorous scientific evaluation. The ongoing engagement of LTER scientists, the new questions they raise, and the consistent databases they collect, allow broad assessment of changes in the social and ecological systems. Thus, LTER creates the opportunity to learn by doing. This learning offers important contributions to future policies for managing nonpoint pollution and other freshwater problems. Moreover, our basic knowledge has grown concerning integrated social-ecological systems through interdisciplinary studies of ecosystems dominated by the activities and decisions of people.

Summary

The history of the Madison lakes is intimately linked to the history of the lands around them. Lake Mendota, the largest of the lakes, has changed substantially since European settlement of the catchment. The lake was eutrophied by excessive phosphorus inputs from sewage and agriculture. The present state of eutrophication is sustained by phosphorus runoff from agriculture, construction sites, and urban areas. Wetland habitat around the lake was reduced by half, and the extent of lake bottom colonized by macrophytes has decreased. Species diversity of macrophytes has decreased. Profundal benthic invertebrates have declined substantially. Fish species richness is roughly constant, but many ancestral species were lost and replaced by introduced species.

Management of Lake Mendota has focused on eutrophication and sport fishing. Phosphorus inputs from sewage were diverted from the lake. Nonpoint inputs of phosphorus remain high. Policies aimed at changing practices of individual landowners have not succeeded in reducing nonpoint pollution at the time of writing, although there is hope that current or future programs may succeed.

Change in Lake Mendota is ongoing. The social context is shifting as the urban population expands and agricultural land is converted to other human uses. Hydrology of the watershed is changing as groundwater tables drop and urbanization causes surface flows to become more flashy. The lake is vulnerable to species invasions, and invaders like the zebra mussel could change the ecosystem in major ways. A changing climate will impact the ecosystem, but the exact nature of these impacts is unknown. The North Temperate Lakes LTER program will observe, understand, and learn from these events as they unfold.

Part III

Developing and Implementing Long-Term Ecological Research

At the onset we were challenged to initiate a new kind of research that required innovation, activities, and infrastructure that were not a part of conventional science at the time. As a member of one of our early external advisory committees stated, one aspect of the challenge was to learn how to do long-term ecology, today. Another challenge was to develop an information management system and make it an integral part of our research program (Chapter 13). Our understanding of how to do long-term ecological research evolved as we learned to meet these and other challenges (Chapter 14). Scientific and technological advances and the generation of the long-term dataset catalyzed our learning and our adaptations, for example, by expanding our research to include a regional component and human interactions. Here we document our progress, history, and thoughts in the orchestration of long-term ecological research.

13

Breaking the Data Barrier: Research Facilitation through Information Management

Barbara J. Benson
Paul C. Hanson
Jonathan W. Chipman
Carl J. Bowser

Where is the wisdom we have lost in knowledge? Where is the knowledge we have lost in information?
—T. S. Eliot (1934, p. 7)

Complex questions, interdisciplinary collaboration, analyses over larger spatial and temporal scales, and pressing issues for policy makers have generated a need for sophisticated information management for ecological research projects. The Long-Term Ecological Research (LTER) Program funded by the National Science Foundation represents a paradigm shift for information management (Fig. 13.1). Under the old practices, a scientist's data might reside in poorly documented files in paper or electronic form that often would become unusable upon his or her death or departure. The magnitude of effort required to assemble data sets such as these across investigators, disciplines, and scales often has meant that access to the data was, and in many cases still is, a major barrier to synthetic research.

Under the LTER paradigm, information management has been an integral part of the research process. This shift in the importance of information management was driven by the scope of the research and by the recognition that LTER data would be an invaluable resource for scientists in the future if the data were managed well. Significant advances in computer and information technology accelerated this shift. The network of LTER information managers has been playing a central role in defining for the ecological community what constitutes good information management (Ingersoll et al. 1997, Michener et al. 1998, Michener and Brunt 2000). Today's in-

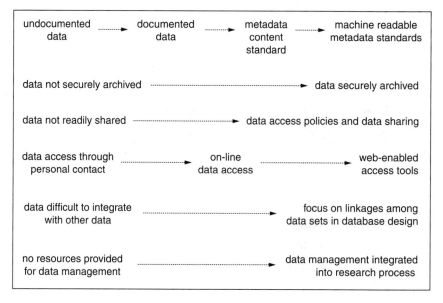

Figure 13.1. Important components of the evolution in data management under LTER information management at North Temperate Lakes LTER.

formation manager deals with research and information system design, data entry, quality control and assurance, documentation, data access and integration, computer networking, Web publishing, database security and backups, and the long-term archival of data sets and sample collections. In addition, the information manager is often the person who purchases, installs, networks, and maintains a lab's computers.

This chapter represents the authors' reflections on the information system developed since 1981 at the North Temperate Lakes LTER site and provides information on the system and its use. We organized this chapter around three questions to focus the discussion on the relationship between science and information management and on how the information management system at our site has evolved in the context of that relationship: (1) How have information technology and management changed the way we do science?, (2) What have been the pivotal decisions and principles that have shaped information management?, and (3) What challenges does the future appear to hold for information management at North Temperate Lakes LTER?

Changing the Way We Do Science

Initial Conceptualization of Information Management

How we conceptualize and implement information management has evolved from 1981 to 2004 (Table 13.1). When the project started in 1980, we knew that we needed to create a data legacy by committing to make the LTER data sets well documented

Table 13.1. Timeline of selected highlights for the North Temperate Lakes LTER program

---

1981—Began data collection.

1982—Incorporated physical and chemical limnology and fish data into a hierarchical database located on a campus mainframe.

1984—Sent data descriptions for 15 data sets to the Jornada LTER site. The transfer was our first attempt at communication between LTER sites via modem and was done to contribute to an intersite data catalog.

1988—Acquired first satellite remote sensing images (with concurrent water quality data for 20 lakes) for the Trout Lake area and hardware and software for Geographic Information Systems. Tom Lillesand, Director of the Environmental Remote Sensing Center, joined the project as a Principal Investigator.

Established the meteorological station at the Woodruff airport and used a data logger to record data.

1989—Implemented a Local Area Network for the Center for Limnology (CFL) and acquired a connection to the campus network.

1993—Moved the database into an Oracle database on a Unix server at CFL.

Entered the bibliography into a bibliographic database, EndNote Plus. Records from this database were made available for the LTER all-site bibliography at the Network Office.

1994—Made the home page on the World Wide Web (WWW) public. The Web site contained a site description, text and figures of proposals and links to the personnel directory, bibliography, and on-line data.

Acquired end-user query software that gives researchers direct access to the database.

1995—Acquired a 56-Kbaud dedicated phone line to provide Trout Lake Station, the field station for the northern study lakes, with an Internet connection.

1997—Created dynamic database queries through the WWW for the meteorological data, the bibliography and the personnel directory.

Developed data entry software for recording fish data in the field.

1999—Acquired a high-speed T1 connection to the Internet for Trout Lake Station, a buoy with sensors on Trout Lake connected via spread spectrum wireless communication, and video conferencing equipment.

Created a Web interface to allow direct data entry into the Oracle database.

2000—Deployed a spatially distributed array of whole-lake metabolism buoys.

2001—Created a Web interface that provides direct querying of the Oracle database using a query engine driven by metadata in the database.

---

and secure. In the early years of the project, research that used data from the 1920s and 1930s on lakes in our study area highlighted the importance of the historical legacy (Bowser 1986). The scientific question of interest was whether lakes in northern Wisconsin had changed chemically, in particular in pH, from the Birge and Juday era, 1925–1941, to the time of our current LTER study. The availability of the early data was necessary to address the question. In addition, documentation (now called metadata) that dealt with methods was crucial to the valid comparison of data from the two different eras. In some cases, the exact details on methods were never published. For example, whether pH was measured in the field or a few hours later in the lab was not recorded, even though this difference in methodology affects pH values. Fortunately, an oral history was still accessible from some of the staff, technicians, and students who worked with Birge and Juday. A conference was held to record the oral history, and personal discussion resolved the pH methodology question. To resolve methodological issues about conductivity, the fortuitous archiving of the working conductivity meter used by Birge and Juday allowed direct comparison of new and old methods. The archiving of equipment, in addition to samples and data, has not been addressed within the LTER program but should be. Part of the LTER mandate is that 50 or even 100 years from now, researchers who want to ask similar questions will have not only the LTER data but also the associated electronic metadata to handle data comparability issues.

In addition to creating a long-term legacy, we decided to design an information management system that would facilitate interdisciplinary research. Many scientists engaged in interdisciplinary research have experienced the less facile process of collaboration, including negotiations with colleagues to obtain data sets, transferring of data files, wrestling with multiple file and data structures, and considerable cutting and pasting and other data manipulations to produce the data set to be analyzed. We designed the data collection and management so that different data sets would be comparable and easily linked (Kratz et al., 1986, Appendix 13.1). Field sampling of different components of the lakes' ecosystems was coordinated to facilitate analysis of relationships among these components. A relational database supported linking of diverse data tables. Certainly, a host of additional integration issues are involved in making data comparable across disciplines. For example, census tract boundaries are not ecologically meaningful boundaries, while watersheds and ecoregions are not used widely in the social sciences.

New Technology Expands the Vision

Developments in computer and network technology since 1980 were fundamental to achieving our information management goals and have led to an expansion of those goals. However, the application of new technology often brings challenges to information management in adapting the technology to function in a research environment or developing custom-built applications. Today we take for granted computing and communication capabilities that were only in rudimentary form two decades ago.

The founding of North Temperate Lakes LTER coincided with the dawn of the microcomputer revolution. In our early years (Fig. 13.2, top), we exchanged data with the field station by hand-carrying or mailing disks and depended on campus mainframes for database applications. Since then, we have changed how we work by incorporating some of the rapid advances in computer hardware and software, networking, user interface standards, and other components of the technological environment (Table 13.2). In some cases, the advances in computer technology have made possible research with extensive and complex requirements for data and analysis.

Some of the most important advances are being made in data access (Fig. 13.2 bottom). When the project first established its database, researchers would access

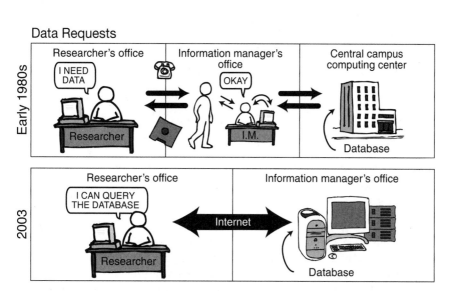

Figure 13.2. Contrasts between the early 1980s and 2003 for the North Temperate Lakes LTER program in respect to data flow (top) and data retrieval (bottom). PDA = personal digital assistant, ERSC = Environmental Remote Sensing Center at the University of Wisconsin-Madison.

Table 13.2. North Temperate Lakes LTER information management infrastructure as of 2005.

Local Area Nets (LAN)	Ethernet LAN supporting Windows, Macintosh, and Linux-based computers and Sun workstations at the Laboratory of Limnology in Madison and Trout Lake Station; laser and color printers; scanner; copiers [a]
Connection to Wide Area Networks (WAN)	Laboratory of Limnology in Madison LAN linked by a 10Mbit/s connection to the campus-wide WAN; Trout Lake Station connected via a T1 (1.544 Mbit/s) land line
Web Server	Apache Web server running on the Sun Sparc Ultra2 workstation (Solaris operating system)
Database	Oracle RDBMS installed on the Sunblade 2000 workstation (Solaris operating system)
Data Collection	Compaq iPaqs used for field data entry
Data Retrieval	Oracle browser application installed on networked Windows computers; ODBC connections on networked computers; dynamic query of the database through the Web site
System Backup	Daily backups of the Sun workstations to DAT tape; daily Oracle database exports; regular backups of data on computers to tape; backups of spatial data to CD-ROM
Wireless Communication	Digital spread spectrum, serial to field equipment/sensors; Ethernet for computers and some sensor systems
Video Conferencing Equipment	Microsoft NetMeeting software, video cameras, computer projectors, Polycom ViaVideo units
Remote Sensing	Space Science and Engineering Center (SSEC) X-band satellite receiving antenna and ingest capability

a Computer facilities at the Center for Limnology are described at limnology.wisc.edu/Lake_Mendota_Lab/computerroom.htm

Note: RDBMS = relational database management system, ODBC = Open DataBase Connectivity, DAT = Digital Audio Tape

data by submitting a request to the information management staff. Retrievals from the database required knowledge of a special computer language. Today we want researchers to be able to access data sets through powerful yet "user-friendly" environments. We provide several avenues for direct access to the database through the use of browsing software. Researchers can select precisely and retrieve data of interest through a point-and-click interface. They can join different data tables, select subsets, and define functions to aggregate the data. Having this simple and direct access to the database has supported a much greater level of exploratory data analysis. NTL information managers regularly conduct training sessions in database access for new users (Fig. 13.3). The use of Open DataBase Connectivity software has improved the flexibility of database access, permitting direct connections to the database by a wide range of software applications, including other commercial database management systems, spreadsheets, scripted Web interfaces, and custom-developed software.

Figure 13.3. Information management staff conducting training in database browsing and querying for researchers at the Laboratory of Limnology on the Madison campus. From left to right: Paul C. Hanson (network administrator), Brian M. Roth (graduate student), David Balsiger (information management), Catherine L. Hein (graduate student), Barbara Benson (information management), John Magnuson (lead principal investigator). (Center for Limnology)

In addition, data availability has been extended through the World Wide Web. In 1985, the National Science Foundation laid the basis for the modern Internet by establishing NSFNet. In 1993, development of the World Wide Web on the Internet exploded with the introduction of the first graphical browser, Mosaic (National Center for Supercomputing Applications). Information managers in the LTER research network quickly realized the importance of the Web, and by 1995 almost all LTER sites had a home page providing a diverse array of information (Benson 1998).

On the North Temperate Lakes LTER home page (lter.limnology.wisc.edu), one can find a searchable bibliography, facilities descriptions, a personnel directory, biodiversity information, grant proposals, major research findings, on-line data and metadata, outreach and education activities, a picture gallery, and links to related sites. The Web provided a way to extend easy and direct data access beyond our local research group. We enabled data browsing capability by implementing dynamic queries of the database through the Web (Stubbs and Benson 1996, Smith et al. 2002). The goal is for a user to be able to choose the data table to access, specify the variables of interest, and limit the data extracted by filtering on fields such as lake and sample date. The application we developed generates a Web form (Fig. 13.4) that is driven by metadata stored in the database and retrieves data in a chosen format based on the user's input to that form (Smith et al. 2002). We have developed graphical displays returned from dynamic database queries. With such functionality, field technicians can view time series of data-logger data directly from queries on the Web. This display can help them troubleshoot equipment and perform quality control.

In the early years of the project, moving data and documents between different computer platforms was not straightforward. The evolution in platform independence, together with communication via the Internet, has reduced or eliminated many of the barriers to scientific collaboration at our site. Developments in the area of groupware and video conferencing are enhancing collaboration among researchers, reducing the need for time-consuming trips, and allowing us to interact together on electronic documents. Discussion group software accessed through the Web site enables threaded document sharing and dialogues.

Data collection and entry are becoming more automated as ecologists make use of data loggers and handheld or laptop computers. For the first time, during the summer of 1997 the crew sampling fish no longer recorded data on paper field sheets but used new data entry software developed by P. Hanson to enter the measurements directly into a computer. In 1997, the field crew used cellular phones to communicate the measurements to a data entry person at the lab; starting in 1998, the field crew used a laptop computer for data entry on the boat. In 2002, the software was compiled to run on a personal digital assistant (see Fig. 14.6k). The device is lightweight, has low power consumption, has a weather-resistant protective case, and is inexpensive when compared to a laptop. Importantly, the screen is viewed easily in direct sunlight. Nonetheless, data entry continues to be a bottleneck in the sampling process. Adequate speech recognition software, or some other keyboardless data entry system, may promote dramatic improvements.

The Internet has become an integral part of the process of providing information management support for research in very pragmatic ways. The whole process has been accelerated from obtaining specifications on gear, placing orders for materials, communicating with colleagues, and upgrading software to diagnosing hardware and software problems. For example, one morning, a field technician at the field site was having difficulty with a fish-crew database file from the Personal Data Assistant device. Within 30 minutes, the computer support staff in Madison downloaded the needed connection software from the Compaq Web site and placed the software on the field station server. The field technician then installed the software on several computers, thus allowing a number of connectivity options, and uploaded the problematic database files to the file server. The programmer in Madison downloaded the database files, thus enabling software debugging, and returned the upgraded software to the field technician. This kind of adaptability and rapid turnaround was impossible even in the early 1990s.

Spatial data play an important role as research expands from the study of individual lakes to watersheds, lake districts, regions, and global scales. Many types of spatial data have been compiled for our LTER site (Appendix 13.1), ranging from digital elevation data to wetlands and soil maps. A series of land use/land cover layers over time (1930s, 1960s, 1990s) for watersheds and riparian zones has been

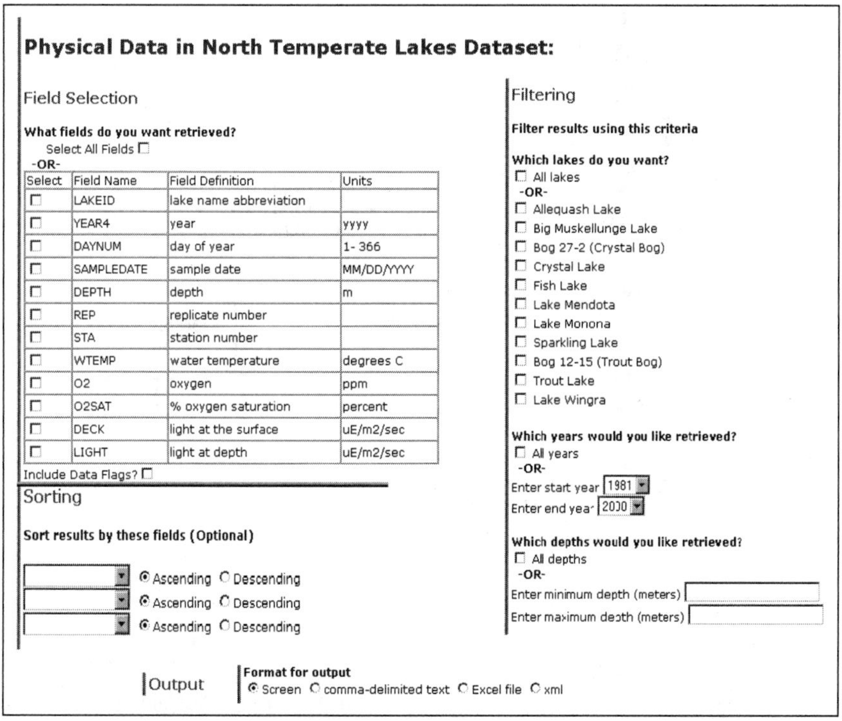

Physical Data in North Temperate Lakes Dataset:

Field Selection

What fields do you want retrieved?
Select All Fields ☐
-OR-

Select	Field Name	Field Definition	Units
☐	LAKEID	lake name abbreviation	
☐	YEAR4	year	yyyy
☐	DAYNUM	day of year	1- 366
☐	SAMPLEDATE	sample date	MM/DD/YYYY
☐	DEPTH	depth	m
☐	REP	replicate number	
☐	STA	station number	
☐	WTEMP	water temperature	degrees C
☐	O2	oxygen	ppm
☐	O2SAT	% oxygen saturation	percent
☐	DECK	light at the surface	uE/m2/sec
☐	LIGHT	light at depth	uE/m2/sec

Include Data Flags? ☐

Sorting

Sort results by these fields (Optional)

[▼] ⦿ Ascending ○ Descending
[▼] ⦿ Ascending ○ Descending
[▼] ⦿ Ascending ○ Descending

Filtering

Filter results using this criteria

Which lakes do you want?
☐ All lakes
-OR-
☐ Allequash Lake
☐ Big Muskellunge Lake
☐ Bog 27-2 (Crystal Bog)
☐ Crystal Lake
☐ Fish Lake
☐ Lake Mendota
☐ Lake Monona
☐ Sparkling Lake
☐ Bog 12-15 (Trout Bog)
☐ Trout Lake
☐ Lake Wingra

Which years would you like retrieved?
☐ All years
-OR-
Enter start year [1981 ▼]
Enter end year [2000 ▼]

Which depths would you like retrieved?
☐ All depths
-OR-
Enter minimum depth (meters) [   ]
Enter maximum depth (meters) [   ]

Output   Format for output
⦿ Screen ○ comma-delimited text ○ Excel file ○ xml

Figure 13.4. With this user interface screen for dynamic database access on the North Temperate Lakes LTER website, researchers can select variables of interest, control sorting variables, filter on specific variables such as lake or year, and select an output format.

produced and will be an important component in understanding the interactions between a lake and its surrounding landscape.

One of the great successes of spatial information management at the North Temperate Lakes LTER has been the development of an extensive, continually growing archive of multiresolution satellite imagery for both the northern and southern lake districts and the use of these data in a wide range of research projects. These activities have resulted from the involvement of researchers associated with the University of Wisconsin's Environmental Remote Sensing Center. Satellite imagery has been used in varied applications on our project (Fig. 13.5) from remote estimates of lake turbidity, water clarity, and surface temperature (Lathrop et al. 1991, Chipman et al. 2004) to vegetation classifications of landscapes (Chapter 6).

Spatial data present some unique information management challenges. Data volumes are typically large, particularly in the case of remote sensing imagery, as the result of both the volume of single scenes, typically 300 to 800 MB, and the increasing use of time-series of imagery. An extreme example of the latter is our automated system for downloading, processing, and archiving twice daily the Moderate-Resolution Imaging Spectro-radiometer images from the National Aeronautics and Space

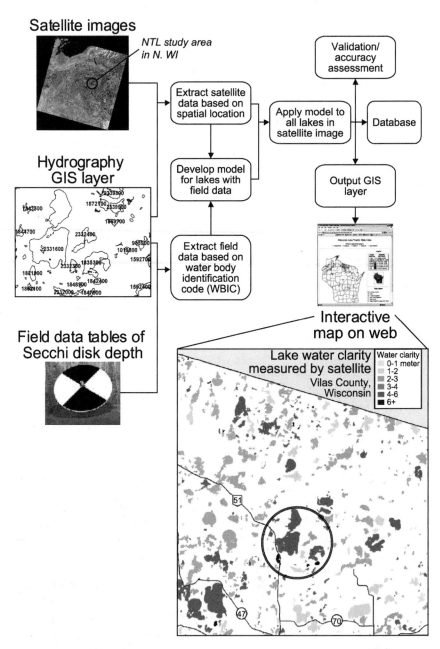

Satellite images

NTL study area in N. WI

Hydrography GIS layer

Field data tables of Secchi disk depth

Extract satellite data based on spatial location

Develop model for lakes with field data

Extract field data based on water body identification code (WBIC)

Apply model to all lakes in satellite image

Validation/ accuracy assessment

Database

Output GIS layer

Interactive map on web

Lake water clarity measured by satellite
Vilas County, Wisconsin

Water clarity
0-1 meter
1-2
2-3
3-4
4-6
6+

Figure 13.5. Schematic flow for development and Web distribution of statewide water clarity data for lakes based on satellite imagery available to LTER researchers through the Environmental Remote Sensing Center at the University of Wisconsin-Madison. The interactive water clarity map can be found at http://www.ersc.wisc.edu (see link to lakesat).

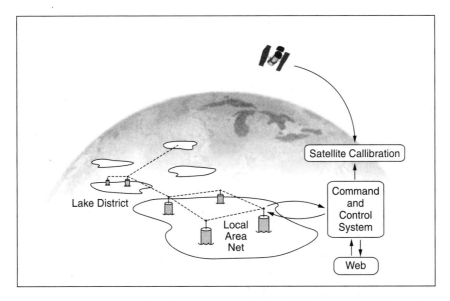

Figure 13.6. Integrated system of lake sensors on buoys, wireless communication, and satellite remote sensing proposed by the North Temperate Lakes LTER project. The command and control system manages communication with buoys, the Web, and the satellite calibration software.

Administration's Terra and Aqua satellites via the Earth Observing System's direct-broadcast system. Other challenges in spatial data management result from the evolution of data structures and metadata. Except for satellite imagery, we are migrating all spatial data at our LTER site from freestanding ESRI shapefiles and coverages to a unified geodatabase, implemented using an Oracle database and ESRI's Spatial Data Engine. One of the main factors in choosing this approach has been the desire for improved access to spatial data and metadata via the Internet.

Satellite imagery represents only one kind of remotely sensed data. Acoustic data for fish have been part of our fish data collected annually. We have begun using spread-spectrum wireless communication to access data from buoys on lakes. Sensors on these buoys measure data such as carbon dioxide concentrations and water temperature at time intervals as frequent as 1 min, and the data are transmitted in near-real time. Experience with the technical and information challenges using these buoys and the integration and dissemination of data in real time will provide us with a knowledge base for deploying an array of buoys that can remotely sense the environment (Fig. 13.6).

Breaking the Data Barrier

Information management has made possible the analysis of large, complex data sets for long-term regional ecology. The effort required to integrate data from multiple research areas or programs would have represented a substantial barrier in the past. Comparison of two multisite workshops held at our site makes the point. In 1988, our site hosted an LTER intersite workshop on variability in North American ecosystems (Chapter 6) (Kratz et al. 1991b, Magnuson et al. 1991, Kratz et al. 1995, Magnuson and Kratz 2000). Each participating LTER site provided data sets on physical, chemical, and biological variables for multiple stations and years to address the question of whether patterns in the variability in space and time were consistent at the 12 sites. Because data management existed at each site, data sets were available for this analysis, but acquiring and processing the data prior to analysis was a daunting (but possible) task. In some cases, a researcher from our site actually traveled to the site to explain the variability project and to acquire data sets.

In 1996, we hosted an international workshop on lake-ice phenology related to climate change and variability (Magnuson et al. 2000a). Prior to the workshop, participants were asked to submit various data sets. Most of the data were provided via the Internet, and by the date of the workshop, a considerable portion of the data was in our LTER database, with useful data and graphical views available to participants. Postworkshop data access was provided to all participants through dynamic queries of the database launched from the Web. The database contains ice phenologies for 749 water bodies throughout the northern hemisphere, including 29 water bodies with time series greater than 100 years. Over the past 150 years, from 1846 to 1995, consistent patterns of increasingly late freeze dates and increasingly early breakup dates are seen across the Northern Hemisphere (Chapter 7) (Magnuson et al. 2000b). The magnitude of data that were brought together rapidly would not have been possible without the sophistication of our information management.

Another research area that illustrates the large amount of data involved in analysis is investigating whether a suite of variables demonstrates temporal coherence across lakes (Chapter 5) (Magnuson et al. 1990, Kratz et al. 1998, Baines et al. 2000). In this instance, we compared patterns of interannual variation among lakes for physical, chemical, and biological variables. We were able to address questions that involve analyzing an extensive amount of data because our information system makes retrieving data relatively straightforward. In addition, our information managers have assisted researchers by writing programs that perform some of the required data analysis.

Having the ability to analyze large, complex data sets rapidly has facilitated our ability to conduct short-term, high-risk studies. The end products of these studies are papers and theses based on data of relatively short-term duration; however, in some cases, these studies have been the proving ground for application of new techniques to the long-term research mission. For example, a short-term project of 3 years' duration that used carbon dioxide and dissolved oxygen sondes (a probe for measuring the environment) to assess lake metabolism (Hanson et al. 2003) led to our changing from bottle measures of metabolism to sonde measures.

In summary, the North Temperate Lakes LTER emphasis on information management and the particular way we have implemented our information management system have led to some significant changes in the way we do science. The information management system is designed explicitly to support a long-term data legacy and interdisciplinary studies. Use of database software and experience with facilitating the research process have increased substantially the ability of researchers to integrate and analyze complex synthetic data sets. We are progressing in the integration of multiple sources of data, for example, field measurements, Geographic Information System layers, and models. Software that permits dynamic database access supports direct queries of the database, enhancing the analysis environment for researchers and for model calibration and validation. The availability of data and data browsing has been extended to the entire ecological community and to the public. This open data policy is expected to increase the number of collaborative studies.

An Information System for the LTER Network

The 26 sites within the LTER research network represent diverse ecosystems, ranging from forests to deserts, Arctic tundra to urban settings. Each site has developed an information system to manage and disseminate its data. The Web provided the technological impetus for building an information system for the entire LTER Network (Baker et al. 2000). Using the Web, a single point of entry can be created to access site databases that are distributed geographically. One prototype design allows individual sites to maintain their own systems for information management while enabling a centralized database to be updated regularly using Web harvesting (Henshaw et al. 1998). The LTER information management community is engaged in constructing the LTER Network Information System envisioned as network-level datasets and generic tools to facilitate data discovery, access, integration, and visualization.

The current modules for the Network Information System include a data catalog, an intersite climate database, an intersite net primary productivity dataset, an intersite bibliography, a site description database (Baker et al. 2002), and a personnel directory. Researchers who have experienced the challenges of gathering data from multiple sites for an intersite research project welcome these developments. Crucial to the success of the prototypes has been an ongoing dialogue between information managers and project researchers (Benson and Olson 2002). The latest innovations in network-level information management have been focused around a standard for metadata description and format called Ecological Metadata Language (EML). This language (Jones et al. 2001, knb.ecoinformatics.org/software/ eml) is being developed by the National Center for Ecological Analysis and Synthesis, the Central Arizona-Phoenix LTER site, the Partnership for Interdisciplinary Studies of the Coastal Oceans at the University of California at Santa Barbara, the Jones Ecological Research Center, the Kellogg Biological Station LTER site, and the LTER Network Office.

For many years, LTER information managers have collaborated on defining what constitutes adequate metadata for a data set. We realize that the data legacy we are creating for the future will require sufficient documentation for tomorrow's scientific use of the data and that metadata are a key component of the data integration required by synthetic research. Other groups have been working in the area of metadata content standards (Scurlock et al. 2002b). The Ecological Society of America's Committee on the Future of Long-Term Ecological Data Sets made recommendations on what constitutes adequate metadata (Michener et al.1997) for the ecological community. The U. S. Geological Survey has coordinated the effort to define metadata for spatial data sets, and the Federal Geographic Data Committee has produced an established standard (Federal Geographic Data Committee 1998, www.fgdc.gov/metadata/contstan.html). The National Biological Information Infrastructure has extended the committee's standards to a Biological Data Profile (www.nbii.gov/datainfo/metadata/standards/).

The adoption of Ecological Metadata Language as a metadata standard for the LTER sites raises exciting possibilities for constructing an advanced information system across diverse sites. EML is an XML-based standard (Extensible Markup Language, www.w3.org/XML), so, in addition to specifying a content standard, the metadata conform to a machine-readable structure. EML-based data discovery software allows users to search, browse, and locate data stored in heterogeneous systems at multiple sites. Ecological Metadata Language sets the stage for generic software development that can produce sophisticated tools for metadata and data management, data integration engines, and online analytical engines (McCartney and Jones 2002, Porter 2002).

The wide availability of information via the Internet has produced an expanded set of expectations regarding the availability of LTER data online. In turn, increasing the availability of data has required the articulation of a data access policy. The LTER research network has developed guidelines that each site must follow in creating a data access policy (lternet.edu; search for data policy). The North Temperate Lakes LTER data access policy is displayed on our Web site as a part of the online data access (Appendix 13.2).

The terms that the LTER information managers use to describe and articulate their roles have evolved. The work now is viewed as information management, as opposed to data management. At our site, we seek to construct an interface to data that provides the relevant information for specific research questions, rather than merely serving up data sets. Raw data may be integrated with other data and distributed as value-added data products to provide useful knowledge to broader communities, including the public and policy makers (Baker et al. 2000). For example, water temperature profiles from four lake districts were aggregated into a database; the summary data were extracted, analyzed, and interpreted, and the results were published (Fig. 13.7).

LTER information managers have developed and published their evolving views on what constitutes good information management for ecological data (Michener 1986, Briggs and Su 1994, Michener et al. 1994, Veen et al. 1994, Baker 1996, Benson 1996, Porter et al. 1996, Spycher et al. 1996, Wasser 1996, Ingersoll et al. 1997, Benson and Stubbs 1998, Michener and Brunt 2000, Henshaw et al. 2002, Meléndez-Colom and Baker 2002). This process is not done in isolation at each site but benefits greatly from the intersite activities of the LTER information managers, including annual meetings, special symposia, publications, working groups, and email groups. As the LTER network has grown and expanded its research agenda, the information management group has self-organized to meet the challenges. The information managers are now a standing committee within the LTER network, and this committee has a steering committee.

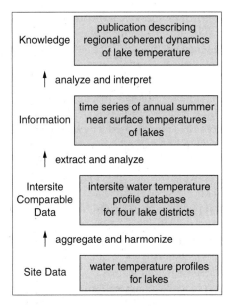

Figure 13.7. Data are transformed from site-specific data to useful knowledge through processes of aggregation, harmonization, extraction, analysis, and interpretation. To answer a particular question, relevant data are extracted and analyzed to produce a derived data set containing the information appropriate for the question. The derived data are then analyzed and interpreted, and the results are conveyed to the science community through publication of manuscripts.

The LTER Network interacts with a wider information management community within environmental sciences and contributes to broader developments and products. LTER sites are listed in the Global Change Master Directory (gcmd.gsfc.nasa.gov), which describes earth science data sets relevant to global change research. Other groups are developing environmental data repositories (Scurlock et al. 2002a), such as the Oak Ridge National Laboratory's Distributed Active Archive Center (www-eosdis.ornl.gov) for biogeochemical and ecological data. Interactions are taking place within the ecological community, both nationally and internationally, through which information technology expertise within the LTER community is being shared. The LTER community will continue to benefit from collaboration with computer scientists who are developing the next generation of technology.

Pivotal Decisions and Principles

Each LTER site developed its information management system in a semiautonomous manner, and, for each site, decisions were made about project organization that constrained how the information management system would be designed. The historical context, characteristics of the user community, and scientific goals were important factors that influenced the design and implementation (Baker et al. 2000). Priorities for resource allocation had effects at each site. The information systems that LTER sites have created are diverse, but all deal with common challenges and functional requirements.

An important characteristic of North Temperate Lakes LTER information management is the centralized collection and management of core data, the data collected over the long term. A crucial step early in the project's history was the decision of principal investigators to support core data collection as a site. Some sites follow a different strategy that has individual investigators collecting, entering and running quality control on data sets before submitting them to a central archive. At our site, project staff, students, and faculty engage in centralized activities of collecting and analyzing samples, performing data entry and some quality control, and submitting data for incorporation into a central database.

This centralization in data collection and processing has had many positive effects on the North Temperate Lakes LTER program. Having considerable data collection performed by technicians who have been with the project a long time has provided a strong consistency in methods over the years. The data are readily available to everyone on the project, as opposed to being viewed as the property of a particular researcher, even though individual investigators play a large role in the design and supervision of data collection. This structure promoted data sharing and interdisciplinary research and may have been a component in creating cohesiveness within the research group, especially in the early years of the project. The commitment to technician support has meant, though, that fewer funds are available to allocate to individual faculty members and graduate students, certainly a facet of project management that has required a balancing effort over the years.

Information management goals and priorities guide the selection and development of appropriate technology and the allocation of resources. At our site, two primary goals for information system design were (1) to create a powerful and accessible environment for the retrieval of information that facilitates linkages among diverse data sets, and (2) to maintain database integrity. Database design was driven by the research agenda. The priority given to supporting ongoing research determined which data sets to process first and which retrieval requests to turn around quickly.

In 1982, our site made a commitment to use database software. The first database package was supported by the central campus computing facility and ran on a Sperry 1100. This hierarchical database was not fully relational. We currently use Oracle as the database management system, run locally on a Sun UltraSparc. The choice of Oracle was influenced by the fact that our use of Oracle is covered by the University of Wisconsin-Madison Oracle license. This makes upgrades less costly and provides a local pool of Oracle expertise. Use of Oracle has allowed us to develop a powerful user query interface. However, Oracle software has a somewhat steep learning curve and requires skills in database administration. The tradeoff is between the increased flexibility and power and the increased time required for system upkeep and design. Certainly, archiving text files is a simpler system to maintain. The centralization of our information management and the early introduction of database management software promoted our decision to invest in relational database technology. Other LTER sites with excellent information systems have chosen to design their systems with less reliance on high-end database management software.

We built flexibility and extensibility into the information system in anticipation of technological change. Because technology changes rapidly and these changes can require significant investment of staff time, our project wanted to control the frequency of upgrades or migrations to new software or a new platform. When we were still dependent on the central campus computer, we were forced to migrate twice to different database management software packages when the central computer facility decided to drop support of a currently used package. Now that our database is on a locally controlled computer, we can plan for and pace such transitions. Software/hardware vendors push the rate of migration, and information managers balance the advantages and disadvantages of installing new versions and establish timelines for upgrades.

Information managers frequently need to decide whether to look for solutions by purchasing off-the-shelf software or by writing (or contracting for writing) in-house programs. Our general philosophy has been to use off-the-shelf software whenever possible. The biggest problem with in-house software is that a sustained commitment to maintenance and upgrades is required. Programming support staff, undergraduates in most cases, changes every few years, and a great deal of knowledge is lost when they depart. On the other hand, commercial software often is designed for the business environment rather than the scientific environment, and available off-the-shelf software may not have the necessary features. Sometimes we have customized a commercial package to meet our needs; for example, in the past we have used highly customized Excel spreadsheets for data entry (Briggs et al. 1998).

In our efforts to increase efficiency, we came to realize that the information management staff was spending too much time dealing with data errors and inconsistencies when loading data. For example, nutrient data come from several sources, and, while loading data, we discovered date inconsistencies among sources. Thus, we are developing more data checks for the early stages of data flow. The second version of the automated fish sampling software contains considerably more error checking than the first version did. Warnings are issued immediately, while crews are still operating in the field. A person with the fish still in hand can easily check an inconsistent length and weight, whereas updates made later in the data flow are time consuming and can result in having to discard or flag the inconsistent measurements. Development of software for field-based quality control is an area that needs further effort. Wireless communication can provide the opportunity to have people in the field make checks against a remote database.

Sometimes tradeoffs exist between making life easier for the end user of the database and spending time in information management activities. Consequently, we have designed a number of Oracle tables to contain some redundant information; in technical jargon, these tables are not normalized completely. The main reason for this design was that end users would have an easier time retrieving data; they would not have to link as many tables in their retrievals, though database administrators can facilitate retrievals by constructing views on the database with predetermined joins. Archived files that were created by dumping database tables to text files are easier to use with additional identifying information in the table. This redundancy does require that information management staff make the updates to more than one field when updating data; however, data updates are infrequent after initial quality assurance screening.

Another decision made early in the project was to support a heterogeneous computer environment. We currently have researchers using Windows XP, Macintosh OS, and Unix. This diversity provides more options for the researchers and has allowed us to use a larger set of software. On the down side, network administration is more complex. We have limited our software selection in some cases, for example, bibliographic software, to products available on both Macintosh and Windows operating systems.

Another important decision was to hire the information manager in a dual role as both an ecologist and an information manager. The advantages of having a scientist in this position include greater sensitivity to research facilitation, enhanced recognition of communication challenges, greater integration of the research agenda with information management, and a higher status given to information management. Our information manager has been a principal investigator on grants, regularly participates in the monthly principal investigator meetings, and frequently is an author on ecological papers, as well as information management papers.

All four authors of this chapter feel that this blend of ecology and information management expertise has been useful in their project activities. The science activities of the information manager contributed to professional development and hence to continuity of staffing in a field noted for rapid turnover. Having people who are comfortable crossing the boundaries between science and technology in both the site and intersite arenas often has been advantageous.

As the information management workload increased and the technological requirements grew and became more specialized, an assistant with formal computer science training was hired in 1985. The technical background necessary for this second position expanded as we moved the database off the central campus computer and onto a local Unix workstation in 1993 and continues to expand. Administration of the Center for Limnology computer facilities is provided by a position shared between the LTER program and the Center for Limnology. This system support releases the LTER information management staff from some of the computer and network troubleshooting that information managers at some LTER sites find quite time consuming.

Future Challenges

Our research agenda has expanded to encompass the study of human interactions with lake ecosystems. Researchers now include social scientists, who are generating types of data new to our site. The new information management challenges include training a new group of researchers in the LTER information management paradigm and helping to create the links between the natural and the social science data sets that will facilitate these new interdisciplinary efforts. Examples of the new types of data are extensive census and property tax records and survey data based on responses of property owners to issues regarding their use of lakes and lakefront property.

We anticipate that in the future we will need to deal with far greater volumes of data than in the past. This increased volume will occur especially with data from satellite-based sensors. As an example of the increased data volume, 1 year of satellite image data from National Aeronautics and Space Administration's Earth Observing System is equivalent in storage requirements to 25 years of Landsat data. A decade ago, the Environmental Remote Sensing Center on our campus might acquire no more than six Landsat images per year over our study areas in northern and southern Wisconsin. Today, while continuing to collect Landsat imagery, the Center also downloads and archives Moderate-Resolution Imaging Spectroradiometer images on a daily basis for the entire state and the adjacent Great Lakes, using the Earth Observing System's direct broadcast system. At the same time, our researchers have access to data from new types of airborne and spaceborne sensors, including multispectral and thermal systems, imaging spectrometers (hyperspectral sensors), and imaging radar systems (Appendix 13.1).

As another example of the increase in data volume, the North Temperate Lakes LTER lake buoys provide high temporal resolution data in near real time. These data provide opportunities to study lake dynamics at a wider range of temporal scales than has been undertaken before. The sonde equipment for measuring lake metabolism will allow us to estimate metabolism on all seven northern Wisconsin study lakes on an hourly basis, compared with the bottle metabolism measurements, which have been limited logistically to three lakes at biweekly intervals. The real-time data collection challenges our information systems in areas of data capture, quality control and assurance, and distribution. Data from buoys are moving directly from the sampling gear into the database, and they are being moved at rates and volumes that exceed the ability of humans to visually inspect the raw data. Thus, we plan to develop robust computer algorithms for quality control and assurance. The creation of software that implements the full spectrum of data validation, is configured easily for new measurements, and scales as the number of buoys increases will make an important contribution to data processing for sensor arrays. Rapid data transfer and screening offer advantages to the field crews by allowing near-immediate feedback. This feedback about the current status of the system allows crews to troubleshoot equipment and scientists to adjust sampling protocols if necessary, minimizing both down time and data errors and providing the possibility of changing the sampling protocol in response to driving events such as storms.

Synthetic, data-intensive research places greater demands on information management. The LTER network has designated the years 2001–2010 as the Decade of Synthesis and will use long-term data resources and networking to integrate information at higher levels. The LTER information managers are developing new partnerships and infrastructure to support these initiatives. The new partnerships and infrastructure will build on existing projects that have been developed by the LTER information managers in collaboration with the LTER Network Office and ecoinformatics research groups at the National Center for Ecological Analysis and

Synthesis and the San Diego Supercomputer Center. Collaboration among the information managers across sites will be critical to making the Decade of Synthesis a success, as will ongoing dialogues with the researchers who are pursuing synthetic research as the new infrastructure is designed and implemented.

The technical background and training needed by information management staff will continue to expand. Information managers across the LTER Network have presented the view that a new educational curriculum should be developed in environmental information management (Stafford et al. 1996). As envisioned by Stafford and colleagues, such programs for information managers will link technology to science and will include internships on ecological projects. The curriculum would cover a wide range of topics in computer science and statistics as well as ecology and might include information science, database development, geographic information systems, modeling, programming and statistical languages, data visualization, experimental design, and statistical techniques. Some training modules have been developed and presented by LTER information managers in national and international workshops (Vanderbilt 2001, Michener 2002, Michener and Bonito 2003). As technology continues to evolve at a rapid pace, training is needed both at the novice and advanced levels. The Eco-Informatics Consortium (Brunt et al. 2002) has been formed to create a vehicle for communication and coordination among developers of information systems and technology for ecology and environmental science. The Consortium has established a Web site at ecoinformatics.org to support exchange of informatics tools and information.

Innovations in computer and network technology will continue to expand both our capabilities and our vision. The World Wide Web already has had a significant impact on the way scientists access and share information. An extension to the current Web, called the Semantic Web, promises to vastly improve the ability to discover and integrate data and information by allowing machines to "understand" the information that currently they can only display (Berners-Lee et al. 2001). Technologies to support this new functionality are being developed now. For example, with XML (www.w3.org/XML) people can create tags in a Web document that includes information about meaning that a computer can then use for processing.

The future promises a technologically rich environment for data exploration and analysis. Within the LTER network and locally at North Temperate Lakes LTER, initial important steps have been made in breaking the data barrier. Many data sets are available readily over the Internet and are well documented. Online data access tools have been created that permit researchers to construct a database query and to receive the data in formats compatible with analysis packages of their choice. We now have a wealth of online data, and the next new challenges are data discovery and integration.

A significant advance in the area of ecological data discovery was the development of Ecological Metadata Language, discussed earlier, a content standard for ecological data specified in XML (www.w3.org/XML) Schemata (McCartney and Jones 2002). As descriptions of data sets are made available in this standardized content and format, networked metadata catalogs can be searched to locate data resources. Once data sets are located, another large barrier is integration of data. Comparability issues may derive from differences in data formats, scaling, measurement protocols, or semantics. Standardized, machine-readable metadata can provide a basis for automating some aspects of the data integration challenge. Other aspects of making data comparable, called semantic mediation, will require the development of formal ways to represent ecological concepts and their relationships.

The development of infrastructure for data discovery, integration, and analysis represents current challenges in ecoinformatics research. In the future, discovery of relevant data sets will become easier, as will extraction of the information pertaining to the questions of interest, and on-line visualization and analysis tools will facilitate pattern examination. We are progressing toward a research environment where access to data will no longer be the major barrier to synthetic research. The vision for information technology infrastructure for environmental science research in the future is a global information network where researchers can discover and access data easily and obtain help with interpretation and integration of data sets across comparability barriers.

Summary

Information management has been an integral part of the research process of the North Temperate Lakes LTER program. From its initiation, a strong commitment was made to information management at both the local and the national levels, a commitment that continues today. The creation of a data legacy of secure and well-documented data and the facilitation of the research process through the design of data collection and the information management system are part of that process. The incorporation of technological advances in computers, networking, and software has greatly increased the functionality of our information system.

Access to data is provided through dynamic database queries on the program's Web site. Data collection and entry have become more automated through the use of handheld computers with custom-built software and instrumented buoys streaming data in near-real time. Spatial data sets play an important role in understanding the interactions between lakes and landscapes. Analysis of large, complex data sets has been enhanced significantly, and in some cases made possible, by the investment in information management.

Certain decisions were pivotal in the design of the information system: the centralized collection and management of core data, the emphasis on accessibility of data by providing advanced retrieval capability to researchers, and the use of relational database software. At the network level, the LTER program has been developing a Network Information System through the partnerships of LTER scientists and information managers, the LTER Network Office, and ecoinformatics collaborators; these collaborations are leading to advances in data discovery, access, and integration.

Future challenges for information management for the North Temperate Lakes LTER program include the creation of new data types from the study of human interactions with lakes, the processing of greater volumes of data from satellite and instrumented buoys, increasing demands for infrastructure support for synthetic, data-intensive research, and the ever-increasing need for new technical expertise.

14

Origin, Operation, Evolution, and Challenges

John J. Magnuson
Barbara J. Benson
Timothy K. Kratz
David E. Armstrong
Carl J. Bowser
Alison C. C. Colby
Timothy W. Meinke
Pamela K. Montz
Katherine E. Webster

The best way to predict the future is to invent it.
—Alan Kay (1971)

The objectives, scope, and realization of the Long-Term Ecological Research (LTER) Program of the National Science Foundation (NSF) evolved as the activities played out at the various sites through the 1980s and 1990s and into the twenty-first century. At the onset of the program, each site had its own historical perspective and a core of individual participants who responded to NSF's challenge to initiate long-term, site-based ecological research. Here we describe and discuss the origin, operation, history, evolution, and challenges of the LTER program from the perspective of the North Temperate Lakes site. This is our view of what happened and how the ecological science community set out in new directions, adapted, and evolved (Magnuson et al. 1984, Kratz et al. 1986, Magnuson and Bowser 1990, Magnuson et al. 1997a). We hope that this autobiographical history of science will be useful in understanding consequences of the multidirectional interactions among the researchers at the North Temperate Lakes, investigators at other LTER sites, and the National Science Foundation.

Before the Origin

At Wisconsin, E. A. Birge and C. Juday had led a comparative long-term investigation in lakes of the Northern Highlands Lake District for about 17 years, from 1925 through 1941. This tradition for long-term research ended with the Second World War and the deaths of Juday in 1944 and Birge in 1950 at the age of 99 (Beckel 1987, Magnuson 2002b). The next generation of limnological study at the University of Wisconsin-Madison was led by A. D. Hasler and focused on experiments both in the field and in the laboratory, fish physiology and behavior, process studies, and the application of science to environmental problems such as lake eutrophication. The tradition of long-term, comparative lake research established by Birge and Juday did not emerge again until the 1980s with the establishment of the NSF-supported North Temperate Lakes LTER.

Limnological science played a key role in the genesis of ecological concepts and approaches before the challenges embodied in long-term ecological research were addressed (Chapter 1). During the late 1970s, ecological scientists both outside and inside the National Science Foundation recognized the need for long-term ecological research at field sites in the United States. The National Science Foundation supported workshops (Callahan 1984, Franklin et al. 1990) that addressed the issues and the opportunities; these efforts resulted in a call for proposals from NSF to establish six LTER sites. North Temperate Lakes was one of the first six sites to be supported, along with H. J. Andrews Experimental Forest, Coweeta Hydrological Laboratory, Konza Prairie, Niwot Ridge, and North Inlet Estuary. The resulting NSF program in long-term ecological research was driven by the recognition (Marzolf 1982) that:

1. Investigations of ecological phenomena occurring at time scales of decades and centuries were not normally supported by NSF funding in ecology.
2. Ecological experiments were conducted with little recognition of the high interannual variability in the studied systems.
3. Long-term trends were not being systematically monitored in ecological systems with the consequence that unidirectional changes could not be distinguished from more cyclic variation.
4. The absence of a coordinated network of ecological research sites inhibited comparative research and its benefits.
5. Natural ecosystems where research was being conducted were being lost to other uses.
6. Ecological research was often done on only a selected component of the system, and multilevel, integrated data were not available at intensive research sites.

The call for proposals from NSF was made in 1979. The call encouraged each site to include five core areas in their research. These were and still are (1) pattern and control of primary production, (2) spatial and temporal distribution of populations selected to represent trophic structure, (3) pattern and control of organic matter accumulation in surface layers and sediments, (4) patterns of inorganic inputs and movements of nutrients through soils, groundwater, and surface waters, and (5) patterns and frequency of disturbance to the research site.

At Wisconsin, the response to NSF's call for long-term ecological research came from a diverse group of aquatic faculty who already were networking through the Oceanography and Limnology Graduate Program at the University of Wisconsin-Madison (Appendix 14.1). Included in this group were Mary P. Anderson (hydrogeology), David E. Armstrong (chemical limnology), Carl J. Bowser (geochemistry and hydrogeology), Thomas D. Brock (microbiology and limnology), John J. Magnuson (fish ecology and limnology), and Robert A. Ragotzkie (atmospheric science and physical oceanography). We submitted a proposal entitled "Long-Term Ecological Research on Lake Ecosystems." The proposal was funded, starting on October 1, 1980, in one-year increments for five years at an average level of $250,000 per year under the title "Comparative Studies of a Suite of Lakes in Wisconsin." The University of Wisconsin Graduate School provided startup funds for the summer of 1980.

The Beginning and the Challenge

The scientific work actually began as we wrote that first proposal and the requested addendum where we thought about the what, how, who, where, and why of doing such science. Even at the onset, we were pushed by NSF to shape and sharpen long-term research with ecological questions and testable hypotheses, to specify the management plan for this new type of research, and to recruit a project review committee. We were asked to address the continuity of leadership for project management and intellectual commitment and other issues to reassure NSF that we were serious and had a high potential to successfully conduct long-term ecological research at the Wisconsin site. Long-term ecological research needed to be distinguished from monitoring at that time. Monitoring was not the program's purpose; conducting long-term, question driven research and establishing the long-term databases required to do this were the purposes. In retrospect, these early tough questions touched on critical issues on which success of long-term ecological research programs would depend.

We believed that the core investigative areas provided a useful template for insuring the depth, breadth, and intellectual scope of the proposed research but that the core areas did not provide the conceptual structure for the questions we wished to address. How did the core areas influence us? They provided a comfortable platform for our group of limnologists and oceanographers to design what, where, and when to measure. Clearly, the ecosystem context covered trophic structure, primary production, organic matter accumulation, and biogeochemical cycling in the hydrosphere. Clearly, a population biology context considered long-term dynamics of populations. Clearly, an emphasis on temporal scale was indicated with long-term measurements and event driven dynamics of disturbance. And, clearly, the seed was present to study multiple systems, rather than a single lake; the single-lake focus had characterized previous long-term lake studies (Chapter 1).

We gathered at the Trout Lake Station in northern Wisconsin for several days in the summer of 1979 to establish a list of major research goals and to discuss which lakes to include and how to organize our effort. These critical early decisions set the stage for our future directions and were, in themselves, decisions for the long

term. We chose a heterogeneous set of lakes from a collection of previously studied lakes near the Trout Lake Station. We chose a common set of measurements
and to have a common data repository. We ended up choosing the lakes of the
Northern Highlands rather than the Madison area lakes in southern Wisconsin because the northern lakes seemed more insulated from human activity and modification, spanned gradients representing important classes of lakes, had data sets
collected annually from the Birge and Juday era in the Northern Highlands from
1925 through 1941, were located near the University of Wisconsin-Madison's Trout
Lake Station, which provided a base of operations, and were of research interest to
one or more of us. We chose not to do a whole host of things for various reasons.
For examples, we decided not to begin with paleolimnology because the sedimentary record would be available to us in the future, and we decided not to measure
the parameters from phytoplankton to fishes as frequently as specialists would wish
because such intensities would limit the breadth of measurements needed to integrate the dynamics of the entire network of lakes.

The Advisory Committee's Challenges

At the onset of our first field season, in June 1981, four scientists external to the project
(Nicholas C. Collins, David G. Frey, Gene E. Likens, and Thomas C. Winter (Appendix 14.2) met with North Temperate Lakes LTER researchers at the Trout Lake
Station to review our plans and to offer advice (Fig. 14.1). A few highlights of their
advice are quoted.

The hydrologist Thomas C. Winter recommended that we shift our lake choices
so that every lake was located in the same groundwater flow system.

> I think the key to success is to locate an area that has the greatest variety of lakes in
> the smallest area. The area should comprise a physically defined unit, and because
> many processes in an ecosystem, especially an aquatic ecosystem, are dependent on
> water, a hydrologic unit is the logical choice . . . the ultimate unit for your studies, in
> my opinion, should be a well-defined ground-water system. (10 June 1981, letter,
> Limnology Archives at UW-Madison)

Nicholas C. Collins provided many insights on the calibration of simple measures
as indices and challenged us to embrace the effort required for data management.

> . . . I'll reiterate my concern that data management . . . is likely to be a time-consum-
> ing job that will require very frequent communication among [the data manager] and
> the PI's. . . . In my experience, thorough, ongoing documentation of the data will be
> much more difficult to maintain than the numbers themselves. (9 June 1981, letter,
> Limnology Archives at UW-Madison)

David G. Frey added to the challenge, spurring us to become interdisciplinary
rather than remain multidisciplinary.

> My overall impression of the program is that thus far even the outline of the elephant
> is perceived somewhat dimly. The seven principal investigators . . . need to [do] more
> than merely intensify their "thing." Some common theme or Leitfaden is needed
> to bind them all together and give them a sense of participating in a program larger
> than their own restricted disciplines but in which their disciplines are integral and

Figure 14.1. First meeting of the Scientific Advisory Committee (Appendix 14.2) for the North Temperate Lakes LTER, held in June 1981 at the Trout Lake Station. The roles of Wisconsin LTER personnel are listed in Appendix 14.1, 14.3, and 14.5. Top, left to right: Robert A. Ragotskie, Thomas D. Brock, Paula Kuscmarski, Brad Price, Nicholas C. Collins (Advisory Committee), David G. Frey (Advisory Committee), John J. Magnuson, Thomas C. Winter (Advisory Committee), Gene E. Likens (Advisory Committee), Carl J. Bowser, John D. Lyons, Timothy K. Kratz, Timothy W. Meinke, and William H. Horns. (Center for Limnology) Bottom, left to right: Thomas A. Brock, Thomas C. Winter, Timothy K. Kratz, Paul W. Rasmussen, unidentified, Paula Kuczmarski, Carl J. Bowser, Vicki Watson, William H. Horns, Robert A. Ragotzkie, Nicholas C. Collins, John J. Magnuson, Timothy W. Meinke, Gene E. Likens, David E. Armstrong, David Frey, Brad Price, and Galen J. Kenoyer. (Center for Limnology)

indispensable parts. Study of carefully selected lakes over their entire ontogeny to the present and on into the future could be one way to accomplish this. (3 June 1981, letter, Limnology Archives at UW-Madison)

Gene E. Likens chose to challenge us to develop and initiate a new science.

... I would urge you not to use this grant to do more of the same thing. You have a great opportunity to do integrative ecosystem science on a long-term basis. You have made a reasonable start, but some of the questions ... did not appear to be uniquely long term. I think it is fair to state that most of us don't know how to develop rigorous and clever long-term ecological studies (I certainly don't). But you have the opportunity. I think a sizeable proportion of your time and brain power should be spent in trying to develop and initiate such studies. (23 June 1981, letter, Limnology Archives at UW-Madison)

All four of these initial advisers shaped our early decisions, but, more important, they challenged our founding group to reach out from the scientific, cultural, and infrastructural legacies of the past (Chapter 1). Their messages were an exciting, almost intoxicating, introduction to the years ahead. In the next 20 years, we benefited from one early NSF review by Tom Callahan, two more external advisory committee meetings, and three site visits (Fig. 14.2) (Appendix 14.2).

Setting Up the Site

Administrative Organization

The initial principal investigators were interdisciplinary, aquatic scientists comfortable with joint cruises or expeditions with physical, chemical, and biological colleagues. The broad disciplines, limnology and oceanography, from which we came were inherently and historically interdisciplinary. We knew and believed that the whole was not just the aggregation of the pieces and that the physics, chemistry, and biology, even at their finest partition, were inexplicable except through their relations to each other. This history led to an operational model of shared visions, resources, planning, and responsibilities.

Many basic structures and procedures established by the initial investigators have been retained to the present. This consistency has been the case for direction and policy, proposal preparation, operation and management, graduate student mentoring, and data collection, management, and use. Program management (Fig. 14.3) emphasized shared governance, communication, and research, and empowered leadership to focus the vision and to resolve conflict. This general structure evolved somewhat to become more articulated as the size of the group and the complexity of our interactions grew. We write what follows in the past tense, but many of these features persist to the present and will likely continue.

The program was administered in the Center for Limnology in the College of Letters and Sciences at the University of Wisconsin-Madison. The lead investigators, John J. Magnuson with Carl J. Bowser from 1980 to 2000, and Steve R. Carpenter from 2000 to this time of writing, provided direction (Fig. 14.4). Investigators wanted to participate in governance and not simply let the lead investigator determine their

Figure 14.2. First site visit (Appendix 14.2) of the North Temperate Lakes LTER program by the National Science Foundation at the Trout Lake Station in 1989. Roles of Wisconsin LTER personnel are listed in Appendix 14.1, 14.3, and 14.5. The top photo was taken during a planning session in February; bottom was taken during the site visit in May. Top, left to right around the table: Barbara J. Benson, Ann S. McLain, John R. Vande Castle, Thomas M. Frost, Timothy K. Kratz, John J. Magnuson, Carl J. Bowser, Stith T. Gower, David E. Armstrong, Mark McKenzie, and Dale M. Robertson. (Center for Limnology) Bottom, left to right regardless of row: Milton G. Ward, Jonathan J. Cole, James E. Schindler, and Monica G. Turner on the Site Visit Team, and John J. Magnuson, Carl J. Bowser, Thomas M. Frost, James T. Callahan (National Science Foundation), Timothy K. Kratz, Thomas S. Lillesand, and David E. Armstrong. (Center for Limnology)

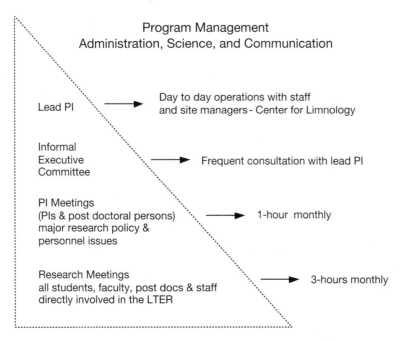

Program Management
Administration, Science, and Communication

Lead PI ⟶ Day to day operations with staff
and site managers - Center for Limnology

Informal
Executive ⟶ Frequent consultation with lead PI
Committee

PI Meetings
(PIs & post doctoral persons) ⟶ 1-hour monthly
major research policy &
personnel issues

Research Meetings
all students, faculty, post docs & staff ⟶ 3-hours monthly
directly involved in the LTER

Figure 14.3. The organizational structure of the North Temperate Lakes LTER program as established early in the program and as presented at the 1999 NSF Site Visit (Appendix 14.2). The structure emphasizes shared governance, communication, research, and leadership.

Figure 14.4. Lead investigators of the North Temperate Lakes LTER. From 1980 to 2000, John J. Magnuson (center) was lead principal investigator (Luquillo LTER Site, May 1990, C. Simenstad), and Carl J. Bowser (left) was the associate principal investigator (Trout Lake Station, March 2000, J. Magnuson); Stephen R. Carpenter (right) became lead principal investigator in 2000. (Trout Lake area, June 2001, M. Turner)

collective future. Principal investigators from across the campus met monthly for one hour to plan and advise on program, personnel, and budget decisions. Usually the group of principal investigators reached consensus on decisions, but when this did not occur, the lead investigator made the decisions. The process was not contentious; revolutions did not occur. We shared the same perspectives, the table was flat and stable, and we were realistic about the immense scientific challenge relative to the magnitude of our financial resources.

The lead investigator directed the administrative and technical staff. Two site managers oversaw data collection activities and use of field resources: Timothy K. Kratz, from 1981 to 2002; Jim Rusak, from 2002 to the time of writing for the northern Wisconsin lakes near the Trout Lake Station; and Dick Lathrop, from 1994 to the time of writing for the southern Wisconsin lakes near the Madison campus. These individuals evolved to become principal investigators in the program, as did the North Temperate Lakes LTER information manager Barbara Benson from 1983 to the time of writing and the geographic information systems/remote sensing specialists Mark MacKenzie, from 1988 to 1995, David Bolgrien, from 1995 to 1998, Joan Riera, from 1998 to 2000, and Jonathan Chipman, from 2001 to the time of writing. These participants and others are detailed in Appendix 14.3. Technical staff consisted of three field technicians and a water chemistry laboratory technician, data managers and programmers, and short-term student employees who assisted in the field and the laboratories.

Core data were managed (Chapter 13) as a project-wide resource rather than being owned by individual principal investigators. From the design of data collection to incorporation in the centralized database to analyses, our intent was to facilitate investigation of linkages among the components of the systems studied. Data from thesis research that were not part of our core data were incorporated into the database along with their associated metadata when the data were judged by the principal investigators to be likely to contribute to future LTER research, such as regional lake surveys.

Graduate students and their mentoring were key components of the program. Often our most in-depth analyses came from the contributions of our younger colleagues. The principal investigators discussed opportunities for new students collectively, and study areas were identified. Usually, a single faculty member mentored each student, but many students were comentored either formally or informally. Students interacted with each other through LTER activities at the Trout Lake Station especially during summers, and at centers, departments, and laboratories across the Madison campus.

The entire group of principal investigators, research staff, and graduate students met monthly for several hours to discuss concepts, to present research results, and to plan. Information and presentations from these meetings often were placed on our Web page as that tool became available. These sessions played an important role in keeping our group cohesive and knowledgeable across an increasingly broad set of disciplines that made up the North Temperate Lakes LTER program. Initially, interdisciplinary science was for us chemical, physical, and biological limnology; Bowser and Magnuson dreamt of causally connecting groundwater and the fishes

Interdisciplinary Approach

Commitment and Opportunity

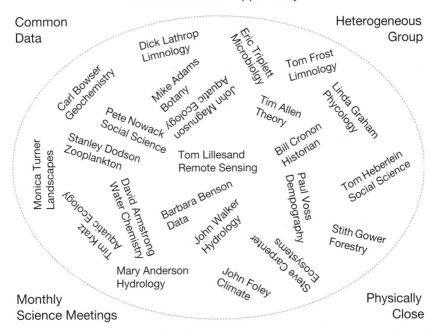

Figure 14.5. Diagram of the structure to catalyze interdisciplinary science at the North Temperate Lakes LTER presented at the 1999 site visit. A largely nonhierarchical system encouraged the exercise of commitment and opportunity. This universe of interactions is porous to interactions outside the system.

through the physical-biological system they visualized at that time. By 2000, inter-disciplinary study included additional aspects of limnology, terrestrial ecology, rural sociology, and economics.

The principal investigators and their students related as an interactive and largely nonhierarchical community within which to be creative in interdisciplinary planning and research (Fig. 14.5). Interactions came from individual initiatives or temporary problem-oriented working groups to identify and develop potential activities or directions. Associations formed and dissolved in the doing of related science rather than by assignment or by implied and real power structures.

Collaboration was encouraged outside the LTER grant proper. We encouraged strong linkages between collaborative projects and the North Temperate Lakes LTER program. Key among those in the first 20 years were the Little Rock Acidification Experiment (Chapter 9), the training grants from NSF on Synthesis of Lake and Stream Ecology (Research Training Grant) and Integration of Social and Natural Sciences (Integrative Graduate Education and Research Traineeship), and programs within agencies such as the U.S. Geological Service's Water, Energy, Biogeochemical Budgets (WEBB) Program at Trout Lake and the Wisconsin Department of Natural Resources on the Madison lakes. In the case of the Little Rock Lake project, the initial impetus came from outside our program through John Eaton, of the U.S. Environmental Protection Agency, in Duluth, who saw advantages of doing the research in conjunction with an LTER site. Significant and mutually beneficial sharing occurred in funding and doing the research. In the case of the Little Rock Lake project, researchers from our LTER program and a number of state and federal groups were brought together (Chapter 9).

Core Data and Sampling Frequency

Our goals for the North Temperate Lakes LTER were to detect long-term trends and to investigate long-term dynamics and interactions in a suite of lake ecosystems. These goals required establishing a long-term database of parameters characterizing multiple components of the ecosystems (Magnuson et al. 1984, Kratz et al. 1986). The selection of parameters to measure (Appendix 13.1) was comprehensive within the constraints of the budget, reflected the research interests of the investigators, and ranged across physical, chemical, and biological measurements suitable for studies spanning population, community, and ecosystem levels (Figs. 14.6a–l). We were influenced in our choices of variables and methods by the measurements made by E. A. Birge and C. Juday and their colleagues in the early 1900s, even though the methods in many cases had changed. In some cases, for example, electrical conductivity of water as a measure of ion concentrations, the same principle of measurement as in the Birge and Juday era, was used, and it was measured in essentially the same way. In other cases, for example, specific ion concentrations such as chloride, sodium, and calcium, the measurement techniques and solute detection limits were substantially different. The kind and range of measurements expanded through the years. Many of the measurements we made were not possible in the Birge and Juday era.

Sampling frequencies were tuned to match the dynamics of individual parameters. For example, zooplankton was sampled at biweekly intervals as zooplankton characteristically peak in abundance over a four- to six-week period. Parameters that vary over longer time scales, such as fishes, macrophytes, groundwater chemistry, and benthic invertebrates, were measured annually. We sampled chemistry most intensively at four limnological seasons: spring overturn, maximum stratification in summer, fall overturn, and winter stratification. These periods are important chemically because differences between spring and fall overturns indicate a net gain or removal of chemical species from the water column, and the midsummer and winter periods are when the maximum seasonal differences between surface and bottom waters occur. We coordinated the lake sampling of physical limnology, water chemistry, and plankton. Our overall sampling program allowed comparisons of parameters and processes among seasons, among years, and among lakes, facilitated the study of linkages between system components, and positioned us to detect invading exotic species in our primary lakes.

Figure 14.6a. Physical and chemical measurements. Top left: Sensor measures photosynthetically active radiation beneath the snow and ice. The sensor is near Daniel Helsel's left shoulder. The framework is inserted down a hole in the ice in a straightened position and then is pulled into the right-angle position shown in photo for light measurements at the bottom of the ice and at increments through the water column. (Trout Lake, winter 1985, J. Magnuson) Top right: Secchi disc being lowered into the water by Richard C. Lathrop to determine water clarity. (Lake Monona, late spring 1980s, Dane County Regional Planning Commission) Bottom left: Yellow Springs instrument to measure temperature and dissolved oxygen in the water column. (Trout Lake, about 1990) Bottom right: Montedoro-Whitney thermometer model TC-5C being lowered by Dale M. Robertson to measure temperature profiles. (Lake Mendota, late 1980s, C. Bowser)

Figure 14.6b. Automated weather and physical and chemical measurements. Top left: Raft on Sparkling Lake (May 1989 to May 1999) used to estimate evaporative components of the lake hydrologic budget from temperature difference between surface water and air, the relative humidity, and the boundary layer properties from wind speed at 1, 2, and 3 m above the water. A thermistor array below the raft measured the water temperature profile. Power came from car batteries and data were saved on a data logger. (C. Bowser) Top right: An improved raft after May 1999 being maintained by Tim Meinke. The same measurements were made, an oxygen sensor was added, the power source was changed to solar, and data were transmitted to the Trout Lake Station via 900 Mhz spread spectrum radios. (C. Bowser) Bottom left: Trout Lake buoy (Apprise Technologies, Minnesota), tested in summer 2000 to estimate evaporation, photosynthetically active radiation (being checked by Paul Hanson), and limnological data. A 2.4 Ghz antenna transmitted information to the Trout Lake Station; a buoyancy-controlled vertical profiler sensed water temperature, conductivity, photosynthetically active radiation, pH, chlorophyll, total dissolved gas, and turbidity at various depths. Solar panels provided the power. (J. Miller, University of Wisconsin-Madison University Communications) Bottom right: LTER weather station at the Noble F. Lee Airport, Arbor Vitae, June 1989. Data are transmitted by modem and phone line to the Trout Lake Station. Instruments from front to back are: precipitation gauge; telephone pedestal; instrument mast with anemometer, photosynthetically active radiation sensor, air temperature and relative humidity sensor, and data logger; and at the back fence a longwave radiation sensor (left) and a shortwave radiation sensor (right). (C. Bowser)

Figure 14.6c. LTER base crews measuring physical, chemical, and plankton variables. Top: Debi Fisk is measuring the volume pumped through a filter used to measure total particulate matter in the water column, and Joe Gressens is holding the Wisconsin net after taking a vertical tow for zooplankton. The elongated box to his left contains a 2-m-long Schindler-Patalas-style plankton trap that we designed and built. (Allequash Lake, July 2000, J. Miller, University of Wisconsin-Madison University Communications) Bottom left: Sampling by Paula Kuczmarski and Alan G. Barbian. (Crystal Bog, summer 1982, J. Magnuson) Bottom right: Base limnological sampling with vertical plankton tow just completed by Ted M. Cummings and Vicki S. Schwantes filtering water for chlorophyll analyses. (Lake Mendota, summer 2003, J. Miller, University of Wisconsin-Madison University Communications)

Figure 14.6d. Winter sampling. Top: Ice drills, tent, snowmobile, and a toboggan for gear facilitated sampling by the winter base crew. (Trout Lake, January 1982, P. Barbian) Middle: A vertical temperature and dissolved oxygen profile was obtained by Dale Robertson with a Yellow Springs temperature oxygen meter model 57 and a Montedoro-Whitney thermometer model TC-5C. The ice drill and snowmobile are close at hand. (Trout Lake, mid-1980s, P. Jacobsen) Bottom left: Base measurements of physical, chemical, and biological limnology being made by Pam Montz. A peristaltic pump on her right was used to collect the water for chemistry samples and was powered by the battery at her feet. Tubing coiled on top of the pump goes down into the lake to the sampling depths. The insulated box contained the Yellow Springs temperature/oxygen meter. (northern LTER lake, late winter about 1992, C. Bowser) Bottom right: Winter base crew measured physical, chemical, and biological limnology. Participants were Barbara Reinecke (on the left, with graduated cylinder and meter stick), Tim Meinke (with a Wisconsin net for vertical zooplankton tows), and Tim Kratz (on the right with the 2–m Schindler-Patalas-style zooplankton trap). (Crystal Bog, about 2000, J. Magnuson)

Figure 14.6e. Primary production measurement and phytoplankton and zooplankton sampling. Top left: Opaque tubes were used to sample phytoplankton for measurements of C_{14} primary production. Tubes were assembled to the correct length to remove water from the epilimnion, the metalimnion, or the hypolimnion. Tim Meinke and Andy Juele are draining the water from the tube into a thermal cooler that excluded light and air. (Crystal Lake, summer 1987, University of Wisconsin Archives) Top right: Incubation tank for C_{14} primary production. Tim Meinke is placing light and dark bottles into the three channels that are at the temperatures of the epilimnion, the metalimnion, and the hypolimnion. Each bottle contains lake water from one of those layers and is spiked with C_{14}. Differences in uptake between the light and dark bottles at different temperatures and light levels are used to calculate primary production for the entire lake water column. (Trout Lake Station, 2004, T. Kratz) Bottom left: The 2-m Schindler-Palalas-style trap is being removed by Tim Kratz. Water is draining from the device through a plankton net hidden from view. (Crystal Bog, winter 2000, J. Magnuson) Bottom right: Zooplankton were sampled in a vertical tow from the bottom to the surface with a Wisconsin net by Paula Kuczmarski. Water is draining from the net, and the sample will be removed from the cup on the bottom of the net. (Trout Bog, January 1982, A. Barbian)

Figure 14.6f. Among the related projects that enriched and were augmented by the North Temperate Lakes LTER were (top) the Little Rock Lake experimental acidification and (bottom) the Microbial Observatory. Top left: Janet M. Fischer sampling zooplankton with a Schindler-Patalas sampler from the mesocosms alongside the boat to test effects of acidification. (Little Rock Lake, July 2001, J. Klug) Top right: Zooplankton feeding rates were measured by Michael E. Sierszen using a Haney chamber and radioactively labeled algae. (Little Rock Lake, 1987, A. Barbian) Bottom left: Bacterial samples were collected by Angela D. Kent from discrete depths to describe bacterial communities associated with the mercury methylation rates measured at each depth by other researchers. The filtration manifold concentrated the bacteria onto 0.2 µm filters. (C. Watras) Bottom right: Anthony C. Yannarell preparing to sample bacteria from 0 to 5 m. The sample tube and pump is used to collect planktonic organisms, and the water will be filtered through a 0.2 µm filter to concentrate the bacteria. (Firefly Lake, October 2002, A. Kent)

Figure 14.6g. Bottom-related studies. Top left: The sedimentation sampler was suspended for weeks or months in the water column and collected settling particulates and dead organisms. The transparent chamber is reflecting the shoreline behind the photographer. This early version was later replaced by one made of opaque plastic. (C. Bowser) Top right: Lake sediments being cored to obtain information on sponge spicules and the past biogeochemistry of the lake. Tim Kratz is on the left; Tim Meinke is setting the drive rod of the piston corer; John Morrice is on the right. (Trout Lake area, winter 1989, D. Schneider) Bottom left: Boat is being prepared to be backed into the lake. Amina Pollard, on the left, studied the benthic invertebrates from a landscape perspective and is being helped by Kelli Melville. (Little Crooked Lake, summer 1999, J. Magnuson) Bottom right above: Mayfly nymph. (Trout Lake, June 1975, D. Stamm) Bottom right below: A Dendy sampler served as a colonization sampler for benthic insects, snails, and other invertebrates. (2004, T. Kratz)

Figure 14.6h. The Department of Geology and Geophysics was an important contributor to LTER research, especially on groundwater, water budgets, and bedrock geology. Top left: David Krabbenhoft is transporting the first Sparkling Lake raft from Madison to the Trout Lake Station. (spring 1989, C. Bowser) Top right: Sledgehammer seismology is used to estimate the depth of the glacial till to the bedrock layer in the Trout Lake groundwater shed. Oemeka E. Okwueze is adjusting the instrument that measures the time for the return echo from bedrock, and Carol Ptacek is pounding a metal plate to generate the source signal. (near Crystal Lake, 1981, J. Magnuson) Lower left: Installation of groundwater observation wells (piezometers). Galen J. Kenoyer, on the left, is aided by Carol Ptacek and Barbara Bickford, on the right. Auger flights being added to the well string during drilling. The drilling rig was owned and operated by the Department of Geology and Geophysics. Well depths varied from a few to more than 20 m in length. (near Crystal Lake, summer 1981, C. Bowser) Bottom right: Two views of the groundwater well heads showing the pipes for sampling three depths and the lowering of a weighted tape to measure the groundwater level in the pipe. (isthmus between Crystal and Big Muskellunge Lakes, early 1980s, J. Magnuson)

Figure 14.6i. Fish sampling with sonar and fyke nets. Top left: Abundance and sizes of pelagic fishes were estimated with sonar. Lars G. Rudstam prepares the instrument for another transect. The transducer is suspended off the bow about 0.5 m below the surface in a tow body, here shown on the front deck, a car battery provided the power, and the 70 kHz Simrad EY/M instrument with a roll of recording carbon paper is in the metal box, together with a cassette player for analog data recording and a small portable oscilloscope. (Trout Lake, summer 1981, A. Barbian) Top right: A subsample of fish caught in the various kinds of nets were weighed with a scaled set of Pesola spring balances, here by Theodore V. Willis. (Trout Lake, summer 1998, J. Magnuson) Bottom left: Fyke net was lifted and emptied of fish. Paisarn Sithigorngul is lifting the net showing the hoop areas where the fish are trapped. The hoops are connected to a net lead set perpendicularly from the shore. Weerawan Chulukasem is in the middle to measure the fish, and John J. Magnuson is adjusting the boat's position. (Trout Lake, 1982, Magnuson) Bottom right: A fyke net is stacked on the bow before being set from shore by John Lyons. (Sparkling Lake early 1980s, A. Barbian)

Figure 14.6j. Vertical gill nets sampled pelagic fishes in the deepest part of the lakes. Top left: Gill net was spooled off the net float by Ann S. McLain to check for windows, holes that needed to be mended before setting the nets. (Trout Lake Station, about 1989, J. Magnuson) Top right: Vertical gill nets rolled on floats being taken to the boat landing after being emptied of fishes. Tim Kratz was at the motor, Weerawan Chulakasem was on the bow, and Paisarn Sithigorngul was amidship. Racks on which the nets were placed for lifting are visible on the starboard bow and stern. (Big Muskellunge Lake, summer 1982, J. Magnuson) Bottom left: Set of seven vertical gill nets, each with a different mesh size. Nets hang from the rollers to the lake bottom. They are held in place by anchors at each end of the set. (summer 1982, J. Magnuson) Bottom right: Fish being removed by Paul T. Jacobson from a gill net as it was being rolled up onto the float. The float was suspended on an iron rack from both ends as it was being lifted. (summer 1985, D. Robertson)

Figure 14.6k. Fishes often were sampled at night with beach seines and boat-mounted electro-shockers. For most other fish-sampling gear, such as the crayfish trap, the greatest catches occurred during the night, even though the gear was serviced during the day. Top: Seine catches often were processed on the boat. Vicki S. Schwantes, on the left, measured fish, while Kelly K. Dunn recorded data into a weatherproof, handheld data logger (Ipaq 3600 personal digital assistant, Hewlett-Packard Company) programmed by Owen C. Langman for easy, prompted data entry of fish species, length, weight, and fish scale information. Fish were held in the metal live well or waterfilled buckets be-tween capture and processing and then returned to the lake. (Fish Lake, 2002, T. Cummings) Middle: Fish were captured with a boat-mounted electroshocker at night. Fish were dip-netted between and outside the electrodes that hang into the water at the front of the boat. (Bass Lake, 2003, M. Woodford) Bottom left: Night photograph of a yellow perch. (Sparkling Lake, July 1975, D. Stamm) Bottom right: Crayfish traps were modified minnow traps with the openings at each end increased in size to 7.6 cm. They were baited with chunks of thawed beef liver. (Sparkling Lake, 2001, B. Roth)

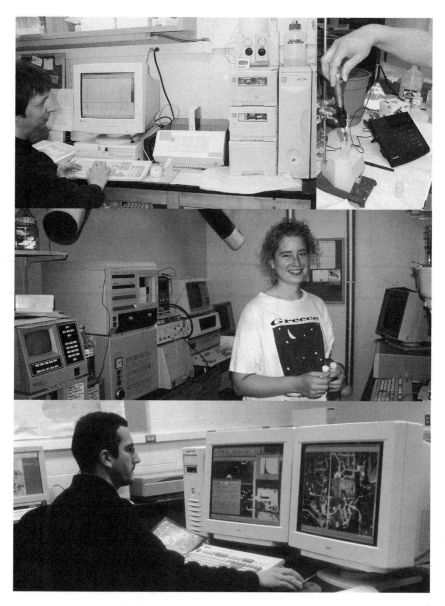

Figure 14.6l. Modern technologies were available to researchers. Top: On the left, an ion chromatograph is used by James A. Thoyre to measure chloride, nitrate, and sulfate, while, on the right, an acid titration is being used to estimate alkalinity. Other principal analytical equipment was in the chemical laboratory in Geology and Geophysics from 1981 but was moved to Limnology when Carl Bowser retired in 2000. (Limnology Laboratory on Lake Mendota, April 2004, J. Magnuson) Middle: High-performance liquid chromatograph and other equipment. Here, Beth L. Sanderson is analyzing phytoplankton samples for pigment composition. (Trout Lake Station, summer 1991, Center for Limnology) Bottom: Remote sensing and spatial imaging and analysis tools. Here, Joan L. Riera is examining changes in land use and cover at the North Temperate Lakes LTER site using ERDAS (Earth Resource Data Analysis System) software on a dual monitor computer. (Environmental Remote Sensing Center on the Madison campus, November 1998, J. Riera)

Lake Choice

A crucial shift in the choice of study lakes can be traced to our first external advisory committee meeting. In the initial proposal to NSF (1980, Proposal, Center for Limnology Archives, UW-Madison), we selected study lakes (Table 1.1, Fig. 1.4) on the basis of the existence of historic data, representation of the heterogeneity among lakes in the lake district, and the research interests of the investigators. Thomas C. Winter's advice—to maximize heterogeneity but within a single groundwater flow system—was taken (see earlier quotation), and the new criteria prepared us to compare our lakes along the gradient of their positions in the groundwater flow system (Chapters 2, 3). Carl J. Bowser, a North Temperate Lakes LTER principal investigator and hydrogeologist, strongly supported the idea of using the integration of the groundwater system as an organizational framework.

Evolution of Objectives

The primary objectives of our program evolved from 1980 to 2002 (Table 14.1). The original three objectives—perceiving long-term change; understanding interactions of physical, chemical, and biological processes; and understanding the role of disturbance—persisted through the first four grant cycles. These themes were the threads that tied our project together and provided a long-term perspective.

Later proposals added five new themes (Table 14.1). Three facets of the program's evolution are of particular interest: (1) intersite science, (2) regional science, and (3) an increase in disciplinary breadth. Each facet becomes apparent in the detailed evolution of our program objectives (Appendix 14.4).

Intersite Science

A challenge faced by each LTER site was to find an appropriate balance between site-focused and intersite science. Although the National Science Foundation had

Table 14.1. Chart of the evolution of the major objectives from each North Temperate Lakes LTER proposal, 1980–2002.*

Objectives	1980–85	1986–90	1991–96	1996–2002
Long-Term Change	X	X	X	X
Internal-External Processes	X	X	X	X
Disturbance & Stress	X	X	X	X
Time & Space Interaction		X	X	
Concept Development		X		
Regional Context			X	X
Organization of Lake Districts				X
Terrestrial & Social Science				X

*The list and wording of more detailed objectives from each proposal are in Appendix 14.4.

Figure 14.7. Several LTER site leaders at dinner following the founding meeting of the LTER network in October 1980 in Washington, D.C. Full faces, from left to right: Nelson Caine (Niwot Ridge), Carl J. Bowser (North Temperate Lakes), Pat Webber (Niwot Ridge), Dennis M. Allen (North Inlet Estuary), G. Richard Marzolf (Konza Prairie), John J. Magnuson (North Temperate Lakes). (G. R. Marzolf)

intended that the individual LTER sites would be part of a coordinated network, each of the sites competed on its individual merits to become a LTER site. Therefore, most site presentations in the early competitions focused primarily on local site issues and questions. Indeed, the six sites funded in the first round of competition in 1980 were reminded that they constituted a network in October 1980, soon after the arrival of their awards (15 October 1980, minutes of the Organizational Meeting of the LTER Directorate at NSF, Limnology Archives at UW-Madison). The ultimatum received from NSF to cooperate and be a network was a source of tension at later intersite meetings and was intensified when an additional five sites were added to the interactions in 1982. The issue of promoting and delivering intersite science has continued into the 2000s as indicated in the 10- and the 20-year reviews of the LTER program at NSF (Risser and Lubchenco et al. 1993, Harris and Krishtalka et al. 2002).

Our 1980 founding meeting for the network (Fig. 14.7) was convened by James T. (Tom) Callahan, from the National Science Foundation (see Memorial pages v–vi). Richard (Dick) Marzolf, from Konza Prairie, was chosen to chair the "LTER Network Directorate" with Richard H. Waring, from Andrews Forest, as vice chair; the first meeting was set up for December 1980 at Konza Prairie to initiate a proposal to NSF for funding to support the activities of the network, especially study groups in subject areas such as meteorology/atmospheric chemistry, analytical chemistry, hydrology, below-ground biology, archiving, primary production, consumers, and data management. NSF mandated annual reports from the network. All sites were asked to send a list of participants with addresses and phone numbers (no email yet) to Dick Marzolf. Colleagues (Fig. 14.8) who led the LTER network as chairs of the LTER Coordinating Committee during the 1980s and 1990s were Richard Marzolf, from 1980 to 1983; Jerry E. Franklin, from 1984 to 1994;

Figure 14.8. Three LTER investigators chaired the LTER Network Coordinating Committee. From left to right: G. Richard Marzolf, 1980–83 (Konza, 1981, J. Magnuson); Jerry E. Franklin, 1984–94 (Olympic Peninsula, 1991, C. Bowser); and James R. Gosz, 1994–present (Sevilleta, 1998, Sevilleta LTER program).

and James R. Gosz, from 1994 to the present. Robert B. Waide has served as executive director of the LTER Network Office from 1997 to present.

The LTER network was launched with Tom Callahan pushing and guiding the early voyage with a thoughtful essay or carrot entitled "The Science of Long-Term Ecological Research LTER" and a set of expectations or stick entitled "Guidelines for Evaluating Progress of LTER Projects—Year One" (October 1980, documents, Center for Limnology Archives at UW-Madison). Tom had great expectations for LTER network science, as revealed in the ending paragraph of his essay or charge.

> The LTER network of research projects . . . and the Foundation are entering into an experiment. The results of this experiment can promote an advance in ecosystem science which will cause the field to change from a largely descriptive discipline to a predictive science.

The importance that NSF placed on intersite science was evident throughout the first several years of the LTER program but became crystal clear in November 1986 when John L. Brooks (Director, Division of Environmental Biology at NSF) (see Memorial pages v–vi) attended an LTER Network Coordinating Committee meeting in Denver (8–9 November 1986, minutes of LTER Coordinating Committee in Denver, Limnology Archives at UW-Madison). He stated that the opportunities for new science lay in LTER network science, not just in the individual site science, and that expectations for network science would be part of NSF evaluations of LTER. His comments were a pep talk on the one hand but on the other hand expressed the concern that by 1990 NSF would have spent $15 million on 11 sites and would be asking, "What have we gotten from it?" "By 1990 we need to incorporate greater time spans and demonstrate comparative capability. . . . In 1990 the whole program needs to be assessed to see if it is worthwhile and we are getting unique results that could not have been obtained in another way." He implied rather directly that if we did not do network level science, we would revert to individual ecosystem sites as far as NSF was concerned. To foster intersite LTER science, NSF was willing to commit $60,000 each year to the Network Coordinating Committee to conduct research at a network level.

At our North Temperate Lakes LTER site, the Brooks message catalyzed the cross-site comparative studies reported in Chapter 6. One challenge that intersite science posed for our lake-oriented site was that we were one of only a few sites with an aquatic focus and the only site at the time with a lake focus. This meant that for us to participate in meaningful intersite comparisons, the comparisons would have to deal with general properties of ecological systems, rather than specific properties of systems such as forests, deserts, grasslands, or estuaries. Our response to this challenge is documented more thoroughly in Chapter 6.

At about this time, we were developing ideas for site-based science that provided a basis for our first intersite project. We added the objective to study interaction between spatial heterogeneity and temporal variability, as well as the intent to develop concepts related to lake ecosystems and landscape ecology (Appendix 14.4). In one early approach, we contrasted the degree to which the variance in a series of limnological parameters was partitioned into variability among lakes, among years, or some interaction between lakes and years (Chapter 6). We related differences in interlake and interyear variability to life history characteristics of planktonic organisms. For example, variability in abundance of the short-lived rotifers was associated more with years than was the variability of the somewhat longer-lived copepods and cladocerans (Kratz et al. 1987a). Interestingly, here we used data from the Birge and Juday era to obtain a sufficient number of years and lakes for the analyses; at that time we had collected about five years of LTER data.

A logical and exciting extension of this work was to ask similar, but generalized, intersite questions. For example, what types of parameters showed the most overall variability? How did variability among years compare in magnitude with variability among locations within LTER sites? Did some types of ecological variables exhibit strong variability among years and others among locations within an LTER site? Were these patterns similar for different types of systems such as lakes and streams and forests and grasslands? We viewed variability as an ecological attribute, an attribute that system variables at all sites had regardless of biome type (Chapter 6).

To explore these ideas, we convened a meeting of representatives from 12 existing LTER sites at the Trout Lake Station in April 1988, with funding from an LTER network grant. In the year leading up to the workshop, we collated data from each site. Some sites transferred their data via email, some via diskettes sent through the U.S. mail, and some via diskettes hand-carried from site visits made by John J. Magnuson. We found that by visiting many of the sites ahead of time, we could explain the rationale of the project, discuss the available data with local site representatives, and encourage participation by interested individuals. This activity preceded by more than 10 years the high degree of online data availability that existed in the network by 2000. At the workshop, master data sets containing among-year and among-location variability for a total of 448 variables from 12 LTER sites were available on desktop computers. Compiling these data and computing the variability was a nontrivial task that took significant staff time. To us, the activity high-lighted the importance of information management systems across the network and of the uneven development that existed at that time.

Results from the workshop are reported in Chapter 6. One illuminating result was that we realized that meaningful intersite science was possible even from disparate sites measuring different variables. The excitement and success of this workshop led directly to several other intersite activities organized by North Temperate Lakes LTER scientists.

Other intersite projects included using Landsat imagery to assess and explain the spatial heterogeneity of greenness indices from regions representing 13 LTER sites (Chapter 6) (Riera et al. 1998); assessing the extent and cause of carbon dioxide supersaturation in aquatic systems (Chapter 10) (Cole et al. 1994), and understanding the importance of landscape position in influencing lake properties and dynamics in lake districts around the northern hemisphere (Chapter 3) (Kratz and Frost 2000). As our intersite research program evolved, more and more non-LTER sites came to be included in our comparative intersite studies.

Regional Science

In writing the initial LTER site proposal, we were faced with two important decisions. The first was whether we should focus our long-term research on intensive study of a single lake or on less intensive study of multiple lakes spanning the diversity of north temperate lakes. We felt that a multiple-lake perspective would allow for a richer set of dynamics and interactions to be discovered. Thus, we had the beginning of a regional perspective in our first proposal. The second decision was whether we should center the LTER program on the Madison area lakes or on the lakes in the Northern Highlands Lake District, where the Trout Lake Station is located. Both areas had strong attributes for an LTER site. Lakes of the Madison area, especially Lake Mendota and Lake Wingra, had a rich history of study (Frey 1966, Bauman et al. 1974, Loucks and Odum 1978, Adams and Prentki 1982, Brock 1985, Kitchell 1992, Lathrop et al. 1992); both lakes are located conveniently near the Madison campus and its limnological research facilities (Fig. 1.5, bottom). The Northern Highlands lakes near the Trout Lake Station (Fig. 1.5, top) also had a legacy of data from the studies of Birge and Juday (Beckel 1987) and the additional attributes of being plentiful, diverse, and less influenced by human development. In the end, the decision was for the northern lakes, but we continued to look for ways to extend studies to the Madison area lakes. We designated Lake Mendota as a secondary LTER study lake from the very start; for example, the LTER crew sampled fish in the lake each year as part of the core data collection. In a sense, we were prepared for the more regional emphasis that occurred in the LTER network in the early to mid-1990s.

Influenced by our 1988 intersite workshop on temporal and spatial variability, we started thinking about a similar but lake-focused comparative study that would include data from other well-studied lake sites with long-term data. Our goal was to extend and test concepts generated on the Wisconsin LTER lakes to lakes across a range of climatic, geologic, and hydrologic settings and to identify large-scale patterns. In our 1991 proposal, we included an objective to expand our understanding from individual lakes to lake districts to broader regions and, in particular, to the Upper Great Lakes region (Appendix 14.4).

Our successful proposal to the National Science Foundation's call for proposals, entitled "LTER Project Augmentation for Comprehensive Site Histories and Increased Interdisciplinary Breadth," allowed us, in 1994, to initiate studies at larger spatial scales and to bring social scientists into the program. We proposed to regionalize our program by including the Madison area lakes (Fig. 1.6) as primary lakes, thus allowing a contrast of agricultural and urban land use with the forested landscapes of the northern study lakes as well as including eutrophic lakes, a lake type not included in the our northern study lakes. We proposed to increase our level of interaction with two Canadian sites renowned for long-term research, the Ontario Ministry of the Environment's Dorset Research Centre, in eastern Ontario, and the Canadian Department of Fisheries and Oceans's Experimental Lakes Area (ELA), in western Ontario.

The augmentation provided opportunities to extend the spatial scale of analysis to inland lakes of the Upper Great Lakes region (Chapters 1, 4, 5, 7). A workshop in January 1995 at the Trout Lake Station brought together researchers (Fig. 14.9) and data from these two Canadian sites and from the northern and southern Wisconsin LTER sites. These four lake districts allowed comparisons across gradients of climate, geological substrate and till thickness, watershed vegetation, coldwater to warmwater biota, and the intensity and type of human influence.

The availability of satellite remote sensing data provided another source of information for expanding analyses to the Upper Great Lakes region. Wynne (1995) used Landsat data to generate estimates of lake ice freeze and thaw dates for lakes over large parts of the Laurentian Shield and to analyze relations to explanatory variables and spatial patterns of interyear coherence (Chapters 5, 7).

Ice phenology data exist for lakes and rivers around the Northern Hemisphere and in some cases provide records extending back 150 years or more. Funding opportunities from NSF for cross-site research through its 1994 LTER Special Competition for Cross-Site/International Research were used to bring 25 scientists from seven countries to Trout Lake Station in October 1996 for a workshop on patterns and trends in lake and river ice phenology. Data contributed by participants included 689 lakes and rivers for which 27 time series were longer than 100 years. These data formed the basis for a series of papers examining trends and dynamics and their relation to large-scale drivers of climate change and variability (Chapter 7). This extensive database now resides at the National Oceanic and Atmospheric Administration's National Snow and Ice Data Center within the NSIDC Data Catalog (http://www.nsidc.org; search for lake ice).

Funding from the Cross-Site/International Research program allowed us to test the applicability of the landscape position concept (Chapter 3) to other lake districts in the Northern Hemisphere. Twenty scientists, representing 13 lake districts from the Antarctic, Canada, the Czech Republic, England, Ireland, Russia, and the United States, attended a workshop held at the Trout Lake Station in October 1997. Analyses from the workshop were published in a special issue of Freshwater Biology in 2000.

Figure 14.9. Several Canadian participants attended the North Temperate Lakes LTER workshop to stimulate collaborative research and synthesis with lake research sites in Ontario held at the Trout Lake Station in January 1998. Top: Everett J. Fee, Robert E. Hecky, and Susan E. M. Kasian, from the Experimental Lakes Area, Canadian Department of Fisheries and Oceans. Bottom: Peter J. Dillon, Martyn N. Futter, and Keith M. Somers, from the Dorset Environmental Science Centre, Ontario Ministry of the Environment. (J. Magnuson)

Disciplinary Breadth

At the same time that our program was considering ways to increase spatial extent, we recognized that human activities are a dominant force both for the Madison lakes (Chapter 12) and for the northern lakes (Chapter 11). Any comprehensive understanding of dynamics would require analyses of the complex interrelationships among humans, landscapes, and lakes. We needed to include sociological and economic researchers in the North Temperate Lakes LTER program. NSF's augmentation in 1994 provided us that mechanism. Consistent with the work proposed in our augmentation, we included a new objective in our 1996 renewal proposal: to "understand the way human, hydrologic, and biogeochemical processes interact within the terrestrial landscape to affect lakes and the way lakes, in turn, influence these interactions" (Table 14.1, Appendix 14.4).

Scientists in the Rural Sociology Department had an ongoing graduate training program, called Science, Technology, Agriculture, Resources, and the Environment (STARE), that focused on interdisciplinary studies starting from a sociology disciplinary base. In November 1994, we held a joint meeting to discuss possible partnerships in an augmented LTER program that would include the social sciences. This meeting produced both mutual excitement and uncertainty in the participants and was held in an atmosphere of courtship rather than of comfortable familiarity. Several graduate-level seminars further developed ideas and ways to integrate social and natural sciences into a cohesive program both at the southern and the northern sites. The results of these initial explorations were fruitful for both the northern (Chapter 11) and the southern (Chapter 12) Wisconsin LTER sites.

The social sciences were not the only area in which we needed to expand our breadth. A common theme of NSF site reviews in the 1990s was for us to include more microbial study. Tom Brock, a microbial ecologist, was one of the initial investigators and participated in the 1980s. Eric Triplett, in the Agronomy Department, initiated discussions and research with us that led to the funding of a North Temperate Lakes Microbial Observatory in 2000 under his leadership. Microbial diversity in aquatic systems became an active component of our LTER program.

Summary of Evolution of Objectives

The North Temperate Lakes LTER program evolved to include new objectives but retained the perception and understanding of ecological change and dynamics as its core goal. The number of investigators in the North Temperate Lakes LTER increased, as did the number of departments and organizations represented. The number of investigators listed in the North Temperate Lakes proposals increased from 7 in the 1981 proposal, to 12 in the 1986 proposal, to 21 in the 1996 proposal, to 25 in the 2002 proposal (Appendix 14.1). Five of the original seven investigators were still in the 1996 proposal. The disciplinary mix has expanded from physical, chemical, and biological limnology to include remote sensing, forest ecology, rural sociology, and economics. In the next section we discuss the evolution of the products of the North Temperate Lakes LTER with respect to the nature of our publications and the involvement of graduate students from 1981 to 2000.

Evolution of Research Products and Graduate Students

From the beginning, we recognized that long-term ecological research was a new way to do science, and we were excited by it. Our first Scientific Advisory Committee, and Tom Callahan and John L. Brooks at NSF, were challenging us to expand creatively the temporal and spatial scales of ecological science (see earlier discussion). Yet, in year one, 1981, we were an LTER site with only an archived set of Birge and Juday era data on paper for the years from 1925 through 1941 and our 1980 data on a heterogeneous set of seven lakes near the Trout Lake Station. The 1980 measurements were made possible by a supportive startup grant from the

University of Wisconsin Graduate School and included most, but not all, of the lakes chosen to be the primary LTER study lakes.

So what does one do at this juncture in terms of initiating scientific publications and topics for graduate student theses? Starting conditions were a constraint. An early step was to evaluate the spatial and temporal extent and the quality of the Birge and Juday data with partners in the Wisconsin Department of Natural Resources (Lehner 1980). Another step was to initiate our measurement system and to get graduate students working on various short-term process studies appropriate for theses. Our analysis of the nature of the publications reveals the evolution that occurred over the first 20 years, assessed in successive 5-year intervals, or pentads.

Publications

Over the first two decades at our site, the rate of publication increased (Fig. 14.10, top) and the nature of the authorship changed to include more authors per paper and a greater percentage of authors from outside the University of Wisconsin (Table 14.2). In the first five years, we averaged only seven papers per year; this average increased to 22 per year in the second pentad and to 34 and 31 per year in the last two pentads. Lower initial publication rates resulted from constraints of startup with respect to data, participants, funding, and the demands of simply setting up the site. The number of participants, funding, and the volume and availability of the data increased from 1980 to 2000. A lesson to be learned from this apparent slow start

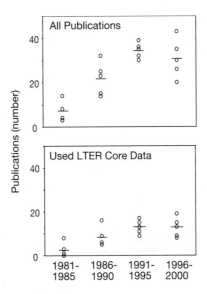

Figure 14.10. Total numbers of publications per year (top) and those using LTER core data (bottom) grouped by pentad from 1981 through 2000. Each time period has five data points (one per year) and the average of these.

Table 14.2. Number of publications by the North Temperate Lakes LTER program by pentad, 1981–2000, by type of authorship.*

Number (%) of Publications	1981–85	1986–90	1991–95	1996–2000
Single Author	16 (43)	39 (36)	45 (26)	30 (19)
Two-Three Authors	18 (49)	50 (46)	90 (52)	72 (47)
Four or More Authors	3 (8)	20 (18)	37 (22)	52 (34)
Total Publications	37	109	172	154
Non-UW Authors	6 (16)	35 (32)	76 (44)	76 (49)

*Number (percent) of papers with different numbers of authors and non-University of Wisconsin-Madison authors.

Note: The first three rows in each column add to 100%.

is that low productivity in early years of a new LTER site may not predict the future productivity or quality of scientific publications. The increase in publications was accompanied by increases in the number of single-authored, two- to three-authored, and greater-than-four-authored papers. However, the percentage of single-authored papers declined from 43% to 19% of the total output, while the percentage of greater-than-four-authored papers increased from 8% to 34%. These changes represent a general trend in publication practices in ecology toward multiauthored papers, as well as an increase in interdisciplinary papers. The increasing percentage of authors from outside the University, from 16% to 49%, resulted from the increase in intersite papers and from the participation of scientists from the U.S. Geological Survey, the Wisconsin Department of Natural Resources, the U.S. Environmental Protection Agency, and other agencies.

In addition to the increase in total publications (Fig. 14.10, top), the number of papers using core LTER data increased from none in the first years of the first pentad to about 14 papers per year in the last two pentads, or almost half of all papers published (Fig. 14.10, bottom). The low initial numbers clearly reflect the short length of time-series data on the LTER lakes and the difficulty of accessing the Birge and Juday historic data. Core data began to be an important resource for publication after about 10 years.

Consistent with the increase in use of core data were significant increases in the number of years and lakes included in single papers (Fig. 14.11). These increases, as well as increases in the geographic extent of studies, reflect the evolution of our objectives (Table 14.1) to take a regional approach and to extend our perspective to global phenomena in some of our comparative studies. The number of lakes and the length of time series included in a single paper began to increase after 10 years. This trend became more pronounced in the last pentad. The number of years averaged about 20 per paper and the number of lakes about 30 per paper in the last pentad. The variation was large; papers in each pentad included some with one or a few years or lakes included in the analyses. One paper extended analysis to more than 500 years, with data from about 600 lakes. These authors were reaching beyond the LTER lakes and time span and in some cases used paleolimnological data. The

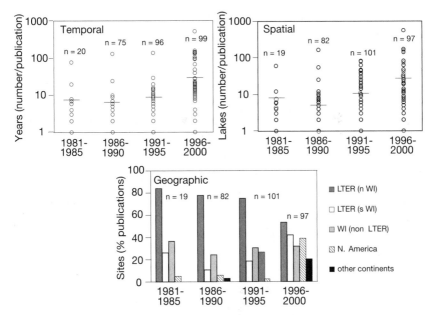

Figure 14.11. Changes in the temporal and spatial extent of data used in research papers published by the North Temperate Lakes LTER program by pentad from 1981 through 2000. Top left: Number of years per paper. Top right: Number of lakes per paper. Bottom: Geographic extent within papers. Notes: Top: Each point represents one paper, numbers above each pentad are the number of papers, and the horizontal bar is the average. Not all papers are included; for example we excluded review papers, data management papers, and papers describing the program. Bottom: Percentages of papers that include the northern Wisconsin LTER site, the southern Wisconsin LTER site, other Wisconsin lakes, other North American lakes, and lakes in other continents. The same paper may appear in more than one category (e.g., if the paper included the northern LTER lakes and lakes from other continents).

frequent occurrence of papers with just less than 20 years of data during the last pentad demonstrates the use of the North Temperate Lakes LTER time series, which was approaching 20 years. Similarly, the frequency of lakes just below 10 lakes suggests considerable use of the seven northern primary lakes plus the reference and treatment basins of the secondary lake, Little Rock Lake, discussed in Chapter 9. The percentage of papers examining lakes at the northern Wisconsin LTER site declined, while the percentage that included lakes at the southern Wisconsin LTER site, or outside of Wisconsin in North America, or on other continents increased as objectives expanded the temporal and spatial scale of our study. The increase in analyses at a global scale is apparent as our intersite workshops and resulting publications included international participation from sites and colleagues outside of North America. Few of these international colleagues or sites were part of the international LTER initiatives of NSF.

The changing objectives and funding opportunities, along with the arrival and retirement of key faculty participants, led to evolution in the topical aspects of

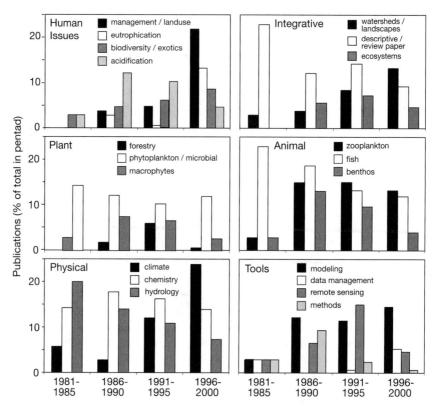

Figure 14.12. Changes in the topic of the papers from the North Temperate Lakes LTER program by pentad from 1981 through 2000. Twenty categories of paper were placed into six groups (human, integrative, plant, animal, physical, and tools) with a figure panel for each group.

our publications (Fig. 14.12). For a coarse resolution, we classified papers into the following categories: human, integrative, plant, animal, physical, and tool topics. Human-related papers and tool-related papers increased over the 20 years. Additional revealing information on how we evolved is apparent in the finer resolution of subcategories. For example, in the human category, the percentage of papers related to lake management and eutrophication increased markedly, and those on biodiversity and exotic organisms increased steadily. In contrast, the percentage of papers concerning acidification of lakes peaked in the second and third pentads, when funding of our Little Rock Lake experimental acidification was at high levels (Chapter 9). Similarly, in the tools category, changes in our use and development of modeling, data management, remote sensing, and method development occurred. Data management received special attention in the first pentad and again in the last pentad, when LTER information managers provided perspectives and training on information management to the broader ecological research community. Method papers peaked in the second pentad, as we explained the methods for our core measurements. Modeling papers have remained high since the second pentad, but their objects have changed. Remote sensing papers peaked in the third pentad, following the first use of these techniques in the first pentad. Other changes are apparent. In the integrative category (Fig. 14.12), the percentage of descriptive review papers of the program or site has decreased, while the percentage of papers increased for the ecosystem and the watershed/landscape levels.

The changes in the plant and animal categories were highly responsive to changes in our principal investigator pool. For example, the percentage of studies on forestry peaked with the brief participation of Tom Gower, and macrophyte studies declined as Mike Adams became less active in the program. Papers on phytoplankton/microbial have remained rather steady since the second pentad through the contributions of Tom Frost and the Little Rock Lake acidification experiment. Their contribution was replaced in part by papers on Lake Mendota and by the Microbial Observatory in the last pentad. The percentage of studies on fishes has declined slowly to percentage levels comparable to those for zooplankton, which have remained high since the second pentad. The high percentages of papers on fishes early in the program were associated with the productivity of two graduate students, John Lyons and Lars Rudstam, and the percentages declined with the increasing diversity of new participants later in the program. Benthic invertebrate studies peaked in the second and third pentad with the growth and decline in the Little Rock Lake acidification experiment.

In the physical category, the percentages of hydrologic papers steadily declined, climate-related papers increased, and papers on chemical topics have remained stable. Several of the original graduate students in hydrology, David Krabbenhoft and Galen Kenoyer, were especially productive, and the percentage of hydrological papers was later decreased by the increasing diversity of the total pool of investigators.

Graduate Student Evolution

Eighty-one students (Appendix 14.5) received one or more graduate degrees between 1983 and 2003 after working with the LTER or with LTER-associated programs such as the Little Rock Lake acidification experiment and the U.S. Geological Survey's Water, Energy, and Biogeochemical Budgets (WEBB). Forty-nine M.S. and 48 Ph.D. degrees were awarded from five universities. Ten postdoctoral persons participated.

The experiences of the graduate students and postdoctoral participants form part of the project's history (Fig. 14.13). We asked a subset of graduate student and postdoctoral participants to respond to a survey about their LTER experiences to characterize the research and attitudes of graduate students during evolution of the program. We divided the 24 respondents into two groups representing the early decade and the later decade. Each was asked to rank 19 types of research from 1 (high) to 3 (low) in terms of its importance to their work with us. We calculated the change in mean rank given to each type of research by subtracting the mean rank for the first decade from the mean rank for the second decade.

Figure 14.13. Photos of graduate students and postdoctoral persons; many others are shown in Figure 14.6, and all are listed in Appendix 14.5. Top, left to right: David B. Lewis, Thomas R. Hrabik, Katherine E. Webster, and Patricia A. Soranno. (Lake Mendota Laboratory, late winter 1996, J. Magnuson) Middle, left to right: Shelley E. Arnott at the workshop with Canadians (Figure 14.9) (winter 1998, J. Magnuson); Harry L. Boston (Trout Lake Station, summer 1982, J. Magnuson); and María J. González (Little Rock Lake, summer 1987, T. Frost). Bottom, left to right: James P. Hurley (Little Rock Lake, about 1990, T. Hoffman); Galen J. Kenoyer (Trout Lake site, summer 1981, C. Bowser); and Liao Guozhang, from the Pearl River Fisheries Research Institute, Guangzhou City, Peoples Republic of China (Trout Lake Station, 1982).

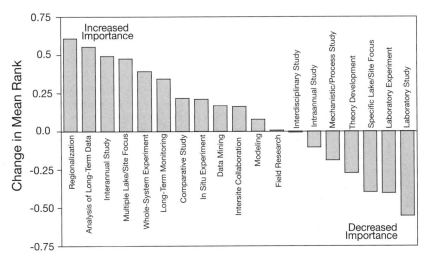

Figure 14.14. Differences in the attitudes of graduate students during the first decade and the second decade of the North Temperate Lakes LTER program. The y-axis is the difference in mean rank between the two groups. Note: The early decade includes 10 graduate students who graduated between 1983 and 1993, plus one postdoctoral person; the later decade includes 10 graduate students who graduated between 1996 and 2001, plus three postdoctoral persons.

Differences in attitude suggested that students from the second decade reflected the vision of LTER types of research more than did students from the first decade (Fig. 14.14). Regionalization, analysis of long-term data, interannual study, multiple lake/site focus, whole-system experiment, and long-term monitoring ranked higher in the second decade. Laboratory study, laboratory experiment, specific lake/site focus, and theory development ranked higher in the first decade. Interestingly, field research, interdisciplinary research, intraannual study, modeling, intersite collaboration, data mining, and process study ranked similarly in both decades. The type of research the students were conducting evolved over the 20 years, while retaining several core elements.

Collectively the two groups of students ranked interdisciplinary, long-term monitoring, and field research the highest (Fig. 14.15). Only slightly below these, came interannual study, analysis of long-term data, comparative study, mechanistic/process study, multiple lake/site focus, theory development, and intraannual study. Progressively lower overall ranks were given to modeling, data mining, specific lake/site focus, regionalization, laboratory study, in situ experiment, whole-system experiment, intersite collaboration, and laboratory experiment. Students in both decades held some common opinions about science but had not yet given high priority to some of the directions that our LTER research program was moving toward, such as regionalization and intersite collaboration. Regionalization ranked higher in the second decade. But, even with that change, the rank for regionalization

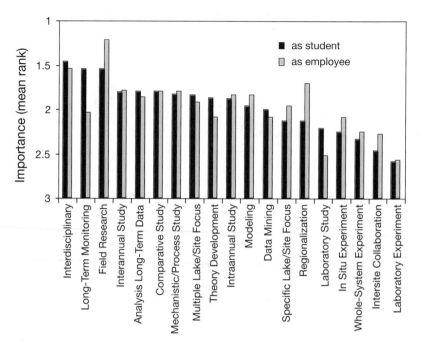

Figure 14.15. Rankings of the importance of 19 topics by 24 graduate students or postdoctoral persons associated with the North Temperate Lakes LTER program and during their present employment. The survey was answered in 2001.

was low. Intersite collaboration seemed unimportant to the students, and the change between decades was small.

The survey suggests that the type of research conducted with the North Temperate Lakes LTER program continued to be important in the professional careers of these students and postdoctoral participants (Fig. 14.16). The mean importance of a type of research to each person as a LTER student or postdoctorate was strongly related to that person's importance ranking of that type of research in his or her professional position. The students and post doctorates in this survey received degrees or worked with nine faculty members in nine departments or programs. Their professional positions at the time of the survey were dominated by teaching and research, with research positions more common for those in the first decade and teaching positions for those in the second decade.

More decades will come, and education is important even before college and graduate programs. North Temperate Lakes LTER has been providing educational experiences for young students at the Laboratory of Limnology on Lake Mendota and at Trout Lake Station (Fig. 14.17). Many graduate students and personnel associated with the LTER program are involved in the educational experiences for these young people. These grade school students learn from shipboard and winter study of limnology. These experiences occur annually and began in 1999 at Trout Lake and in 1989 on Lake Mendota.

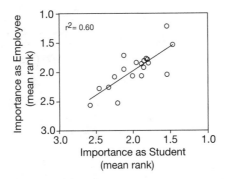

Figure 14.16. Relation between the importance rankings of 19 topics by participants when they were students associated with the North Temperate Lakes LTER program and in their employment in 2001.

Summary of Evolution of Publications and Students

On a personal level, we are pleased to confirm in these publication and student analyses the evolution to a new kind of ecological science. The dynamics and character of our publications and our graduate students over the first two decades of study are rather striking. The results indicate that we have indeed become a functioning long-term, site-based research group, extensively using the time series of core data available on our suites of lakes in northern and southern Wisconsin. We have expanded the interdisciplinary breadth of our group. We have taken on the challenge of regional and global science. We have reached out to and welcomed colleagues in other agencies, universities, nations, and continents. We have flourished despite turnover in graduate students and colleagues. We think that John Brooks and Tom Callahan would be pleased to see the increase in the number of years, the broader spatial extent, and the synthesis included in the analyses during the 1990s.

Challenges for the Future

Several dynamic tensions must be balanced in an LTER program. We comment next on four of these tensions that we believe are both general and important.

A first set of tensions relates to the need to maintain the old while embracing the new. These tensions include the need to collect and manage high-quality, consistent, long-term data versus the need to study new processes and phenomena; the need for consistent measurements versus the need to integrate new technologies; the need for intellectual commitment of new scientists with new ideas versus the need for a long-term perspective first articulated by those perhaps no longer with the program; the need to continually synthesize site and intersite science while maintaining cutting-edge research in our specialties. If these tensions are not balanced properly, an LTER program will either become stagnant by not incorporat-

Figure 14.17. Education and the next generation. Many graduate students and personnel associated with the North Temperate Lakes LTER program enriched the education of
younger students by teaching in the College for Kids program for 10- and 11-year-olds
organized on the Madison campus by the School of Education; and in the Schoolyard LTER
for third- to seventh-grade students organized at the Trout Lake Station by the Center for
Biology Education. Top: Students on board the RV Limnos operated by the Center for
Limnology; on the left, Captain David P. Harring waits to take the group out, (Lake
Mendota, summer 2001, G. Wagner) center and, on the right, students being helped with
their limnological sampling by Anthony C. Yannarell. (Lake Mendota, summer 2003,
G. Wagner) Middle: A beach seine is used by Karen A. Wilson and David B. Lewis to
collect fishes for the College for Kids class. (Lake Mendota, summer 1994, J. Miller, Uni-
versity of Wisconsin-Madison University Communications) Bottom: In the Schoolyard
LTER, students sample through the ice and learn about winter limnology from Tim Kratz.
(Trout Lake, about 2000, M. Woodford)

ing new ideas or will lose its long-term focus by not maintaining long-term research grounded in long-term core data.

A second set of tensions involves the balance between project manageability and growth. As our LTER program grew in numbers of investigators, disciplines, and spatial extent, organizational complexity increased. When our NTL program first started with six principal investigators, we considered ourselves to be interdisciplinary because we studied the interrelationships among hydrology, lake chemistry, and lake biota. In the 2002–2008 proposal, which included 25 principal investigators, our disciplinary breadth included physical, chemical, and biological limnology plus sociology, economics, remote sensing, and terrestrial ecology. The trend toward expanded interdisciplinary studies likely will continue, and the set of represented disciplines will continue to increase. What disciplines will be included in our LTER program 10 and 20 years from now? Will participating investigators all be from the sciences? If so, what sciences? Will disciplines from the arts and letters be included, as well? Growth in size, disciplinary breadth, and organizational complexity will continue to challenge us to create an environment that maximizes the potential for creativity and integration within and across our program and the entire LTER network.

A third set of tensions involves the issues of retaining the continuity and integrity of long-term research. Clearly, acquiring the first 10 years of data was important, allowing us to address ecological phenomena from a multipleyear perspective. Will the third decade of data be as important? How about the sixth or the fifteen decade? One would certainly not argue that we should stop collecting meteorological data because we already have 100 years of daily data from the Trout Lake area. But are the reasons for doing long-term research that were articulated at the beginning of the LTER program, and listed at the start of this chapter, still valid as we enter our third decade? We think the rationale is more compelling than ever. Understanding long-term ecological phenomena, having the appropriate long-term context for shorter-term observations, distinguishing among trends and cycles in long-term phenomena, having a coordinated network of sites to allow greater generality and synthesis of results, maintaining study areas for long-term research, and having comprehensive measurements at multiple levels of organization and multiple temporal and spatial scales will be as important in the future as when LTER was just beginning.

A fourth set of tensions involves the need to continue to synthesize even as we make more site- or system-specific advances to science. Synthesis in LTER comes in many forms. For our program, some of these are development of general concepts; analyses or syntheses among years and among lakes or both; interdisciplinary analyses or syntheses between scientific disciplines, including those between the natural and the social sciences; analyses and syntheses between the fundamental findings and their applications; analyses or syntheses among lake systems across the LTER network and other lake sites around the world; analyses and syntheses among disparate LTER ecological systems, such as deserts and lakes; and finally, even the preparation of this book. As the longer time series became available, we soon found ourselves analyzing the time series and combining that analysis with among-lakes analyses. As we did these analyses, we were stimulated to seek new concepts and develop existing ones. Our site has been committed to synthesis at all of these levels, and the earlier analysis of publications evidences that fact. The only major conflict between these synthetic activities and more disciplinary, site-specific analyses is one of allocation of time and funding.

Meeting the challenges of balancing old versus new, manageability versus growth, maintaining the continuity and integrity of long-term research, and ongoing synthesis will continue to require flexible and creative leadership and reliance on the creativity and generosity of the participating scientists. We look forward to the next 20 years of research at the North Temperate Lakes LTER site.

Summary

This chapter on the origin, operation, history, evolution, and challenges of the North Temperate Lakes LTER describes and discusses the response of one group to the challenges of doing long-term regional ecology. The founding group of investigators was interdisciplinary in the sense of physical, chemical, and biological limnology and oceanography. We designed a program of common goals and objectives with a philosophy of shared resources and management both of data collection and management. Early advisory committees and later site visit teams continued to challenge and help shape our evolving program. Principal investigators and their students related in an interactive community with frequent group meetings. Initially seven lakes in northern Wisconsin were chosen; later, four lakes from southern Wisconsin were added. The initial major objectives that tied the program together were to perceive long-term change, to understand interactions of physical, chemical, and biological processes, and to understand the role of disturbance. Evolution of the program added objectives of which those associated with intersite science, regional science, and incorporation of the social sciences were key. We began with a primordial vision of an LTER and evolved into a long-term regional ecology program over the first 20 years, as evidenced by the change in the nature of our publications and the views of our students.

As in the beginning, significant challenges face an LTER site. Those most apparent to us are how best to balance the need to maintain the old while embracing the new in terms of science, measurements, and technologies; how to balance project manageability and program growth in terms of numbers of investigators and disciplines and spatial extent of the research; how to retain the continuity and integrity of truly long-term study; and how to synthesize at many levels across time, space, disciplines, and ecosystems. These are welcome challenges; we look back with a certain pleasure and forward with commitment and excitement.

Part IV

Summary and Synthesis

15

Synthetic Summary
and Prospective

Barbara J. Benson
Timothy K. Kratz
John J. Magnuson

> Life is no brief candle to me. It is a sort of splendid torch which I have got a hold of for the moment, and I want to make it burn as brightly as possible before handing it on to future generations.
> —George Bernard Shaw (Henderson 1911, p. 512)

We are in the third decade of the North Temperate Lakes LTER program. The challenge LTER researchers undertook was to understand the dynamics of temperate lakes in a landscape context and over time scales ranging from years to decades or longer. We used the idea of the invisible present and the invisible place (Chapter 1) to visualize why broader spatial extents and longer time series were essential to understanding lake ecosystems. A point in time and space, out of context with its past and its surroundings, is destined to be misunderstood and misinterpreted. In many ways, we were attempting to do a new kind of science. From the vantage of more than two decades of LTER research, we summarize the gain in scientific understanding from the North Temperate Lakes LTER program, our contributions to a changing infrastructure for ecological science, and our reflections on the future.

Scientific Contributions

At least eight scientific themes permeate chapters of this synthesis volume on the North Temperate Lakes LTER program. The hydrologic template in which lakes are embedded provides a powerful framework for studying lake status and dynamics. External drivers are the genesis of many interannual and interdecadal dynamics in lakes. Lake features and internal lake processes filter the expression of external drivers on lake dynamics and alter and complicate these dynamics especially for lake chemistry and biology. Humans value lakes as part of their sense of place, and their interactions with lakes include not only the stresses they impose on lakes but also their responses to the lake status or changes they perceive. Generality can be achieved through ecological comparisons and analogs even on dissimilar systems. Temporal dynamics and spatial heterogeneity are important and interrelated, forming a template for a rich set of dynamic lake-landscape interactions. Variability is an informative ecological property. The landscape ecology of lakes is complementary to the ecosystem ecology of lakes, but it is a different perspective that brings new realizations to contribute to a predictive understanding of lake ecology. These themes are evident in the summary that follows.

The LTER study lakes in northern Wisconsin were selected to be within the same groundwater flow system, the Trout Lake basin. Within this flow system, the percentages of groundwater in the lakes' water budgets differ systematically in character and dynamics from lakes high in the flow system to those low in the flow system (Chapter 2, 3). At the outset, we had thought that groundwater would behave as a steady state system and would not have the dynamics of surface water systems like streams and rivers. We found instead that groundwater contributions to a lake's water budget were dynamic on seasonal, interannual, and interdecadal time scales, with important consequences. In some upland lakes, flows alternated between groundwater recharge and discharge in the same year. History becomes important because the groundwater entering a lake originated from rain and snow that occurred in the past, introducing time lags from a few years to several hundred years. Water levels and chemical constituents of upland lakes with shorter flowpaths were more responsive to drought than they were in lower lakes with longer flow paths. In some upland lakes with little groundwater input, drought reduced groundwater inflow sufficiently to reduce chemical buffering, thus making the lake more responsive to acidic deposition. Even a small amount of groundwater was sufficient to buffer a seemingly sensitive lake from acidic deposition.

The influence of a lake's position in the groundwater flow system on the physical, chemical, biological, and social features of lakes is pervasive (Chapter 3). The landscape position concept is powerful, in part, because the position of a lake in the Northern Highlands landscape is a 10,000-year-old legacy of the receding glaciers. Thus, the influence of landscape position on lake status and dynamics is a remarkably stable feature. We gave lakes a numerical position in the landscape based on their position in the groundwater flow system, much like the stream orders in the stream continuum concept. This simple framework explains a surprising portion of the differences in status and dynamics among lakes. With regard to status, lakes high in the landscape are smaller in area, have more dilute waters, lower fish diversity, and fewer human structures per unit shoreline and differ in many other features. The dynamics of water level and chemical constituents that differ with landscape position were mentioned above. The importance of landscape position emerges in many of the chapters in our book, not as a cause or mechanism, per se, but as a context for a host of ecological and social processes.

Lakes in a biogeographic sense are like islands where colonization and extinction contribute to species richness and dynamics of small, relatively isolated ecosystems (Chapter 4). One might expect relatively stable species composition in insular ecosystems. However, the species structure and richness of the phytoplankton, zooplankton, and fishes, are surprisingly dynamic in ecological time, rather than only in evolutionary time. The frequent arrival and loss of species in the study lakes led to high rates of species turnover, not only in the urban and agricultural lakes in southern Wisconsin but also in the northern Wisconsin LTER lakes and in the Ontario lakes, where human influences would appear to be smaller. As a consequence, species richness estimates increase with the number of years included in the estimates, so that cumulative species richness over a set of years is considerably larger than annual estimates. This increase results, in part, because we have only a sample of the assemblages in each year, but extinction and invasion are real even over the relatively short duration of LTER. In any lake, detecting the arrival or loss of species common to the area is possible only with long-term data. In-lake variables that influence extinction and landscape variables that influence access to the lake are important in determining species structure of fishes in the Northern Highlands. When certain exotic species from other areas are among the new arrivals, the invasion is obvious, and the dynamics take on the character of a disturbance (Chapter 8).

We found it helpful to think of an external driver of lake dynamics as a signal that is filtered by lake features and processes to cause interannual dynamics in lake physical, chemical, and biological variables (Chapter 5). Filtering of the signal by the lake ecosystem can amplify, attenuate, delay, and even extend the in-lake response over the years, owing to the interaction of the external signal, such as average summer air temperature, and the lake ecosystem. Lake variables that respond to external signals with little filtering have coherent or synchronous interannual dynamics across the area, while lake dynamics that are altered strongly by complex internal filters may be incoherent even between adjacent lakes. Not surprisingly, coherence between lakes was greatest for physical variables and least for biological variables. Coherences in dates of ice breakup and summer surface water temperature were high and uniform across the landscape regardless of landscape position. Lake districts with strong surface water connections between lakes had some uniform patterns of coherence across the landscape in water chemistry. Lake districts dominated by groundwater had little coherence in water chemistry high in the landscape where slow groundwater flows dampened interannual variability; they did have some coherence low in the landscape where streams existed. Coherence among many variables such as almost all biological variables was low and unstructured across the landscape. Coherence is strong enough for some variables, even for an occasional biological variable, that predicting the dynamics of a population of lakes from a sample may be possible for those variables.

We began to seek generality and common properties among the disparate set of LTER sites from deserts, to forests, to lakes, to estuaries (Chapter 6). General patterns were apparent across these diverse systems even without a common set of measurements. For comparison, we chose variability as a common property of all ecosystems and explored metrics for cross-ecosystem research. Interannual variability was greater for biological parameters than for chemical parameters, which, in turn, were more variable than physical parameters. The proportion of variance explained by location within an LTER site (usually along an altitudinal gradient) was greater than the proportion explained by interannual dynamics. Thus, spatial variability and extent become apparent as necessary components of studies of temporal dynamics. For some sites, water movement from high to low in the landscape was a common context for understanding spatial patterns in variability. Comparison of disparate systems would appear to be more powerful when the same sensors and properties could be measured. Satellite imagery provided this opportunity. To compare spatial heterogeneity, we used the standard deviation of a vegetation index derived from Landsat images. Topographic relief and land use largely explained the differences in spatial heterogeneity in the vegetation index among landscapes. The greater the relief, the greater was the heterogeneity. For relatively level landscapes with low relief, the greater the percentage of agriculture, the greater was the heterogeneity. For some landscapes such as the North Temperate Lakes LTER in Wisconsin, water was a significant contributor to landscape heterogeneity; thus, removing water from a scene prior to analysis of terrestrial heterogeneity seems shortsighted. These properties of spatial variability were related strongly to the grain of aggregation of the Landsat images.

Interannual and interdecadal variability in climatic variables constitute a strong external signal that drives lake dynamics over large spatial scales (Chapter 7). Air temperatures and solar radiation exhibited some coherent interannual dynamics, with some coherence between northern and southern Wisconsin and Canadian sites in the upper Great Lakes region. Ice-on and ice-off dates similarly exhibited coherent interannual dynamics and had some association with large-scale climatic drivers such as the El Niño/Southern Oscillation and the North Atlantic Oscillation. Most physical variables we analyzed in lakes in relation to lake thermal structure were coherent among the northern Wisconsin LTER lakes, but some, such as bottom water temperatures in summer, were not coherent even between adjacent lakes, owing to the importance of lake specific filters. Interannual dynamics in lake water levels displayed some coherence across Wisconsin and with the Laurentian Great Lakes, but even among the northern Wisconsin LTER lakes, differences in water-level dynamics, associated with landscape position, produced complex landscape patterns of low and high coherences. Several lakes in southern Wisconsin had long-term increases in water level. Ice-on and ice-off records as long as 150 years in duration, including two Wisconsin LTER lakes, provided strong evidence for warming around the Northern Hemisphere. As climate warming continues, we anticipate that ice cover will continue to decline, lakes will warm, and there will be changes in thermal stratification, hydrology, and ecology.

The invasion of several LTER study lakes in northern Wisconsin by two exotic species, rusty crayfish and rainbow smelt, were unintended natural experiments through which we studied the underlying mechanisms for invader success and impacts as revealed by the long-term data (Chapter 8). These exotic invaders were ecological disturbances that caused major transformations of the invaded ecosystems. Declines in abundance and likely extinctions of two native fishes, cisco and yellow perch, were related to smelt predation on the young cisco and competitive interactions between the young of year of yellow perch and smelt. Dramatic losses of macrophyte species and biomass followed the invasion of the rusty crayfish, and other related changes occurred to the lake communities of bottom-living animals. In these relatively small LTER lakes, the impacts of the invasions on native species and communities were revealed over an interdecadal time scale. Other exotics are likely to enter the LTER lakes, with zebra mussels entering the southern lakes most likely early in the 2000s.

Acidic deposition is another strong external driver of lake status and dynamics where biotic interactions strongly filter the acidification signal (Chapter 9). In cooperation with the U.S. Environmental Protection Agency and other institutions, we conducted a whole ecosystem experiment on Little Rock Lake at our northern Wisconsin LTER site. We acidified one half of a two-basin lake divided by a curtain to provide a strong external stimulus to the lake ecosystem in the context of understanding the effects of acidic precipitation on lake ecosystem behavior. Dramatic changes occurred in the physics, chemistry, and biology of the treatment basin. Compensatory dynamics among zooplankton species characterized the responses as some species declined and others became abundant. Surprisingly to us, many lake responses to acidification were driven indirectly through changes in food web dynamics, rather than by direct toxicity. Again, internal lake processes played a large role in how external drivers influenced the interannual dynamics of lake communities.

It becomes increasingly apparent that internal processes in lakes can change the effect of external drivers on lake ecosystems, often with surprising consequences (Chapter 4, 7, 8, 9, 10), and, perhaps more surprisingly, that internal processes can generate and dominate interannual lake dynamics (Chapter 10). Phosphorus cycling through the food web is the product of external loading to the lake, as well as a suite of biogeochemical cycles in the lake, all of which are influenced by the basin morphology, the land use and cover of the watershed, and species present in the lake. The resulting intraannual and interannual dynamics of phosphorus could not be described or understood except as an interactive system of internal and external factors. Biotic communities differ predictably among lakes and form different matrices of interaction and dynamics. Food web relations can exacerbate the influence of limiting nutrients, especially in the dilute waters that characterize many of the northern Wisconsin LTER lakes, as is seen, for example, with sponges (requiring silica) and snails (requiring calcium) being more vulnerable to predation at low nutrient levels. In Crystal Lake, oscillatory dynamics of year-class strengths of yellow perch, the zooplankton and phytoplankton communities, and water clarity were generated from a complex interaction of cannibalism in perch, planktonic food web consumption processes, and perch life histories related to longevity and age of first reproduction. This interannual cyclic behavior with a multiyear period of Crystal Lake physics and biology was largely independent of interannual variability in external drivers, that is, until the exotic smelt (Chapter 8) invaded the lake and drove the perch to near-extinction. The complexity of interactions among organisms and the lake chemistry and physics generates a need for some caution in explaining lake dynamics from external drivers or, more positively, provides an opportunity to understand the interannual dynamics of complex ecosystems as resulting from interactions between external and internal dynamics and processes.

When social scientists joined the team (Chapter 11), the nature and the scope of the LTER research changed. We incorporated new objectives such as understanding the reciprocal interactions between human activities and lake ecosystems (Chapter 11), and we became more aware of the human history and the human dynamics in our systems. Surprising to at least some of us were the rate and distribution of human development in the lake-rich Northern Highlands, where the northern Wisconsin LTER lakes are located. The number of structures has increased dramatically since 1970, and these structures are concentrated along the lakes' shorelines. Most homeowners are state residents, but many are not. Lakes are an important component of people's sense of place about the Northern Highlands, which often is affectionately known as "up north." Surprisingly and inconsistent with prevailing theory, seasonal residents with second homes had a greater sense of attachment to the property and area than did permanent residents. Homeowners often formed a lake management association for collective behavior with respect to their lake. Many features of lakes such as coarse woody habitat, landscape position, and exotic species seem below perception for many shoreline homeowners, and their beliefs were not necessarily consistent with protection of lake ecosystems. Studying the system of beliefs and values by which individuals and groups attribute meaning to lakes is illuminating the relationships between humans and lakes.

Lake Mendota, one of the southern Wisconsin LTER lakes since the North Temperate Lakes LTER program augmentation in 1994, has provided a long-term history of the impacts of human development and activities on lake eutrophication and has been the focus of management activities to improve water clarity over many decades (Chapter 12). Study of the lake brings to bear the joint contributions of social and natural scientists because history, land use, and human perception and actions become critically important to understanding the lake ecosystem and its dynamics. Input of phosphorus from sewage and agriculture began in the late 1800s. Consequently, the lake is excessively eutrophic, at least to the human eye and enjoyment. To judge from their willingness to pay, people put a remarkably high value on cleaning up the lake. Many actions have been taken over the years to reduce nutrient inputs and to manage the consequences of those inputs within the lake. A major recent initiative is a large-scale effort in the watershed to reduce nonpoint phosphorus inputs to Lake Mendota. This management effort constitutes for the LTER research team a major long-term experiment at the whole-lake and watershed levels, as well as a focus for outreach and service to the communities in which we live.

Cultural and Infrastructural Contributions

The cultural contributions of the North Temperate Lakes LTER program include not only the students, publications, and measurement systems that came from the North Temperate Lakes LTER but also the ways we conceptualize ecology, our activities, and our modes of operation. Our LTER program has contributed to changing the view of the greater ecological community on the value of long-term research.

In the beginning of LTER in the early 1980s, it was important for the sites to distinguish LTER measurements from monitoring. Monitoring was viewed as the taking of routine measurements, as opposed to a research-driven activity. The Long-

Term Ecological Research program as initiated by the National Science Foundation played a key part in legitimizing the use of repeated long-term measurements to enable an understanding of ecological processes that play out across the landscape over interannual, interdecadal, and longer time scales. As we believed at the outset and continue to discover, a multitude of important ecological and environmental dynamics occur at these longer time intervals. Signs of the changing view are the fact that the Ecological Society of America established a Long-Term Studies Section and that long-term monitoring is becoming an essential component for the adaptive management of natural systems by government agencies and nations.

During the first 20 years of the North Temperate Lakes LTER, the nature of our research objectives, our publications, and the views of our students evolved (Chapter 14). The objectives expanded to include the social sciences and spatial scales beyond the boundaries of our particular LTER site. In the early 1980s, we viewed ourselves as interdisciplinary because we had a research team that consisted of physical, chemical, and biological limnologists. Now we are learning to do phys-geo-chem-bio-socio research. The publications evolved to include broader spatial and longer temporal analyses, more modeling, and considerations of human-related issues and study. Our graduate students began placing higher values on regionalization, analysis of long-term data, interannual study, and multiple-lake study. The leadership role that will be played by our graduates is evident even in this book; four former graduate students and two former postdoctoral associates led the author teams on 6 of the 15 chapters. Former students and postdocs suggested that their LTER experience has been formative in their research directions in their professional careers.

An important cultural shift occurred in regard to data sharing at our site and across the LTER network (Chapter 13). At the North Temperate Lakes LTER, we now have an open data access policy for our long-term data. Data are published on the site's LTER Web site as soon as they have been entered into the database and passed quality assurance tests. Our experience has been that open data access has stimulated and facilitated collaboration with investigators outside our LTER site.

The infrastructural components of a research program allow the science activities to flourish and maintain themselves. The long-term data we have collected and managed constitute a rich legacy for future generations of researchers. These data provide the context for future research at the North Temperate Lakes LTER site, including thesis research that is truly long term, as well as a source of data to apply to unanticipated future research questions, cross-site synthetic research endeavors, and environmental resource management. Our site, in conjunction with other LTER sites and the LTER Network Office, has generated a knowledge base of best practices for ecological information management. This knowledge has been and will continue to be shared with the broader ecological community through training workshops and publications. The information management practices are designed to ensure that the data legacy is preserved for and interpretable by future generations.

A new technological infrastructure of equipment, software, protocols, and expertise now exists at the North Temperate Lakes LTER (Chapter 13, 14). This infrastructure provides a research platform for our science and profoundly affects the scale of scientific investigations we undertake and the ways in which we collaborate as scientists. Interactions among our researchers, and between researchers and our data, have been facilitated not only across our campuses but across the globe. The LTER information management community has been active in ecoinformatics research and is engaged in developing new technology for data discovery, access, and integration.

The North Temperate Lakes LTER program developed its own administrative framework and organization for interactions (Chapter 14) that provided an infrastructural contribution that is being perpetuated in the third decade. This largely nonhierarchal framework promotes the scientific agenda through regular meetings, facilitates proposal writing, and provides a method for resolving conflicts and making difficult decisions.

The long-term funding base that the National Science Foundation established for the LTER program was crucial to the contributions made by our site. Also crucial was the working relationship between NSF program officers and the leadership in the LTER network of sites in guiding the evolution of a vision of the program and its implementation.

The Future

We anticipate that the value of a long-term study will increase with duration of the study as a progression of stair-step changes (Fig. 15.1), rather than, say, linearly or asymptotically. As the length of the time increases, new kinds of analyses become possible. Qualitatively different questions can be addressed with techniques not applicable to shorter time series. The longer window increases the probability that an infrequent event or disturbance will be captured. Researchers from a new generation already familiar with long-term regional ecology will bring new perspectives and questions to the program. More powerful information systems will develop. Synthesis and intersite network level research will lead to unexpected breakthroughs in this developing science. To flourish in such an upward progression of increasing value, LTER programs will need to creatively balance four sets of tensions (Chapter 14): maintaining the old while embracing the new, balancing project manageability and growth, maintaining project continuity and integrity in spite of future uncertainties, and achieving continuous synthesis while making site-specific or system-specific advances to science.

As when we began, we are interested in the development of a predictive understanding of long-term dynamics of lakes in a landscape setting. We have come to include humans as a key part of this system, an augmentation that substantially improves the odds of significant successes. The time series we have generated are just beginning to be sufficiently long to permit the use of more powerful statistical models such as time-series analyses and prediction. The multifaceted dynamics we have observed and the whole ecosystem experiments we have conducted, and are conducting, are revealing the complexity of the systems whose future we are attempting to visualize. We have a better grasp of the underlying uncertainties in lake ecosystems. What we have understood provides a basis for an increasing use of process models. New initiatives by new leadership are leading the way in developing

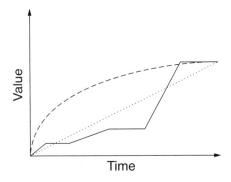

Figure 15.1. Hypothetical relationships between the value of a legacy and time.

scenarios to better understand possible alternative futures. New technologies continue to increase our ability to understand lakes and their human-inhabited landscapes. This book is a synthesis after 20 years of LTER research, but, even though we are proud of what we have learned, we remain modest because many pieces still seem loosely connected.

We believe that long-term ecological research has yet to make its greatest contributions. To achieve a predictive understanding at a reasonable level of success will require the evolving long-term data and broad spatial contexts. The endeavor will need to meet the challenges of conducting synthetic, comparative studies among disparate ecosystems, sometimes at continental or global scales, and of incorporating human systems and beliefs. Scientific advances will be facilitated by new and improved technology and information systems, as well as by a multiplicity of approaches, including whole-ecosystem experiments in the real world and statistical and process modeling. Significant scientific breakthroughs will require interdisciplinary efforts made up of a diversity of talents, and, probably most important, new generations of students and scholars curious about the world in which they live.

Appendices

Appendix 13.1. North Temperate Lakes LTER core parameters.*

Local Measurements

Parameter	Frequency	Location	Method

Weather
  Air Temperature[a]
  Relative Humidity[a]
  Wind Speed[a]
  Precipitation[a]
  Solar Radiation[a]
 	
  Wind Direction[a]
  Soil Temperature[a]
Evaporation
  Air Temperature
  Relative Humidity
  Wind Speed
  Water Temperature
Precipitation Chemistry
Other Weather Data[a]
	Hourly except half-
	  hourly for solar
	  radiation and 5
	  minute for rain
	  gauge during rain
	  events
	
	Weekly
	Daily
		Noble Lee Airport,
		  Sparkling Lake, Trout
		  Lake, and Madison NWS
		
		Noble Lee Airport and
		  Madison NWS
		
		Lake rafts on Sparkling
		  and Trout Lakes
		
		Trout Lake NADP site
		32-km radius
			Thermistor, Campbell
			  HMP 35c probe, 3-cup
			  anemometer
			
			Tipping bucket gauge,
			  Eppley Pyranometer
			  (long and short wave),
			  electronic wind vane,
			  thermistor probes
			
			Thermistor, Campbell
			  HMP 35c probe, 3-cup
			  anemometers,
			  thermistor array
			
			NADP protocols
			Weather Service
			  Stations

Groundwater
  Water Level
  Water Chemistry
    same as Chemical
    Limnology, except no
    total particulate matter
	
	Monthly
	Annually in autumn
		Selected wells near study
		  lakes
			Tape and popper
			Chemistry same as
			  lake samples

Physical Limnology
  Water Temperature[a]
  Vertical Light
    Attenuation
  Secchi Disc Depth[a]
  Dissolved Oxygen[a]
  Ice Thickness[a]
  Snow Depth on Lake[a]
  Ice Duration[a]
	Every two weeks
	  during ice-free season
	
	Every six weeks during
	  ice-covered season
	Annually
		Deepest part of lake, 0.25
		  to 1-m depth intervals
		  depending on lake
		
		Research lakes
			YSI 58, Campbell
			  dataloggers, LiCor
			  Cosine Quantum
			  Sensors, meter stick

Instrumented Buoys
  Water Temperature,
  Dissolved Oxygen,
  Wind Speed, Wind
  Direction, Relative
  Humidity
	10 minutes	Oxygen at surface,
		  temperature at approxi-
		  mately 1-m intervals
			Greenspan oxygen
			  sondes, Apprise
			  Templine thermistor
			  chain, R. M. Young
			  wind sensors

(continued)

Parameter	Frequency	Location	Method
Chemical Limnology		Deepest part of lake, top and	Samples collected with
Total Nitrogen[a]	Every four weeks	bottom of epilimnion,	peristaltic pump and
Total Dissolved	during ice-free	mid thermocline and top,	in-line filtration, N by
Nitrogen	season, every six	middle, and bottom of	$K_2S_2O_8$ digestion and
Nitrate, Ammonia[a]	weeks during ice-	hypolimnion	copper cadmium
Total Phosphorus[a]	covered season		digestion and diazo
Total Dissolved			complex, P by $K_2S_2O_8$
Phosphorus[a]			digestion and phospho-
Total Silica			molybdate complex, Si
Dissolved Reactive			by $NaHCO_3$ digestion
Silica[a]			and silica-molybdate
Field pH[a]			complex, pH with
Air Equilibrated pH[a]			meter, alkalinity by
Total Alkalinity			gran titration, C by
Total Inorganic Carbon			persulfate digestion
Dissolved Inorganic			
Carbon[a]			
Total Organic Carbon			
Dissolved Organic			
Carbon[a]			
Total Particulate Matter[a]			
Major Ions	Quarterly at spring and		Anions by ion chroma-
Chloride[a], Sulfate[a],	fall mixis, summer		tography, cations by
Calcium[a], Magnesium[a],	and winter		atomic absorption
Sodium[a],	stratification		
Potassium[a], Iron[a],			
Manganese[a], Specific			
Conductance[a]			
Biological Limnology			
Chlorophyll a[a]	Same as physical	Deep part of lake, Chl at 2-9	Chl by spectroscopy,
Primary	limnology	discrete depths, Prod by	Prod by C_{14} incuba-
Production		thermal strata for	tion, and analysis of
		euphotic zone	diurnal oxygen and
			pCO_2
Sedimentation Rate			Sediment traps
Macrophyte	Annually in August in	Selected shoreline sites	Permanent line transects
Distribution and	Trout Lake and in		
Biomass[a]	summer in Madison		
	lakes		
Zooplankton	Same as physical	Deep part of lake, 2-9	2-m-long Schindler-Patalas
Biomass[a]	limnology	depths per lake	trap, Wisconsin net
Crayfish	Annually in August	Selected shoreline sites	Cylindrical traps baited
Abundance[a]			with beef liver
Other Benthic	Annually in August	Deep part of lake and	Conical net for *Mysis*
Invertebrates	and September	selected shoreline sites	and *Chaoborus*,
			"Dendy" samplers
Fish[a]	Annually in August	Deep part of lake and	Vertical gill nets, fyke
		selected shoreline sites	nets, trammel nets,
			seines, electroshocker,
			acoustic transects
Human Demography			
Housing Density			
	Each decade since	Northern Highlands Lake	U.S. Census
	1940	District, partial block	
		groups	

(continued)

Spatial Data: Vector Data

Description	Scale or Resolution	Spatial Coverage	Comments
Hydrography	1:24,000	Wisconsin	From Wisconsin DNR
Bathymetry	1:24,000	MLR, TLR	LTER data
Watersheds	1:100,000	Wisconsin	From Wisconsin DNR
Watersheds	1:24,000	TLR	LTER data
Soils	1:250,000	Wisconsin	From USDA
Soils	1:24,000	MLR, TLR	From Dane County and Vilas County
Original Vegetation	1:100,000	Wisconsin	From Wisconsin DNR
Historical Land Use/Land Cover (1930s, 1960s, 1990s)	1:20,000	MLR, TLR	LTER data
Riparian Vegetation	1:24,000	MLR	Mendota watershed only; LTER data
Building Locations	1:30,000	TLR	From Vilas County
Roads	1:31,680	MLR, TLR	From Vilas County and UW Land Information & Computer Graphics Facility
Wetlands	1:24,000	TLR	From Wisconsin DNR
Coarse Woody Debris & Shore Characteristics	variable	TLR	Based on GPS survey; LTER data
Satellite-Derived Water Clarity (1980, 1992, 2000)	variable	MLR, TLR	Based on analysis of Landsat imagery; LTER data.

Spatial Data: Raster Thematic Data

Description	Scale or Resolution	Spatial Coverage	Comments
Digital Elevation Models	30 m	Wisconsin	From USGS
Digital Elevation Models	1 arc-second	Wisconsin	Based on Shuttle Radar Topography Mission, from USGS
Digital Raster Graphs	1:24,000	MLR, TLR	Scanned USGS 7.5' quadrangle maps, from Wisconsin DNR
Land Cover (1992–94)	30 m	Wisconsin	Based on Landsat satellite imagery (WISCLAND), from Wisconsin DNR
Land Economic Inventory (1930s)	variable	MLR	Scanned maps

Spatial Data: Aerial Photography

Film Type[b]	Dates	Scale	Spatial Coverage	Comments
BW	1930s	1:20,000	MLR, TLR	Scanned images and digital orthophotos
BW	1960s	1:20,000	MLR	Scanned images and digital orthophotos
BWIR	1960s	1:15,840	TLR	Scanned images and digital orthophotos
Color & CIR	1986	1:10,000	TLR	Shorelines only, hardcopy film only
BW	1993	1:40,000	MLR	Digital orthophotos
BW	1996	1:31,680	TLR	Digital orthophotos
CIR	1997	1:9,000	MLR	Hardcopy film, some scanned images
CIR	1998	1:9,800	MLR, TLR	Hardcopy film, some scanned images

(continued)

Appendix 13.1. *continued*

Spatial Data: Raster (Digital Images)

Sensor	Type[c]	Resolution	Spatial Coverage	Years
Landsat MSS	M	80	TLR	1972, 1986, 1991
Landsat MSS	M	80	MLR	1975, 1986, 1990
Landsat TM	M,T	30–120	TLR	1984, 1988, 1989, 1991, 1992, 1993, 2000, 2003
Landsat TM	M,T	30–120	MLR	1984, 1986, 1987, 1988, 1989, 1990, 1991, 1992, 1994, 1995, 1999
Landsat ETM+	M,P,T	15–60	TLR	1999
Landsat ETM+	M,P,T	15–60	MLR	1999, 2000, 2001
Terra ASTER	M,P,T	30–90	TLR	2000
Terra/Aqua MODIS	M,T	250–1000	Wisconsin	2000 (various dates), daily since 2001
SPOT HRV	M,P	10–20	TLR	1988
SPOT HRV	M,P	10–20	MLR	1986, 1988, 1989
IKONOS	M,P	1–4	TLR	2000
IKONOS	M,P	1–4	MLR	2000
QuickBird	M,P	0.6–2.4	TLR	2002
QuickBird	M,P	0.6–2.4	MLR	2002
EO-1 ALI	M,P,T	10–30	MLR	2001
EO-1 Hyperion	H	30	TLR	2002, 2003
EO-1 Hyperion	H	30	MLR	2001, 2003
ATLAS	A,M,T	3	TLR	1998
ATLAS	A,M,T	3	MLR	1997, 1998
SIR-C	R	25	TLR	1994
ERS-1	R	25	TLR	1992
ERS-1	R	25	MLR	1992

*More details on measurement protocols are available in the Data Catalog at the project Web site (lter.limnology.wisc.edu).

[a]local measurements at both the Madison Lake area and Trout Lake area; otherwise, Trout Lake area only.

[b]aerial photography film type codes: BW = black and white (panchromatic), BWIR = black and white infrared, CIR = color infrared.

[c]spatial data sensor type codes: A = airborne, H = hyperspectral, M = multispectral, P = panchromatic, R = radar, T = thermal.

Note: Acronym definitions: NWS = National Weather Service; NADP = National Atmospheric Deposition Program; Chl = chlorophyll a; Prod = primary production; MLR = Madison Lake Region; TLR = Trout Lake Region; DNR = Department of Natural Resources; USDA = U.S. Department of Agriculture; USGS = U.S. Geological Survey; DEM = Digital Elevation Model; UW = University of Wisconsin; WISCLAND = Wisconsin Initiative for Statewide Cooperation on Landscape Analysis and Data; GPS = Global Positioning System; MSS = Multispectral Scanner; TM = Thematic Mapper; ETM+ = Enhanced Thematic Mapper Plus; ASTER = Advanced Spaceborne Thermal Emission and Reflection Radiometer; MODIS = Moderate Resolution Imaging Spectro-Radiometer; SPOT = Satellite Pour l'Observation de la Terre; HRV = High Resolution Visible; EO-1 = Earth Orbiter-1; ALI = Advanced Land Imager; ATLAS = Airborne Terrestrial Land Applications Scanner; SIR-C = Shuttle Imaging Radar-C; ERS-1 = European Remote Sensing Satellite-1.

Appendix 13.2. North Temperate Lakes LTER data acess policy.

The North Temperate Lakes (NTL) LTER Data Access Policy is designed to make data from the NTL-LTER database as freely available as possible for academic, research, education, and other professional purposes. Each data set published on the Web page is accompanied by documentation (metadata). We encourage the use of our data sets but ask that users read and agree to our Data Use Agreement.

Data Use Agreement

Permission to download NTL-LTER data sets is granted to the Data User subject to the following terms:

- The Data User must realize that these data sets are being actively used by others for ongoing research and that coordination may be necessary to prevent duplicate publication. The Data User is urged to contact Steve Carpenter, lead Principal Investigator (srcarpen@wisc.edu), to check on other uses of the data. Where appropriate, the Data User may be encouraged to consider collaboration and/or co-authorship with original investigators.
- The Data User must realize that the data may be misinterpreted if taken out of context. We request that you provide Steve Carpenter (ATTN: Data Access, Center for Limnology, University of Wisconsin-Madison, 680 North Park St., Madison, WI 53706) with a copy of any manuscript using the data so that he may review and provide comments on the presentation of our data.
- The Data User must acknowledge use of the data by an appropriate citation (see Citation) of the NTL-LTER database.
- The Data User must send two reprints of any publications resulting from use of the data to the address above. We would like to include such manuscripts in our LTER publications list.
- The Data User must not redistribute original data and documentation without permission from Steve Carpenter, lead Principal Investigator (srcarpen@wisc.edu).

By using these data, the Data User agrees to abide by the terms of this agreement. Thank you for your cooperation.

Citation

Use of the data in publications should acknowledge the North Temperate Lakes LTER project. A generic citation for our databases is: <name of data set>, North Temperate Lakes Long Term Ecological Research program (http://lter.limnology.wisc.edu), NSF, <contact person for data set>, Center for Limnology, University of Wisconsin-Madison. The data set name and contact person for each data set can be found in the metadata header of the online data sets.

Data Availability

Our goal is to release all long term data associated with core research areas within 2 years of collection. These data and accompanying metadata will be available for download from the NTL-LTER Web site.

Disclaimer

While substantial efforts are made to ensure the accuracy of data and documentation, complete accuracy of data sets cannot be guaranteed. All data are made available "as is." The North Temperate Lakes LTER shall not be liable for damages resulting from any use or misinterpretation of data sets. Data users should be aware that we periodically update data sets.

Appendix 14.1. Investigators and staff scientist involved in writing each proposal for the North Temperate Lakes Long-Term Ecological Research program. Most listed began working in or on the program before the proposal on which they are first listed. Many graduate students and center for Limnology staff that participated in the proposal preparation are not listed.

Name	Role	Affiliation	Proposal Years						
			1981–85	1986–90	1991–96	1994–96 Augmentation	1996–2002	1999–2004[b] Microbial Observatory	2002–08
First Proposal									
Adams, Michael S.	Faculty	Botany	X	X	X	X	X		
Anderson, Mary P.	Faculty	Geology & Geophysics	X	X	X	X	X		X
Armstrong, David E.	Faculty	Civil & Environmental Engineering/Water Chemistry	X	X	X	X	X	X	X
Bowser, Carl J.	Faculty, Head LTER Chemistry Laboratory	Geology & Geophysics	X	X	X	X	X		X
Brock, Thomas D.	Faculty	Bacteriology	X	X					
Magnuson, John J.	Faculty, Director Center for Limnology	Zoology/Limnology	X[a]	X[a]	X[a]	X[a]	X[a]		X
Ragotskie, Robert A.	Faculty	Meteorology	X	X					
New 1986–90									
Allen, Timothy F. H.	Faculty	Botany			X	X	X		X
Benson, Barbara J.	Scientist, Data and Information Manager	LTER/Limnology			X	X	X		X
Frost, Thomas M.	Scientist, Trout Lake Station Director	Limnology			X	X	X		
Kratz, Timothy K.	Scientist, Northern Site Manager	LTER/Limnology			X	X	X	X	X
New 1991–96									
Gower, Stith A.	Faculty	Forest Ecology & Management				X	X		

Name	Role	Affiliation				
Lillesand, Thomas S.	Faculty	Institute for Environmental Studies/Environmental Remote Sensing Center		X		X
MacKenzie, Mark D.	Remote Sensing Scientist	LTER/Environmental Remote Sensing Center	X	X		
New 1994–96 Augmentation						
Carpenter, Stephen R.	Faculty	Zoology/Limnology	X	X[a]	X	X[a]
Hanson, Paul C.	Scientist, Laboratory Manager & Computer Networking	Zoology/Limnology	X	X		X
Heberlein, Thomas A.	Faculty	Rural Sociology	X	X		X
Lathrop, Richard C.	Scientist, Southern Site Manager	Wisconsin Department of Natural Resources/Research	X	X		X
Riera, Joan L.	Remote Sensing Scientist	LTER/Limnology	X	X		
Turner, Monica G.	Faculty	Zoology	X	X		X
Voss, Paul R.	Faculty	Rural Sociology	X	X		X
New 1996–2002						
Cronon, William	Faculty	History		X		X
Dodson, Stanley I.	Faculty	Zoology		X		X
Foley, Jonathan A.	Faculty	Institute of Environmental Studies		X		X
Nowak, Peter J.	Faculty	Rural Sociology		X		X
New 1999–2004 Microbial Observatory						
Triplett, Eric W.	Faculty	Agronomy				X[a]
Graham, Linda E.	Faculty	Botany			X[a]	X

(continued)

Appendix 14.1. continued

Name	Role	Affiliation	Proposal Years						
			1981–85	1986–90	1991–96	1994–96 Augmentation	1996–2002	1999–2004[b] Microbial Observatory	2002–08
New 2002–08									
Chipman, Jonathan W.	Remote Sensing Scientist	Environmental Remote Sensing Center							X
Hammer, Roger B.	Faculty	Rural Sociology							X
Stanley, Emily H.	Faculty Head, LTER Chemistry Laboratory	Limnology							X
Vander Zanden, M. Jake	Faculty	Limnology							X
Walker, John F.	Government Scientist	U.S. Geological Survey							X
Wu, Chin H.	Faculty	Civil & Environmental Engineering							X
Summary Numbers									
Total number			7	12	12	19	21	5	25
Number new			7	5	3	7	4	2	6
Number continuing				7	9	12	17	3	19[c]

[a]Lead Principal Investigator
[b]Specialized LTER related proposal
[c]Continued from either the 1996–2002 or the Microbial Observatory Proposal

Appendix 14.2. Chronology of program review and the reviewers of the North Temperate Lakes LTER Program.*

Date	Purpose	Reviewers	Affiliation at Date of Meeting	NSF Staff	Location
1981 June 1–3	External Scientific Advisory Committee	Collins, Nicholas C. Likens, Gene E. Frey, David G. Winter, Thomas C.	U. Toronto Erindale Campus Cornell U. Indiana U. U.S. Geological Survey	None	Trout Lake
1982 May 27	NSF Review	None		Callahan, James T. (Tom)	Trout Lake
1984 Feb. 16–19	External Scientific Advisory Committee	Frey, David G. Goldman, Charles R. O'Neill, Robert V. Winter, Thomas C.	Indiana U. U. California-Davis Oak Ridge National Laboratory U.S. Geological Survey	None	Trout Lake
1986 April 10–11	External Scientific Advisory Committee	Carpenter, Stephen R. D'Angelis, Donald L. Goldman, Charles R. Schindler, David W.	U. Notre Dame Oak Ridge National Laboratory U. California-Davis Freshwater Institute, Winnipeg	None	Trout Lake
1989 May 22–25	Site Visit	Cole, Jonathan J. Schindler, James E. Turner, Monica G. Ward, G. Milton	Institute for Ecosystem Studies Clemson U. Oak Ridge National Laboratory U. Alabama	Callahan, James T. (Tom)	Madison & Trout Lake
1993 July 12–15	Site Visit	Briggs, John M. De Angelis, Donald L. Korterba, Michael T. Power, Mary E. Stanford, Jack A.	Kansas St. U. Oak Ridge National Laboratory U.S. Geological Survey U. California-Berkeley Flathead Lake Biological Station	Callahan, James T. (Tom) Dame, Richard	Madison & Trout Lake
1999 Oct. 12–15	Site Visit	Henshaw, Donald L. McIntyre, Sally Mittelbach, Gary G. Redman, Charles L. Wurtsbaugh, Wayne A.	USDA Forest Ser. Corvallis, OR U. California-Santa Barbara Kellogg Biological Station Arizona St. U. Utah St. U.	Collins, Scott L. Scheiner, Sam	Madison & Trout Lake

*External Review Committees selected by North Temperate Lakes LTER; site visit teams selected by NSF.

Appendix 14.3. North Temperate Lakes LTER staff, 1981–2004.

Activity	Location	Person	Years
Lead Investigator	Programwide	John J. Magnuson with Carl J. Bowser	1981–2000
		Stephen R. Carpenter	2000–
Site Managers	Trout Lake	Timothy K. Kratz	1981–2002
		Jim A. Rusak	2002–
	Madison Lakes	Richard C. Lathrop	1994–
Data and Information Managers	Programwide	Paul W. Rasmussen	1980–83
		Barbara J. Benson	1983–
Data Management Support	Programwide	Joyce M. Tynan	1985–95
		Maryan Stubbs	1995–99
		David Balsiger	2000–
Remote Sensing Specialists	Programwide	Mark D. MacKenzie	1988–95
		John R. Vande Castle	1988–89
		David W. Bolgrien	1995–98
		Joan L. Riera	1998–2000
		Jonathan W. Chipman	2001–
Computer/Network Manager	Centerwide	Paul C. Hanson	1992–
Chemical Laboratory Technicians		Brad Price	1981–82
		Doug Lindloff	1983–89
		Sue Holloway	1989–94
		James A. Thoyre	1994–
Field Technicians	Trout Lake	Paula Barbian	1981–91
		Daniel R. Helsel	1983–87
		Timothy W. Meinke	1984–
		Pamela K. Montz	1984–
	Madison Lakes	Martha J. Wessels	1995–98
		Nicholas Voichick	1999–2000
		Ted M. Cummings	2002–

Appendix 14.4. Detailed objectives of the North Temperate Lakes LTER proposal, 1981–2002.

1980–85: Long-term ecological research on lake ecosystems.
  (a) The perception and description of long-term trends in physical, chemical, and biological properties of lake ecosystems.
  (b) Detection of linkages or interrelationships among the physical, chemical, and biological properties of lakes and their relationships to the climatic and hydrologic environment.
  (c) Determination of lake features that influence long-term ecosystem stability and resiliency in face of natural and anthropogenic disturbances.
1986–90: Comparative studies of a suite of lakes in Wisconsin.
  (a) To perceive and describe long-term trends and patterns in physical, chemical, and biological properties of lake ecosystems.
  (b) To understand the dynamics of internal and external processes affecting lake ecosystems.
  (c) To analyze the temporal responses of lake ecosystems to disturbance and stress.
  (d) To evaluate the interaction between spatial heterogeneity and temporal variability of lake ecosystems.
  (e) To develop and test concepts and theories, that relate temporal and spatial variability of lake ecosystems at scales relevant to long-term landscape ecology.
1991–96: Comparative studies of a suite of lakes in Wisconsin.
  (a) To perceive long-term trends in physical, chemical, and biological properties of lake ecosystems.
  (b) To understand the dynamics of internal and external processes affecting lake ecosystems.
  (c) To analyze the temporal responses of lake ecosystems to disturbance and stress.
  (d) To evaluate the interaction between spatial heterogeneity and temporal variability of lake ecosystems.
  (e) To expand our understanding of lake-ecosystem properties to a broader, regional context.
1994–96: Augmentation: upper midwest lakes and their landscapes: 1800–2100.
  (a) Do the dominant factors that control inland lake ecosystems, and the predictability of their effects, vary systematically as spatial scales expand from individual lakes, to watersheds, to lake districts, to broader regions and as temporal scales extend from years, to decades, to centuries?
  (b) How have natural and human-induced changes in the landscape interacted with aquatic ecosystem structure and dynamics in the Upper Great Lakes Region over the past two centuries, and what changes can be expected over the next hundred years?
1996–2002: Comparative study of a suite of lakes in Wisconsin.
  (a) Perceive long-term changes in the physical, chemical, and biological properties of lake ecosystems.
  (b) Understand interactions among physical, chemical, and biological processes within lakes and their influences on lake characteristics and long-term dynamics.
  (c) Develop a regional understanding of lake ecosystems through an analysis of the patterns and processes organizing lake districts.
  (d) Develop a regional understanding of lake ecosystems through integration of atmospheric, hydrologic, and biotic processes.
  (e) Understand the way human, hydrologic, and biogeochemical processes interact within the terrestrial landscape to affect lakes and the way lakes, in turn, influence these interactions.
2002–08: Comparative study of a suite of lakes in Wisconsin.
  (a) Perceive long-term changes in the physical, chemical, and biological properties of lake districts.
  (b) Understand the drivers of temporal variability in lakes and lake districts.
  (c) Understand the interaction of spatial processes with long-term change.
  (d) Understand the causes and predictability of rapid extensive change in ecosystems.
  (e) Build a capacity to forecast the future ecology of lake districts.

Source: North Temperate Lakes LTER proposal, Archives of the Center for Limnology, UW-Madison.

Appendix 14.5. Graduate students and postdoctoral participants in the North Temperate Lakes LTER program, 1981–2003, by graduation year or year of departure.

Name	Thesis Year or Departure	Degree or Status	Adviser/s	Thesis Area
Okwueze, Emeka E.	1983	Ph.D.	Clay	Geophysics
Rudstam, Lars G.	1983	M.S.	Magnuson	Fish Ecology
Watson, Vicki	1983	Postdoc.	Brock	Phytoplankton Ecology
Guozhang, Liao	1983	Postdoc.	Magnuson	Fish Ecology
Attig, John W.	1984	Ph.D.	Michelson	Pleistocene Geology
Boston, Harry L.	1984	Ph.D.	Adams	MacrophytePhysiology
Lyons, John D.	1984	Ph.D.	Magnuson	Fish Ecology
Lay, Bibiana	1985	Ph.D.	Brock	Phytoplankton Ecology
Rudensky, Ksenia	1985	M.S.	Perry, U. Minnesota	Leaf Litter Input & Decomposition
Kenoyer, Galen J.	1986	Ph.D.	Bowser & Anderson	Hydrogeochemistry
Marin, Luis E.	1986	M.S.	Bowser & Kratz	Hydrogeochemistry
Schneider, Daniel W.	1986	M.S., Ph.D.	Frost & Magnuson	Aquatic Invertebrate Ecology
Detenbeck, Naomi E.	1987	Ph.D.	Brezonik, U. Minnesota	Benthic Algae & Nutrients
Newsom, Mhora	1987	M.S.	Perry, U. Minnesota	Leaf Litter Decomposition
Wachtler, John N.	1987	M.S.	Brezonik, U. Minnesota	Primary Production
Hurley, James P.	1988	M.S., Ph.D.	Armstrong	Nutrient Water Chemistry
Knight, Susan E.	1988	Ph.D.	Waller	Aquatic Macrophyte Ecology
Krabbenhoft, David P.	1988	M.S., Ph.D.	Bowser	Groundwater Lake Interaction
Martinez, Neo D.	1988	M.S.	Magnuson	Aquatic Food Web Ecology
Sherman, Leslie A.	1988	M.S.	Brezonik, U. Minnesota	Sediment Water Chemistry
Sierszen, Michael E.	1988	Ph.D.	Frost & Magnuson	Zooplankton Ecology
Tacconi, J. E.	1988	M.S.	Brezonik, U. Minnesota	Biogeochemistry
Urban, Noel R.	1988	Postdoc.	Brezonik, U. Minnesota	Sulfur Biogeochemistry
Baker, Lawrence A.	1989	Postdoc.	Brezonik, U. Minnesota	Internal Alkalinity Generation
Olson, T. Mark	1989	M.S.	Lodge, U. Notre Dame	Crayfish Ecology
Robertson, Dale M.	1989	M.S., Ph.D.	Ragotskie	Physical Limnology & Cryosphere
Weir, Edward P.	1989	M.S.	Brezonik, U. Minnesota	Ion Budgets
Jacobsen, Paul T.	1990	M.S., Ph.D.	Magnuson & Clay	Fish Ecology and Sonar
Crystal-Rose, L. M.	1991	M.S.	Scarpace	Remote Sensing of Wetland Plants
McLain, Ann S.	1991	Ph.D.	Magnuson	Fish Invasion Ecology
Ackerman, J. A.	1992	M.S.	Bowser	Water Budget with Isotopes

(continued)

Appendix 14.5. *continued*

Name	Thesis Year or Departure	Degree or Status	Adviser/s	Thesis Area
González, María J.	1992	M.S., Ph.D.	Frost & Magnuson	Zooplankton Ecology
Kershner, M. W.	1992	M.S.	Lodge, U. Notre Dame	Crayfish Ecology
Mach, Carl E. J.	1992	Ph.D.	Brezonik, U. Minnesota	Geochemistry of Metals
Sampson, Carolyn J.	1992	M.S.	Brezonik, U. Minnesota	Water Chemistry
Segova, Marketa	1992	M.S.	Adams	Benthic Insect Ecology
Asplund, Tim R.	1993	M.S.	Magnuson	Winter Limnology
Cisneros, Rigel 0.	1993	M.S.	Magnuson	Fish Biodiversity
Hoffman, Judith I.	1993	M.S.	Frost & Magnuson	Zooplankton Ecology
Shelley, Brian C. L.	1994	M.S., Ph.D.	Perry, U. Minnesota	Acidification & Leaf Litter Processing
Arancibia-Avila, Patricia E.	1994	Ph.D.	Graham	Algal Physiology
Cheng, Xiangxue	1994	Ph.D.	Anderson	Groundwater Modeling
Hill, Anna M.	1994	Ph.D.	Lodge, U. Notre Dame	Crayfish Ecology
Johnson, Timothy B.	1995	M.S.	Kitchell	Fish Ecology
Keating, Elizabeth H.	1995	Ph.D.	Bahr	Hydrogeochemistry
Michaels, Susannah	1995	M.S.	Bowser	Hydrogeochemistry
Poister, David	1995	M.S., Ph.D.	Armstrong	Nutrient Lake Chemistry
Weaver, Melissa J.	1995	Ph.D.	Magnuson	Fish Ecology
Wynne, Randolph H.	1995	M.S., Ph.D.	Lillesand	Remote Sensing & Cryosphere
Bobo, M. R.	1996	M.S.	Lillesand	Multispectral Image Classification
Fischer, C. S.	1996	M.S.	Lillesand	Landuse Change & Remote Sensing
Gergel, Sarah E.	1996	M.S.	Turner	Landscape Ecology
Jennings, N. P.	1996	M.S.	Lillesand	Data Integration & Texture
Kim, Kanjoo	1996	Ph.D.	Bowser	Hydrogeochemistry
Klosiewski, S. P.	1996	Ph.D.	Stein, Ohio State U.	Fish Ecology
Soranno, Patricia A.	1996	Ph.D., Postdoc.	Carpenter	Nutrients, Lake Mendota & Watershed
Fischer, Janet M.	1997	M.S., Ph.D.	Frost & Magnuson	Zooplankton Ecology
Puth, Linda M.	1997	M.S.	Allen	Crayfish Ecology
Arnott, Shelley E.	1998	Ph.D.	Magnuson & Frost	Zooplankton Biodiversity
Champion, Glen	1998	M.S.	Anderson & Hunt	Groundwater Modeling
Lathrop, Richard C.	1998	Ph.D.	Carpenter	Water Clarity Phosphorus & *Daphnia*
Sanderson, Beth L.	1998	M.S., Ph.D.	Frost & Magnuson	Limnology & Fish Ecology
Webster, Katherine E.	1998	Ph.D.	Magnuson	Geochemistry & Hydrology
Fallah-Moghaddam, Payam	1999	Ph.D.	Shearer, U. Illinois Champaign-Urbana	Fungal Systematics

(continued)

Appendix 14.5. *continued*

Name	Thesis Year or Departure	Degree or Status	Adviser/s	Thesis Area
Foreman, William J.	1999	M.S.	Carpenter	Fish and Phytoplankton Nutrients
Hrabik, Thomas R.	1999	M.S., Ph.D.	Magnuson	Ecology of Fish & Exotic Fish
Jorgensen, Bradley S.	1999	Postdoc.	Heberlein	Social Science
Reed-Andersen, Tara	1999	Ph.D.	Carpenter	Watershed and Nutrients
Greenfield, Ben K.	2000	M.S.	Carpenter	Fish Contaminant Ecology
Lauster, George H.	2000	Ph.D.	Armstrong	Nutrients, Bacteria, & Phytoplankton
Lewis, David B.	2000	M.S., Ph.D.	Magnuson	Benthic Invertebrate Ecology
Riera, Joan L.	2000	Postdoc.	Magnuson & Lillesand	Landscape Ecology of Lakes
Schnaiberg, Jill L.	2000	M.S.	Langston & Turner	Shoreline Development
Stedman, Richard C.	2000	Ph.D.	Heberlein	Social Psychology & Lakes
Wilson, Matthew A.	2000	Ph.D.	Heberlein	Attitudes & Economics
Bennett, Elena M.	2001	M.S., Ph.D.	Carpenter	Landscape Nutrient Patterns
Lenters, John D.	2001	Postdoc.	Foley	Climatology & Hydrology
Wegener, Mark W.	2001	M.S.	Lillesand & Turner	Land Use & Runoff
Bade, Darren L.	2002	M.S.	Carpenter	Chemical Limnology of Carbon
Pint, Christine D.	2002	M.S.	Anderson & Hunt	Groundwater Modeling
Pollard, Amina I.	2002	Ph.D.	Magnuson & Frost	Lake & Stream Interaction Benthos
Wilson, Karen A.	2002	Ph.D.	Magnuson	Crayfish Invasion Ecology
Dripps, Wes	2003	Ph.D.	Anderson & Hunt	Groundwater Recharge Patterns
Hanson, Paul C.	2003	Ph.D.	Carpenter	Lake Metabolism & Buoy Sensors
Schell, Jeffery M.	2003	Ph.D.	Dodson	Aquatic Invertebrate Ecology
Willis, Theodore V.	2003	M.S., Ph.D.	Magnuson	Fish Ecology
Chipman, Jonathan W.	Here	M.S.(2001) Postdoc.	Lillesand	Remote Sensing
Kent, Angela D.	Here	Postdoc.	Triplett	Microbial Observatory
Yannarell, Anthony C.	Here	M.S.	Triplett	Microbial Observatory

Note: Students from other universities, various funding sources, and the Little Rock Lake acidification project and the U.S. Geological Survey's Water Energy and Biogeochemical Budgets project are included.

References

Abbott, I. 1983. The meaning of z in species/area regressions and the study of species turnover in island biogeography. Oikos 41:385–390.

Ackerman, J. 1993. Extending the isotope based ($d^{18}O$) mass budget technique for lakes and comparison with solute based lake budgets. M.S. Thesis. University of Wisconsin-Madison.

Adams, M. S., and R. T. Prentki. 1982. Biology, metabolism and functions of littoral submerged weedbeds of Lake Wingra, Wisconsin, USA: A summary and review. Archive für Hydrobiologie Supplement 62, 3/4:333–409.

Adams, M. S., T. W. Meinke, and T. K. Kratz. 1990. Primary productivity in three northern Wisconsin lakes, 1985–1987. Verhandlungen Internationale Vereigung für Limnologie 24:432–437.

Addis, J. 1992. Policy and practice in UW-WDNR collaborative programs. Pages 7–15 in J. F. Kitchell, editor. Food web management: A case study of Lake Mendota. Springer-Verlag, New York.

Adrian, R., and T. M. Frost. 1992. Comparative feeding ecology of *Tropocyclops prasinus mexicanus* (Copepoda, Cyclopoeda). Journal of Plankton Research 14:1369–1382.

Adrian, R., and T. M. Frost. 1993. Omnivory in cyclopoid copepods: Comparisons of algae and invertebrates as food for three differently sized species. Journal of Plankton Research 15:643–658.

Allen, T. F. H., and T. W. Hoekstra. 1992. Toward a unified ecology. Columbia University Press, New York.

Allen, T. F. H., and T. B. Starr. 1982. Hierarchy: Perspectives for ecological complexity. University of Chicago Press, Chicago, Illinois.

Anderson, J. 1996. Vilas County planning and zoning office annual report.

Anderson, M. P. 2002. Groundwater-lake interaction: Response to climate change, Vilas County, Wisconsin. Groundwater Research Report WRI GRR 02–02, University of Wisconsin Water Resources Institute, Madison.

Anderson, M. P., and C. J. Bowser. 1986. The role of groundwater in delaying lake acidification. Water Resources Research 22:1101–1108.

Anderson, M. P., and X. Cheng. 1993. Long- and short-term transience in a groundwater/lake system in Wisconsin, U.S.A. Journal of Hydrology 145:1–18.

Anderson, M. P., and X. Cheng. 1998. Sensitivity of groundwater/lake systems in the Upper Mississippi River Basin, Wisconsin, U.S.A. to possible effects of climate change. Pages 3–8 in K. Kovar, U. Tappeiner, N. E. Peters, and R. G. Craig, editors. Hydrology, Water Resources and Ecology in Headwaters. Proceedings of the International Association of Hydrologic Sciences. IAHS Press, Wallingford, Oxfordshire.

Anderson, W., D. M. Robertson, and J. J. Magnuson. 1996. Evidence of recent warming and ENSO variation in ice breakup of Wisconsin lakes. Limnology and Oceanography 41:815–821.

Arnott, S. E. 1998. Species turnover and richness of aquatic communities in north temperate lakes. Ph.D. Thesis. University of Wisconsin-Madison.

Arnott, S. E., J. J. Magnuson, and N. D. Yan. 1998. Crustacean zooplankton species richness: Single- and multiple-year estimates. Canadian Journal of Fisheries and Aquatic Sciences 55:1573–1582.

Arnott, S. E., N. D. Yan, J. J. Magnuson, and T. M. Frost. 1999. Inter-annual variability and species turnover of crustacean zooplankton in shield lakes. Canadian Journal of Fisheries and Aquatic Sciences 56:162–172.

Arora, M., and D. Wik. 1988. Feasibility study for resolving the crayfish problem in northern Wisconsin. University of Wisconsin-Stout, Menomonee.

Asplund, T. R. 1993. Patterns and mechanisms of year-to-year variability in winter oxygen depletion rates in ice-covered lakes. M.S. Thesis. University of Wisconsin-Madison.

Assel, R. A., and D. M. Robertson. 1995. Changes in winter air temperatures near Lake Michigan during 1851–1993 as determined from regional lake-ice records. Limnology and Oceanography 40:165–176.

Assel, R., K. Cronk, and D. Norton. 2003. Recent trends in Laurentian Great Lakes ice cover. Climatic Change 57:517–527.

Astor, B. 1982. Brooke Astor on the pleasures of collecting. Architectural Digest 39:34.

Attig, J. W. 1985. Pleistocene geology of Vilas County, Wisconsin. Wisconsin Geological and Natural History Survey, Survey Information Circular 50:32.

Baines, S. B., K. E. Webster, T. K. Kratz, S. R. Carpenter, and J. J. Magnuson. 2000. Synchronous behavior of temperature, calcium, and chlorophyll in lakes of northern Wisconsin. Ecology 81:815–825.

Bajkov, A. D. 1949. Do fish fall from the sky? Science 109:402.

Baker, K. S. 1996. Development of Palmer long-term ecological rearch information management. Pages 725–730 in Eco-Informa '96: Global Networks for Environmental Information. Environmental Research Institute of Michigan, Ann Arbor.

Baker, K. S., J. W. Brunt, and D. Blankman. 2002. Organizational informatics: Site description directories for research networks. Pages 355–360 in vol. VII: Informatics Systems Development II. Proceedings of the 6th World Multiconference on Systemics, Cybernetics and Informatics, Orlando, Florida.

Baker, K. S., B. J. Benson, D. L. Henshaw, D. Blodgett, J. H. Porter, and S. G. Stafford. 2000. Evolution of a multisite network information system: The LTER information management paradigm. BioScience 50:963–978.

Baker, L. A., and P. L. Brezonik. 1988. Dynamic model of internal alkalinity generation: Calibration and application to precipitation-dominated lakes. Water Resources Research 24:65–74.

Barbour, C. D., and J. H. Brown. 1974. Fish species diversity in lakes. American Naturalist 108:473–489.

Baron, J. S., and N. Caine. 2000. The temporal coherence of two alpine lake basins of the Colorado Front Range. Freshwater Biology 43:517–527.

Baron, R. M., and D. A. Kenny. 1986. The moderator-mediator variable distinction in social psychological research. Journal of Applied Psychology 50:212–220.

Baumann, P. C., J. F. Kitchell, J. J. Magnuson, and T. B. Kayes. 1974. Lake Wingra 1837–1973: A case history of human impact. Transactions of the Wisconsin Academy of Sciences, Arts and Letters 62:57–94.

Beckel, A. L. 1987. Breaking new waters: A century of limnology at the University of Wisconsin. Transactions of the Wisconsin Academy of Sciences, Arts, and Letters, Madison, Wisconsin.

Becker, G. C. 1983. Fishes of Wisconsin. University of Wisconsin Press, Madison.

Benndorf, J. 1990. Conditions for effective biomanipulation: Conclusions derived from whole-lake experiments in Europe. Hydrobiologia 200/201:187–203.

Benndorf, J. 1995. Possibilities and limits for controlling eutrophication by biomanipulation. Internationale Revue der Gesamten Hydrobiologie 80:519–534.

Bennett, E. M. 2002. Patterns of soil P: Concentrations and variability across an urbanizing agricultural watershed. Ph. D. Thesis. University of Wisconsin-Madison.

Bennett, E. M., T. Reed-Andersen, J. N. Houser, J. R. Gabriel, and S. R. Carpenter. 1999. A phosphorous budget for the Lake Mendota watershed. Ecosystems 2:69–75.

Benson, B. J. 1996. The North Temperate Lakes LTER research information management system. Pages 719–724 in Eco-Informa '96: Global Networks for Environmental Information. Environmental Research Institute of Michigan, Ann Arbor.

Benson, B. J. 1998. The World Wide Web as a tool for ecological research programs. Pages 59–63 in W. K. Michener, J. H. Porter, and S. G. Stafford, editors. Data and information management in the ecological sciences: A resource guide. Long-Term Ecological Research Network Office, University of New Mexico, Albuquerque.

Benson, B. J., and D. Olson. 2002. Conducting cross-site studies: Lessons learned from partnerships between scientists and information managers. Bulletin of the Ecological Society of America 83:198–200.

Benson, B. J., and M. Stubbs. 1998. Information access and database integrity at the North Temperate Lakes Long-Term Ecological Research Project. Pages 95–97 in W. K. Michener, J. H. Porter, and S. G. Stafford, editors. Data and information management in the ecological sciences: A resource guide. Long-Term Ecological Research Network Office, University of New Mexico, Albuquerque.

Benson, B. J., J. J. Magnuson, R. L. Jacob, and S. L. Fuenger. 2000a. Response of lake ice breakup in the Northern Hemisphere to the 1976 interdecadal shift in the North Pacific. Verhandlungen Internationale Vereinigung für Limnologie 27:2770–2774.

Benson, B. J., J. D. Lenters, J. J. Magnuson, M. Stubbs, T. K. Kratz, P. J. Dillon, R. E. Hecky, and R. C. Lathrop. 2000b. Regional coherence of climatic and lake thermal variables of four lake districts in the Upper Great Lakes Region of North America. Freshwater Biology 43:517–527.

Berners-Lee, T., J. Hendler, and O. Lassila. 2001. The Semantic Web—a new form of web content that is meaningful to computers will unleash a revolution of new web possibilities. Scientific American 284:34–43.

Betz, C. R., editor. 2000. Nonpoint source control plan for the Lake Mendota priority watershed. Wisconsin Department of Natural Resources, Bureau of Watershed Management, Madison.

Bloom, N. S., C. J. Watras, and J. P. Hurley. 1991. Impact of acidification on the methylmercury cycle of remote seepage lakes. Water, Air, and Soil Pollution 56:477–492.

Borman, F. H., and G. E. Likens. 1967. Nutrient cycling. Science 155:424–429.

Boston, H. L., and M. S. Adams. 1983. Evidence of crassulacean acid metabolism in two North American isoetids. Aquatic Botany 15:381–386.

Boston, H. L., and M. S. Adams. 1985. Seasonal diurnal acid rhythms in two aquatic crassulacean acid metabolism plants. Oecologia 65:573–579.

Boston, H. L., and M. S. Adams. 1986. The contribution of crassulacean acid metabolism to the annual productivity of two aquatic vascular plants. Oecologia 68:615–622.

Boston, H. L., M. S. Adams, and T. P. Pienkowski. 1987. Utilization of sediment CO_2 by selected North American isoetids. Annals of Botany 60:485–494.

Bowser, C. J. 1986. Historic data sets: Lessons from the past, lessons for the future; symposium. Pages 155–179 in W. K. Michener, editor. Research data management in the ecological sciences. University of South Carolina Press, Columbia.

Bowser, C. J. 1992. Groundwater pathways for chloride pollution of lakes. Pages 283–301 in F. M. D'Itri, editor. Chemical deicers and the environment. Lewis Publishers, Chelsea, Michigan.

Bowser, C. J., and B. F. Jones. 2002. Mineralogic controls on the composition of natural waters dominated by silicate hydrolysis. American Journal of Science 302:582–662.

Box, G., and G. Tiao. 1975. Intervention analysis with application to economic and environmental problems. Journal of Statistical Association 29:193–204.

Brezonik, P. L., C. E. Mach, and C. J. Sampson. 2003. Geochemical controls for Al, Fe, Mn, Cd, Cu, Pb, and Zn during experimental acidification and recovery of Little Rock Lake, WI, USA. Biogeochemistry 62:119–143.

Brezonik, P. L., L. A. Baker, J. G. Eaton, T. M. Frost, P. J. Garrison, T. K. Kratz, J. J. Magnuson, J. Perry, W. A. Rose, B. Shepherd, W. A. Swenson, C. J. Watras, and K. E. Webster. 1986. Experimental acidification of Little Rock Lake, Wisconsin. Water, Air, and Soil Pollution 31:115–121.

Brezonik, P. L., J. G. Eaton, T. M. Frost, P. J. Garrison, T. K. Kratz, C. E. Mach, J. H. McCormick, J. Perry, W. A. Rose, C. J. Sampson, B. C. L. Shelley, W. A. Swenson, and K. E. Webster. 1993. Experimental acidification of Little Rock, Wisconsin: Chemical and biological changes over the pH range 6.1 to 4.7. Canadian Journal of Fisheries and Aquatic Sciences 50:1101–1121.

Briggs, J. M., and H. Su. 1994. Development and refinement of the Konza Prairie LTER research information management program. Pages 97–100 in W. K. Michener, J. W. Brunt, and S. G. Stafford, editors. Environmental information management and analysis: Ecosystem to global scales. Taylor and Francis, Bristol, Pennsylvania.

Briggs, J. M., B. J. Benson, M. Hartman, and R. C. Ingersoll. 1998. Data entry. Pages 29–31 in W. K. Michener, J. H. Porter, and S. G. Stafford, editors. Data and information management in the ecological sciences: A resource guide. Long-Term Ecological Research Network Office, University of New Mexico, Albuquerque.

Brock, T. D. 1985. A eutrophic lake: Lake Mendota, Wisconsin. Springer-Verlag, New York.

Brönmark, C. 1985. Interactions between macrophytes, epiphytes and herbivores: An experimental approach. Oikos 45:26–30.

Broughton, W. A. 1941. The geology, ground water and lake basin seal of the region south of the Muskellunge Moraine, Vilas County, Wisconsin. Transactions of the Wisconsin Academy of Sciences, Arts and Letters 33:5–20.

Brown, D. J. A. 1982. The effect of pH and calcium on fish and fisheries. Water, Air, and Soil Pollution 18:343–351.

Brown, J. H. 1971. Mammals on mountaintops: Non-equilibrium insular biogeography. American Naturalist 105:467–478.

Browne, R. A. 1981. Lakes as islands: Biogeographic distribution, turnover rates, and species composition in the lakes of central New York. Journal of Biogeography 8:75–83.

Brunt, J. W., P. McCartney, K. Baker, and S. G. Stafford. 2002. The future of ecoinformatics in long-term ecological research. Pages 367–372 in vol. VII: Informatics System Development II. Proceedings of the 6th World Multiconference on Systemics, Cybernetics and Informatics, Orlando, Florida.

Bullen, T. D., D. P. Krabbenhoft, and C. Kendall. 1996. Kinetic and mineralogic controls on the evolution of groundwater chemistry and $^{87}Sr/^{86}Sr$ in a sandy silicate aquifer, northern Wisconsin. Geochemica et Cosmochimics Acta 60:1807–1821.

Burgis, M. J., and P. Morris. 1987. The natural history of lakes. Cambridge University Press, Cambridge, England.

Burnett, A. W., M. E. Kirby, H. T. Mullins, and W. P. Patterson. 2003. Great Lake-effect snowfall during the twentieth century: A regional response to global warming. Journal of Climate 16:3535–3542.

Cáceres, C. E., and D. A. Soluk. 2002. Blowing in the wind: Field test of overland dispersal and colonization by aquatic invertebrates. Oecologia 131:402–408.

Callahan, J. T. 1984. Long-term ecological research. BioScience 34:363–367.

Capelli, G. M. 1975. Distribution, life history, and ecology of crayfish in northern Wisconsin, with emphasis on *Orconectes propinquus* (Girard). Ph.D. Thesis. University of Wisconsin-Madison.

Capelli, G. M. 1982. Displacement of northern Wisconsin crayfish by *Orconectes rusticus* (Girard). Limnology and Oceanography 27:741–745.

Capelli, G. M., and J. J. Magnuson. 1983. Morphoedaphic and biogeographical analysis of crayfish distribution in northern Wisconsin. Journal of Crustacean Biology 3:548–564.

Caraco, N. F., J. J. Cole, and G. E. Likens. 1992. New and recycled primary production in an oligotrophic lake: Insights for summer phosphorus dynamics. Limnology and Oceanography 37:590–602.

Carpenter, S. R. 1980. The decline of *Myriophyllum spicatum* in a eutrophic Wisconsin lake. Canadian Journal of Botany 58:527–535.

Carpenter, S. R. 1999. Acceptance speech, American Society of Limnology and Oceanography 1999, G. Evelyn Hutchinson Award. ASLO Bulletin 8(1):15–16.

Carpenter, S. R., and J. F. Kitchell, editors. 1993. The trophic cascade in lakes. Cambridge University Press, Cambridge, England.

Carpenter, S. R., and R. C. Lathrop. 1999. Lake restoration: Capabilities and needs. Hydrobiologia 395/396:19–28.

Carpenter, S. R., and D. M. Lodge. 1986. Effects of submerged macrophytes on ecosystem processes. Aquatic Botany 26:341–370.

Carpenter, S. R., D. Ludwig, and W. A. Brock. 1999. Management of eutrophication for lakes subject to potentially irreversible change. Ecological Applications 9:751–771.

Carpenter, S. R., T. M. Frost, D. Heisey, and T. K. Kratz. 1989. Randomized intervention analysis and the interpretation of whole-ecosystem experiments. Ecology 70:1142–1152.

Carpenter, S. R., S. W. Chisholm, C. J. Krebs, D. W. Schindler, and R. F. Wright. 1995. Ecosystem experiments. Science 269:324–327.

Carpenter, S. R., D. Bolgrien, R. C. Lathrop, C. A. Stow, T. Reed, and M. A. Wilson. 1998.

Ecological and economic analysis of lake eutrophication by nonpoint pollution. Australian Journal of Ecology 23:68–79.

Carpenter, S. R., J. F. Kitchell, K. L. Cottingham, D. E. Schindler, D. L. Christensen, D. M. Post, and N. Voichick. 1996. Chlorophyll variability, nutrient input, and grazing: Evidence from whole-lake experiments. Ecology 77:725–735.

Carpenter, S. R., T. M. Frost, J. F. Kitchell, T. K. Kratz, D. W. Schindler, J. A. Shearer, W. G. Sprules, M. J. Vanni, and A. P. Zimmerman. 1991. Patterns of primary production and herbivory in 25 North American lake ecoystems. Pages 67–96 in J. Cole, G. Lovett, and S. Findlay, editors. Comparative analyses of ecosystems: Patterns, mechanisms, and theories. Springer-Verlag, New York.

Carpenter, S. R., J. F. Kitchell, J. R. Hodgson, P. A. Cochran, J. J. Elser, M. M. Elser, D. M. Lodge, D. W. Kretchmer, X. He, and C. N. von Ende. 1987. Regulation of lake primary productivity by food web structure. Ecology 68:1863–1876.

Champion, G. S., and M. P. Anderson. 2000. Assessment of impacts on groundwater/lake systems using MODFLOW with a lake package: Application to the Trout Lake Basin, Northern Wisconsin WRI GRR 00-05, Water Resources Institute, University of Wisconsin-Madison.

Chapin, III, F. S., B. H. Walker, R. J. Hobbs, D. U. Hooper, J. H. Lawton, O. E. Sala, and S. Tilman. 1997. Biotic control over the functioning of ecosystems. Science 277:500–504.

Charles, D. F., and S. Christie, editors. 1991. Acidic deposition and aquatic ecosystems: Regional case studies. Springer-Verlag, New York.

Cheng, X., and M. P. Anderson. 1993. Numerical simulation of ground-water interaction with lakes allowing for fluctuating lake levels. Ground Water 31:929–933.

Cheng, X., and M. P. Anderson. 1994. Simulating the influence of lake position on groundwater. Water Resources Research 30:2041–2049.

Chipman, J. W., T. M. Lillesand, J. E. Schmmaltz, J. E. Leale, and M. J. Nordheim. 2004. Mapping lake water clarity with Landsat images in Wisconsin, U.S.A. Canadian Journal of Remote Sensing, Special collection on remote sensing and resource management in nearshore and inland waters.

Christensen, D. L., B. R. Herwig, D. E. Schindler, and S. R. Carpenter. 1996. Impacts of lakeshore residential development on coarse woody debris in north temperate lakes. Ecological Applications 6:1143–1149.

Christie, W. J. 1974. Changes in the fish species composition of the Great Lakes. Journal of the Fisheries Research Board of Canada 31:827–854.

Cisneros, R. O. 1993. Detection of cryptic invasions and local extinctions of fishes using a long–term database. M.S. Thesis. University of Wisconsin-Madison.

Colby, A. C. C., T. M. Frost, and J. M. Fischer. 1999. Sponge distribution and lake chemistry in northern Wisconsin lakes: Minna Jewell's survey revisited. Memoirs of the Queensland Museum 44:93–99.

Cole, J. J., N. F. Caraco, and G. E. Likens. 1990. Short-range atmospheric transport: A significant source of phsophorus to an oligotrophic lake. Limnology and Oceanography 35:1230–1237.

Cole, J. J., G. Lovett, and S. Findlay, editors. 1991. Comparative analyses of ecosystems: Patterns, mechanisms, and theories. Springer-Verlag, New York.

Cole, J. J., N. F. Caraco, G. W. Kling, and T. K. Kratz. 1994. Carbon dioxide supersaturation in the surface waters of lakes. Science 265:1568–1570.

Commoner, B. 1971. The closing circle: Nature, man, and technology. Alfred A. Knopf, New York.

Conde-Porcuna, J. M., and S. Declerck. 1998. Regulation of rotifer species by invertebrate predators in a hypereutrophic lake: Selective predation on egg-bearing females and induction of morphological defences. Journal of Plankton Research 20:604–618.

Connell, J. H. 1980. Diversity and the coevolution of competitors, or the ghost of competition past. Oikos 35:131–138.

Cook, R. E. 1977. Raymond Lindeman and the trophic-dynamic concept in ecology. Science 198:22–26.

Cooke, G. D., E. B. Welch, S. A. Peterson, and P. R. Newroth. 1993. Restoration and management of lakes and reservoirs. Lewis Publishers, Boca Raton, Florida.

Creaser, C. W. 1927. The smelt in Lake Michigan. Science 69:623.

Creaser, E. P. 1932. The decapod crustaceans of Wisconsin. Transactions of the Wisconsin Academy of Sciences, Arts and Letters 27:321–338.

Cronin, G., D. M. Lodge, M. E. Hay, M. Miller, A. M. Hill, T. Horvath, R. C. Bolser, N. Lindquist, and M. Wahla. 2002. Crayfish feeding preferences for freshwater macrophytes: The influence of plant structure and chemistry. Journal of Crustacean Biology 22:708–718.

Curtis, J. T. 1959. The vegetation of Wisconsin; an ordination of plant communities. University of Wisconsin Press, Madison.

Dennis, J. 1993. It's raining frogs and fishes. Harper Collins, UK, London.

Department of Agriculture Trade and Consumer Protection. 1995. Wisconsin 1994. Agricultural Statistics, Madison, Wisconsin.

Deppe, E. R., and R. C. Lathrop. 1993. Recent changes in the aquatic macrophyte community of Lake Mendota. Transactions of the Wisconsin Academy of Sciences, Arts and Letters 81:47–58.

De Stasio Jr., B. T., D. K. Hill, J. M. Kleinhans, N. P. Nibblelink, and J. J. Magnuson. 1996. Potential effects of global climate change on small north-temperate lakes: Physics, fish, and plankton. Limnology and Oceanography 41:1136–1149.

Devito, K. J., I. F. Creed, R. L. Rothwell, and E. E. Prepas. 2000. Landscape controls on the loading of phosporous to boreal lakes following timber harvest: A physical basis for adaptive riparian forest buffer strip strategies. Canadian Journal of Fisheries and Aquatic Sciences 57:1977–1984.

Diamond, J. M. 1969. Avifaunal equilibria and species turnover rates on the Channel Islands of California. Proceedings of the National Academy of Science 64:57–63.

DiDonato, G. T., and D. M. Lodge. 1993. Species replacements among *Orconectes* crayfishes in Wisconsin lakes: The role of predation by fishes. Canadian Journal of Fisheries and Aquatic Sciences 50:1484–1488.

Dillon, P. J., and R. H. Rigler. 1974. The phosphorus-chlorophyll relationship in lakes. Limnology and Oceanography 19:767–773.

Dodson, S. I. 1992. Predicting crustacean zooplankton species richness. Limnology and Oceanography 37:848–856.

Dodson, S. I., S. E. Arnott, and K. L. Cottingham. 2000. The relationship in lake communities between primary productivity and species richness. Ecology 81:2662–2679.

Dorn, N. J., and G. G. Mittelbach. 1999. More than predator and prey: A review of interactions between fish and crayfish. Vie et Milieu 49:229–237.

Dripps, W., M. P. Anderson, and K. W. Potter. 2001. Temporal and spatial variability of natural groundwater recharge WRI GRR 01–07, Water Resources Institute, University of Wisconsin-Madison.

Druse, M. J., and W. B. Neufeld. 1981. Petition to Vilas County Board of Supervisors and Vilas County Zoning, Planning, and Control Committee. Eagle River, Wisconsin.

Duarte, C. M. 1991. Variance and the description of nature. Pages 301–318 in J. Cole, G. Lovett, and S. Findlay, editors. Comparative analyses of ecosystems: Patterns, mechanisms, and theories. Springer-Verlag, New York.

Dymond, J. R. 1944. Spread of smelt, *Osmerus mordax*, in the Canadian waters of the Great Lakes. Canadian Field-Naturalist 58:12–14.

Eadie, J. M., and A. Keast. 1984. Resource heterogeneity and fish species diversity in lakes. Canadian Journal of Zoology 62:1689–1695.

Eadie, J. M., A. Hurly, R. D. Montgomerie, and K. L. Teather. 1986. Lakes and rivers as islands: Species-area relationships in the fish faunas of Ontario. Environmental Biology of Fishes 15:81–89.

Eaton, J. G., W. A. Swenson, J. H. McCormick, T. D. Simonson, and K. M. Jensen. 1992. A field and laboratory investigation of acid effects on largemouth bass, rock bass, black crappie, and yellow perch. Transactions of the American Fisheries Society 121:644–658.

Edgerton, F. N. 1962. The scientific contributions of Francois Alphonse Forel, the founder of limnology. Schweizerische Zeitschrift für Hydrobiologie 34:181–199.

Edmondson, W. T. 1991. The uses of ecology: Lake Washington and beyond. University of Washington Press, Seattle.

Eilers, J. M., D. F. Brakke, and D. H. Landers. 1986. Chemical and physical characteristics of lakes in the Upper Midwest. Environmental Science and Technology 22:164–172.

Eilers, J. M., G. E. Glass, K. E. Webster, and J. A. Rogalla. 1983. Hydrologic control of lake susceptibility to acidification. Canadian Journal of Fisheries and Aquatic Sciences 40:1896–1904.

Elder, J. F., D. P. Krabbenhoft, and J. F. Walker. 1992. Water, energy, and biogeochemical budgets (WEBB) program: Data availability and research at the northern temperate lakes site Wisconsin: U.S. Geological Survey 92–48.

Elder, J. F., N. B. Rybicki, V. Carter, and V. Weintraub. 2000. Sources and yields of dissolved carbon in northern Wisconsin stream catchments with differing amounts of peatland. Wetlands 20:113–125.

Eliot, T. S. 1934. The rock; a pageant play. Faber & Faber, London, England.

Elser, J. J., and R. P. Hassett. 1994. A stoichiometric analysis of the zooplankton-phytoplankton interaction in marine and freshwater ecosystems. Nature 370:211–213.

Elser, J. J., D. R. Doberfuhl, N. A. MacKay, and J. H. Schampel. 1996. Organism size, life history, and N:P stoichiometry. BioScience 46:674–684.

Elster, H. J. 1974. History of Limnology. Mitteilungen Internationale Veriningung für Theoretische und Angewandte Limnologie 20:7–30.

Elton, C. S. 1958. The ecology of invasions by animals and plants. Methuen, London.

Emerson, R. W. 1909. Essay X-Circles. Vol. 5 in C. W. Eliot, editor. The Harvard classics: Essays and English traits. P. F. Collier & Son, New York.

Eriksson, M. O. J., L. Henrikson, B. I. Nilsson, H. G. Oscarson, and A. E. Stenson. 1980. Predator-prey relations important for the biotic changes in acidified lakes. American Biology Teacher 9:248–249.

Evans, D. O., and D. H. Loftus. 1987. Colonization of inland lakes in the Great Lake region by rainbow smelt, *Osmerus mordax*: Their freshwater niche and effects on indigenous fishes. Canadian Journal of Fisheries and Aquatic Sciences 44:249–266.

Fallon, R. D., and T. D. Brock. 1980. Planktonic blue-green algae: Production, sedimentation, and decomposition in Lake Mendota, Wisconsin. Limnology and Oceanography 25:72–88.

Fang, X., and H. G. Stefan. 1998. Potential climate warming effects on ice covers of small lakes in the contiguous U.S. Cold Regions Science and Technology 27: 119–140.

Fassett, N. C. 1957. A manual of aquatic plants. University of Wisconsin Press, Madison.

Federal Geographic Data Committee. 1998. Content standard for digital geospatial metadata (revised June 1998) Federal Geographic Data Committee, Washington, D.C.

Fischer, J. M., and T. M. Frost. 1997. Indirect effects of lake acidification on *Chaoborus* population dynamics: The role of food limitation and predation. Canadian Journal of Fisheries and Aquatic Sciences 54:637–646.

Fischer, J. M., T. M. Frost, and A. R. Ives. 2001a. Compensatory dynamics in zooplankton community responses to acidification: Measurement and mechanisms. Ecological Applications 11:1060–1072.

Fischer, J. M., J. L. Klug, A. R. Ives, and T. M. Frost. 2001b. Ecological history affects zooplankton community responses to acidification. Ecology 82:2984–3000.

Fisher, M. M., J. L. Klug, G. H. Lauster, M. Newton, and E. W. Triplett. 2000. Effects of resources and trophic interactions on freshwater bacterioplankton diversity. Microbial Ecology 40:125–138.

Forbes, S. A. 1887. The lake as a microcosm. Bulletin of the Peoria Science Association, Illinois Natural History Survey Bulletin 15:537–550.

Forbes, S. A. 1912. Forbes Biological Station: The past and the promise. Page 6 in S. P. Havera and K. E. Roat, editors. Illinois Natural History Survey Special Publication 10.

Force, J. E., and G. E. Machlis. 1997. The human ecosystem. Part II: Social indicators in ecosystem management. Society and Natural Resources 10:369–382.

Franklin, J. F., C. S. Bledsoe, and J. T. Callahan. 1990. Contributions of the Long-Term Ecological Research Program. BioScience 40:509–523.

Franzin, W. G., B. A. Barton, R. A. Remnant, D. B. Wain, and S. J. Pagel. 1994. Range extension, present and potential distribution, and possible effects of rainbow smelt in Hudson Bay drainage waters of northwestern Ontario, Manitoba, and Minnesota. North American Journal of Fisheries Management 14:65–76.

Frey, D. G., editor. 1966. Limnology in North America. University of Wisconsin Press, Madison.

Frost, T. M. 1991. Porifera. Pages 95–124 in J. H. Thorp and A. P. Covich, editors. Ecology and classification of North American freshwater invertebrates. Academic Press, San Diego, California.

Frost, T. M., and E. R. Blood. 1996. The role of major research centers in the study of inland aquatic ecosystems. Pages 279–288 in Freshwater ecosystems: Revitalizing educational programs in limnology. National Academy Press, Washington, D.C.

Frost, T. M., and J. E. Elias. 1990. The balance of autotrophy and heterotrophy in three freshwater sponges with algal symbionts. Pages 478–484 in K. Rutzler, editor. New perspectives in sponge biology: Proceedings of the Third International Conference on the Biology of Sponges. Smithsonian Institution Press, Washington, D.C.

Frost, T. M., and P. K. Montz. 1988. Early zooplankton response to experimental acidification in Little Rock Lake, Wisconsin, USA. Verhandlungen Internationale Vereinigung für Limnologie 23:2279–2285.

Frost, T. M., and C. E. Williamson. 1980. In situ determination of the effect of symbiotic algae on the growth of the freshwater sponge *Spongilla lacustris*. Ecology 61:1361–1370.

Frost, T. M., S. R. Carpenter, and T. K. Kratz. 1992. Choosing ecological indicators: Effects of taxonomic aggregation on sensitivity to stress and natural variability. Pages 215–227 in D. H. McKenzie, D. E. Hyatt, and V. J. McDonald, editors. Ecological indicators. Elsevier Applied Science, New York.

Frost, T. M., T. K. Kratz, and J. J. Magnuson. 1999a. Center for Limnology-Trout Lake Station, University of Wisconsin-Madison. Bulletin of the Ecological Society of America 80:70–73.

Frost, T. M., P. K. Montz, and T. K. Kratz. 1998a. Zooplankton community responses during recovery from acidification: Limited persistence by acid-favored species in Little Rock Lake, Wisconsin. Restoration Ecology 6:336–342.

Frost, T. M., S. R. Carpenter, A. R. Ives, and T. K. Kratz. 1995. Species compensation and complimentarity in ecosystem function. Pages 224–239 in C. G. Jones and J. H. Lawton, editors. Linking species and ecosystems. Springer, New York.

Frost, T. M., D. L. DeAngelis, S. M. Bartell, D. J. Hall, and S. H. Hurlbert. 1988. Scale in the design and interpretation of aquatic community research. Pages 229–260 in S. R. Carpenter, editor. Complex interactions in lake communities. Springer-Verlag, New York.

Frost, T. M., P. K. Montz, M. J. Gonzalez, B. L. Sanderson, and S. E. Arnott. 1998b. Rotifer community responses to increased acidity: A comparison of long-term patterns in a whole-lake experiment and two unmanipulated lakes in Wisconsin, USA. Hydrobiologia 387/388: 141–152.

Frost, T. M., L. E. Graham, J. E. Elias, M. J. Haase, D. W. Kretchmer, and J. A. Kranzfelder. 1997. A yellow-green algal symbiont in the freshwater sponge *Corvomeyenia everetti*: Convergent evolution of symbiotic associations. Freshwater Biology 38:395–399.

Frost, T. M., P. K. Montz, T. K. Kratz, T. Badillo, P. L. Brezonik, M. J. Gonzalez, R. G. Rada, C. J. Watras, K. E. Webster, J. G. Wiener, C. E. Williamson, and D. P. Morris. 1999b. Multiple stresses from a single agent: Diverse responses to the experimental acidification of Little Rock Lake, Wisconsin. Limnology and Oceanography 44:784–794.

Galloway, J. N., G. E. Likens, and M. E. Hawley. 1984. Acid precipitation: Natural versus anthropogenic components. Science 226:829–831.

Garrison, P. J., S. R. Greb, D. R. Knauer, D. A. Wentz, J. T. Krohelski, J. G. Bockheim, S. A. Gherini, and C. W. Chen. 1987. Application of the ILWAS model to the northern Great Lake States. Lake and Reservoir Management 3:356–364.

Garvey, J. E., R. A. Stein, and H. M. Thomas. 1994. Assessing how fish predation and interspecific prey competition influence a crayfish assemblage. Ecology 75:532–547.

General Land Office Surveys. 1864. Wisconsin Board of Commissioners of Public Lands.

George, D. G., J. F. Talling, and E. Rigg. 2000. Factors influencing the temporal coherence of five lakes in the English Lake District. Freshwater Biology 43:449–461.

Gergel, S. E., M. A. Turner, and T. K. Kratz. 1999. Dissolved organic carbon as an indicator of the scale of watershed influence on lakes and rivers. Ecological Applications 9:1377–1390.

Gilbert, J. J., and R. Stemberger. 1985. Control of *Keratella* populations by interference from *Daphnia*. Limnology and Oceanography 30:180–118.

Gillett, N. P., H. F. Grat, and T. J. Osborn. 2003. Climate change and the North American Atlantic Oscillation. Pages 193–209 in J. W. Hurrell, Y. Kushnir, G. Otterson, and M. Visbeck, editors. The North Atlantic Oscillation, climate significance and environmental impact. American Geophysical Union, Washington, D.C.

Gilpin, M., and I. Hanski. 1991. Metapopulation dynamics: Empirical and theoretical investigations. Academic Press, San Diego, California.

Gitay, H., S. Brown, W. Easterling, B. Jallow, with J. Antle, M. Apps, R. Beamish, T. Chapin, W. Cramer, J. Frangi, J. Laine, L. Erda, J. Magnuson, I. Noble, J. Price, T. Prowse, T. Root, E. Schulze, O. Sirotenko, B. Sohngen, and J. Soussana, editors. 2001. Ecosystems and their goods and services. Cambridge University Press, Cambridge, England.

Glaser, P. H., J. A. Janssens, and D. I. Siegel. 1990. The response of vegetation to chemical and hydrological gradients in the Lost River peatland, northern Minnesota. Journal of Ecology 78:1021–1048.

Glaser, P. H., D. I. Siegel, E. A. Romanowicz, and Y. P. Shen. 1997. Regional linkages between raised bogs and the climate, groundwater, and landscape of north-western Minnesota. Journal of Ecology 85:3–16.

Goldman, C. R. 2000. Four decades of change in two subalpine lakes. Verhandlungen Internationale Vereinigung für Limnologie 27:7–26.

Golley, F. B. 1993. A history of the ecosystem concept in ecology: More than the sum of the parts. Yale University Press, New Haven, Connecticut.

Gonzalez, M. J. 1992. Effects of experimental acidification on zooplankton populations: A multiple-scale approach. Ph.D. Thesis. University of Wisconsin-Madison.

Gonzalez, M. J., and T. M. Frost. 1992. Food limitation and the seasonal population dynamics of rotifers. Oecologia 89:560–566.

Gonzalez, M. J., and T. M. Frost. 1994. Comparisons of laboratory bioassays and a whole-lake experiment: Rotifer responses to experimental acidification. Ecological Applications 4:69–80.

Gore, A. 1993. Earth in the balance: Ecology and the human spirit. Plume, New York.

Gotelli, N. J., and W. G. Kelley. 1993. A general model of metapopulation dynamics. Oikos 68:36–44.

Gough, R. 1997. Farming the cutover: A social history of northern Wisconsin, 1900–1940. University of Kansas Press, Lawrence.

Graumlich, L. J., L. B. Brubacker, and C. C. Grier. 1989. Long-term trends in forest net primary productivity: Cascade Mountains, Washington. Ecology 70:405–410.

Great Lakes Indian Fish and Wildlife Commission. 2004. Open water spearing in Northern Wisconsin by Chippewa Indians 1985–1998. Biological Services Administration Report. Odanah, Wisconsin. www.glifwc.org.

Gunderson, L. H. 1998. Briefing book for annual meeting of the resilience network. University of Florida, Gainesville.

Gunderson, L. H., C. S. Holling, and S. S. Light. 1995. Barriers and bridges to the renewal of ecosystems and institutions. Columbia University Press, New York.

Gurnell, A. M., K. J. Gregory, and G. E. Petts. 1995. The role of coarse woody debris in forest aquatic habitats: Implications for management. Aquatic Conservation: Marine and Freshwater Ecosystems 5:143–166.

Hagerthey, S. E., and W. C. Kerfoot. 1998. Groundwater flow influences the biomass and nutrient ratios of epibenthic algae in a north temperate seepage lake. Limnology and Oceanography 43:1227–1242.

Hairston Jr., N. G. 1996. Zooplankton egg banks as biotic reservoirs in changing environments. Limnology and Oceanography 41:1087–1092.

Hamed, K. H., and R. A. Rao. 1998. A modified Mann-Kendall trend test for autocorrelated data. Journal of Hydrology 204:182–196.

Hamilton, D. P., C. M. Spillman, K. L. Prescott, T. K. Kratz, and J. J. Magnuson. 2002. Effects of atmospheric nutrient inputs and climate change on the trophic status of Crystal Lake, Wisconsin. Verhandlungen Internationale Vereinigung für Limnologie 28: 467–470.

Hammer, R. B., S. I. Stewart, R. L. Winkler, V. C. Radcloff, and P. R. Voss. 2004. Characterizing dynamic spatial and temporal residential density patterns from 1940–1990 across north central United States. Landscape and Urban Planning 69:185–195.

Hanson, P. C., D. L. Bade, S. R. Carpenter, and T. K. Kratz. 2003. Lake metabolism:relationship with dissolved organic carbon and phosphorus. Limnology and Oceanography 48:1112–1119.

Hansson, L. A., H. Annadotter, E. Bergman, S. F. Hamrin, E. Jeppesen, T. Kairesalo, E. Luokkanen, P. A. Nilsson, M. Sondergaard, and J. Strand. 1998. Biomanipulation as an application of food-chain theory: Constraints, synthesis, and recommendations for temperate lakes. Ecosystems 1:558–574.

Harmon, M. E., J. F. Franklin, F. J. Swanson, P. Sollins, S. V. Gregory, J. D. Lattin, N. H.

Anderson, S. P. Cline, N. G. Aumen, J. R. Sedell, G. W. Lienkaemper, K. Cromack Jr., and K. W. Cummins. 1986. Ecology of coarse woody debris in temperate ecosystems. Advances in Ecological Research 15:133–302.

Harris, F., L. Krishtalka, et al. 2002. Long-Term Ecological Research Program twenty-year review. National Science Foundation. Report to NSF of the twenty-year review committee. Available at LTER Network Web page, http://intranet.lternet.edu/archives/documents/reports/20_yr_review/.

Hasler, A. D. 1947. Eutrophication of lakes by domestic drainage. Ecology 28:383–395.

Hasler, A. D. 1966. Wisconsin 1940–1961. Pages 55–93 in D. G. Frey, editor. Limnology in North America. University of Wisconsin Press, Madison.

Hasler, A. D., editor. 1975. Coupling of land and water systems. Springer-Verlag, New York.

Heberlein, T. A. 1988. Improving interdisciplinary research: Integrating the social and natural sciences. Society and Natural Resources 1:5–16.

Heckrath, G., P. C. Brookes, P. R. Poulton, and K. W. T. Goulding. 1995. Phosphorus leaching from soils containing different phosphorus concentrations in the Broadbalk experiment. Journal of Environmental Quality 24:904–910.

Heinselman, M. L. 1970. Landscape evolution, peatland types, and the environment in the Lake Agassiz Peatlands Natural Area, Minnesota. Ecological Monographs 40:235–261.

Henderson, A. 1911. George Bernard Shaw, his life and works. Stewart and Kidd, Cincinnati, Ohio.

Henshaw, D. L., G. Spycher, and S. M. Remillard. 2002. Transition from a legacy databank to an integrated ecological information system. Pages 373–378 in vol. VII: Informatics Systems Development II. Proceedings of the 6th World Multiconference on Systemics, Cybernetics and Informatics, Orlando, Florida.

Henshaw, D. L., M. Stubbs, B. J. Benson, K. S. Baker, D. Blodgett, and J. Porter. 1998. Climate database project: A strategy for improving information access across research sites. Pages 123–127 in W. K. Michener, J. H. Porter, and S. G. Stafford, editors. Data and information management in the ecological sciences: A resource guide. LTER (Long Term Ecological Research) Network Office, University of New Mexico, Albuquerque.

Hershey, A. E., G. M. Gettel, M. E. McDonald, M. C. Miller, H. Mooers, W. J. O'Brien, J. Paster, C. Richards, and J. A. Schuldt. 1999. A geomorphic-trophic model for landscape control of arctic lake food webs. BioScience 49:887–897.

Hill, A. M. 1994. Competition and predation as mechanisms of replacement of congeneric crayfishes by the exotic crayfish *Orconectes rusticus* (Girard). Ph.D. Thesis. University of Notre Dame.

Hill, A. M., and D. M. Lodge. 1994. Diel changes in resource demand: Competition and predation in species replacement among crayfishes. Ecology 75:2118–2126.

Hill, A. M., and D. M. Lodge. 1999. Replacement of resident crayfishes by an exotic crayfish: The roles of competition and predation. Ecological Applications 9:678–690.

Hillbricht-Ilkowska, A. 2002a. Nutrient loading and retention in lakes of the Jorka River system (Masurian Lakeland, Poland): Seasonal and long-term variation. Polish Journal of Ecology 50:459–474.

Hillbricht-Ilkowska, A. 2002b. River-lake system in a mosaic landscape: Main results and some implications for theory and practice from studies on the River Jorka system (Masurian Lakeland, Poland). Polish Journal of Ecology 50:543–550.

Holling, C. S., editor. 1978. Adaptive environmental assessment and management. John Wiley and Sons, New York.

Hope, D., T. K. Kratz, and J. L. Riera. 1996. Relationship between pCO_2 and dissolved organic carbon in northern Wisconsin lakes. Journal of Environmental Quality 25:1442–1445.

Houghton, J. T., Y. Ding, D. J. Griggs, M. Noguer, P. J. van der Linden, X. Dai, K. Maskell, and C. A. Johnson. 2001. Climate change 2001: The scientific basis. Cambridge University Press, Cambridge, England.

Howarth, R. W. 1991. Comparative responses of aquatic ecosystems to toxic chemical stress. Pages 169–195 in J. Cole, G. Lovett, and S. Findlay, editors. Comparative analyses of ecosystems. Springer-Verlag, New York.

Hrabik, T. R. 1999. Factors influencing fish distribution and condition within lakes and across landscapes. Ph.D. Thesis. University of Wisconsin-Madison.

Hrabik, T. R., and J. J. Magnuson. 1999. Simulated dispersal of exotic rainbow smelt in a northern Wisconsin lake district and implications for management. Canadian Journal of Fisheries and Aquatic Sciences 56:35–42.

Hrabik, T. R., and C. J. Watras. 2002. Recent declines on mercury concentration in a freshwater fishery: Isolating the effects of de-acidification and decreased atmospheric mercury deposition in Little Rock Lake. Science of the Total Environment 297:229–237.

Hrabik, T. R., M. P. Carey, and M. S. Webster. 2001. Interactions between young-of-year exotic rainbow smelt and native yellow perch in a northern temperate lake. Transactions of the American Fisheries Society 130:568–582.

Hrabik, T. R., J. J. Magnuson, and A. S. McLain. 1998. Predicting the effects of rainbow smelt on native fishes in small lakes: Evidence from long-term research on two lakes. Canadian Journal of Fisheries and Aquatic Sciences 55:1364–1371.

Hrabik, T. R., B. K. Greenfield, D. B. Lewis, A. I. Pollard, K. A. Wilson, and T. K. Kratz. 2005. Landscape scale variation in taxonomic diversity in four groups of aquatic organisms: The influence of physical, chemical, and biological properties. Ecosystems 8.

Hunt, R. J., M. P. Anderson, and V. A. Kelson. 1998. Improving a complex finite-difference ground water flow model through the use of an analytic element screening model. Ground Water 36:1011–1017.

Hunt, R. J., C. D. Pint, and M. P. Anderson. 2003. Using diverse data types to calibrate a watershed model of the Trout Lake basin, northern Wisconsin. Pages 600–604 in Proceedings of MODFLOW 2003, IGWMC in Golden, Colorado.

Hunter, R. D., and W. W. Lull. 1977. Physiologic and environmental factors influencing the calcium-to-tissue ratio in populations of three species of freshwater pulmonate snails. Oecologia 29:205–218.

Hunter, T. S., and T. E. Croley. 1993. Great Lakes monthly hydrologic data, NOAA data report ERL GLERL National Oceanic and Atmospheric Administration, Great Lakes Environmental Research Laboratory, Springfield, Virginia, 22161 (http://www.glerl.noaa.gov/data/arc/hydro/mnth–hydro.html).

Hurley, J. P. 1984. Nutrient cycling in three northern Wisconsin lakes. M.S. Thesis. University of Wisconsin-Madison.

Hurley, J. P., and D. E. Armstrong. 1991. Pigment preservation in lake sediments: A comparison of sedimentary environments in Trout Lake, Wisconsin. Canadian Journal of Fisheries and Aquatic Sciences 48:472–486.

Hurley, J. P., and C. J. Watras. 1991. Identification of bacteriochlorophylls in lakes via reverse-phase HPLC. Limnology and Oceanography 36:307–315.

Hurley, J. P., D. E. Armstrong, and A. L. DuVall. 1992. Historical interpretation of pigment stratigraphy in Lake Mendota sediments. Pages 49–68 in J. F. Kitchell, editor. Food web management: A case study of Lake Mendota. Springer-Verlag, New York.

Hurley, J. P., D. E. Armstrong, G. J. Kenoyer, and C. J. Bowser. 1985. Groundwater as a silica source for diatom production in a precipitation-dominated lake. Science 227:1576–1578.

Hurst, J. W. 1964. Law and economic growth: The legal history of the lumber industry in Wisconsin, 1836–1915. M.S. Thesis. Harvard University, Cambridge, Massachusetts.

Imberger, J., and J. C. Patterson. 1990. Physical limnology. Advances in Applied Mechanics 27:303–475.

Ingersoll, R. C., T. R. Seastedt, and M. A. Hartman. 1997. A model information management system for ecological research. BioScience 47:310–316.

Inouye, R. S., and D. Tilman. 1988. Convergence and divergence of old-field plant communities along experimental nitrogen gradients. Ecology 69:1872–1887.

Jackson, D. A., and H. H. Harvey. 1989. Biogeographic associations in fish assemblages: Local vs. regional processes. Ecology 70:1472–1484.

Jacobson, P. T. 1990. Pattern and process in the distribution of cisco, *Coregonus artedii*, in Trout Lake, Wisconsin. Ph.D. Thesis. University of Wisconsin-Madison.

Jävinen, M., M. Rask, J. Ruuhijarvi, and L. Arvola. 2002. Temporal coherence in water temperature and chemistry under the ice of boreal lakes (Finland). Water Research 36: 3949–3956.

Jenkins, D. G. 1995. Dispersal-limited zooplankton distribution and community composition in new ponds. Hydrobiologia 313/314:15–20.

Jensen, H. S., and F. O. Andersen. 1992. Importance of temperature, nitrate, and pH for phosphate release from aerobic sediments of four shallow, eutrophic lakes. Limnology and Oceanography 37:577–589.

Jensen, H. S., P. Kristensen, E. Jeppesen, and A. Skytthe. 1992. Iron:phosphorus ratio in surface sediment as an indicator of phosphate release from aerobic sediments in shallow lakes. Hydrobiologia 235/236:731–743.

Jewell, M. E. 1935. An ecological study of the fresh-water sponges of northern Wisconsin. Ecological Monographs 5:461–504.

John, R., C. D. Pint, M. P. Anderson, and R. J. Hunt. 2003. Effects of potential climate change on lake levels and capture zones. Pages 212–216 in Proceedings of MODFLOW 2003, IGWMC, Golden, Colorado.

Johnson, B. M., and S. R. Carpenter. 1994. Functional and numerical responses: A framework for fish-angler interactions. Ecological Applications 4:808–821.

Johnson, B. M., and M. D. Staggs. 1992. The fishery. Pages 353–375 in J. F. Kitchell, editor. Food web management: A case study of Lake Mendota. Springer-Verlag, New York.

Johnson, B. M., S. J. Gilbert, R. S. Stewart, L. G. Rudstam, Y. Allen, D. M. Fago, and D. Dreikosen. 1992. Piscivores and their prey. Pages 319–351 in J. F. Kitchell, editor. Food web management: A case study of Lake Mendota. Springer-Verlag, New York.

Johnson, L. B., and S. H. Gage. 1996. A landscape perspective for analyzing aquatic ecosystems. Freshwater Biology 36:101–120.

Johnson, T. B., and J. F. Kitchell. 1996. Long-term changes in zooplanktivorous fish community composition: Implications for food webs. Canadian Journal of Fisheries and Aquatic Sciences 53:2792–2803.

Jones, M. B., C. Berkelye, J. Bojilova, and M. Schildhauer. 2001. Managing scientific metadata. IEEE Internet Computing 5:59–68.

Jones, M. S., J. P. Goettl, and S. A. Flickinger. 1994. Changes in walleye food habits and growth following a rainbow smelt introduction. North American Journal of Fisheries Management 14:409–414.

Jorgensen, B. S. 2002. Relating the concepts of Sense of Place and Sense of Community in environmental management. In Eighth Trans-Tasman Community Psychology Conference, Perth, Western Australia.

Jorgensen, B. S., and R. Stedman. 2001. Sense of place as an attitude: Lakeshore property owners' attitudes toward their properties. Journal of Environmental Psychology 21:233–248.

Juday, C., and E. A. Birge. 1933. The transparency, the color and the specific conductance of the lake waters of northeastern Wisconsin. Transactions of the Wisconsin Academy of Sciences, Arts and Letters 28:205–259.

Juday, C., and V. W. Meloche. 1944. Physical and chemical evidence relating to the lake basin seal in certain areas of the Trout Lake region of Wisconsin. Transactions of the Wisconsin Academy of Sciences, Arts and Letters 35:157–174.

Juday, C., E. A. Birge, and V. W. Meloche. 1935. The carbon dioxide and hydrogen ion content in the lake waters of northeastern Wisconsin. Transactions of the Wisconsin Academy of Sciences, Arts and Letters 29:1–82.

Kalff, J. 2002. Limnology; inland water ecosystems. Prentice Hall, Upper Saddle River, New Jersey.

Kawasaki, T. 1993. Why do some pelagic fishes have wide fluctuations in their numbers? Biological basis of fluctuations from the point of view of evolutionary ecology. Food and Agriculture Organization Fisheries Report 291:1065–1080.

Kay, A. 1971. Statement at a meeting of Xerox PARC team. Founder of Xerox Corporation's Palo Alto Research Center and inventor of Small talk.

Keddy, P. A. 1976. Lakes as islands: The distributional ecology of two aquatic plants, *Lemna minor* L. and *L. Trisulca* L. Ecology 57:353–359.

Keeley, J. E. 1983. Crassulacean acid metabolism in the seasonally submerged aquatic *Isoetes howellii*. Oecologia 58:57–62.

Keller, W., N. D. Yan, K. E. Holtze, and J. R. Pitblado. 1990. Inferred effects of lake acidification on *Daphnia galeata mendotae*. Environmental Science and Technology 24:1259–1261.

Kenny, D. A., and L. La Voie. 1985. Separating individual and group effects. Journal of Personality and Social Psychology 48:339–348.

Kenoyer, G. J., and M. P. Anderson. 1989. Groundwater's dynamic role in regulating acidity and chemistry in a precipitation-dominated lake. Journal of Hydrology 109:287–306.

Kenoyer, G. J., and C. J. Bowser. 1992a. Groundwater chemical evolution in a sandy silicate aquifer in northern Wisconsin; 1: Patterns and rates of change. Water Resources Research 28:579–589.

Kenoyer, G. J., and C. J. Bowser. 1992b. Groundwater chemical evolution in a sandy silicate aquifer in northern Wisconsin; 2: Reaction modeling. Water Resources Research 28:591–600.

Kim, K., M. P. Anderson, and C. J. Bowser. 1999. Model calibration with multiple targets: A case study. Ground Water 37:345–351.

Kim, K., M. P. Anderson, and C. J. Bowser. 2000. Enhanced dispersion in groundwater caused by temporal changes in recharge rate and lake levels. Advances in Water Resources 23:625–635.

Kitchell, J. F., editor. 1992. Food web management: A case study of Lake Mendota. Springer-Verlag, New York.

Kitchell, J. F., and S. R. Carpenter. 1993. Variability in lake ecosystems: Complex responses by the apical predator. Pages 111–124 in M. J. McDonnell and S. T. A. Picket, editors. Humans as components of ecosystems. Springer-Verlag, New York.

Kitchell, J. F., and P. R. Sanford. 1992. Paleolimnological evidence of food web dynamics in Lake Mendota. Pages 31–47 in J. F. Kitchell, editor. Food web management: A case study of Lake Mendota. Springer-Verlag, New York.

Klessig, L. L. 1973. Recreational property owners and their institutional alternatives for resource protection: The case of Wisconsin lakes. Ph.D. Thesis. University of Wisconsin-Madison.

Kling, G. W., M. C. Kipphut, and M. C. Miller. 1991. Arctic lakes and streams as conduits to the atmosphere: Implications for tundra carbon budgets. Science 251:298–301.

Kling, G. K., G. W. Kipphut, M. M. Miller, and W. J. O'Brien. 2000. Integration of lakes and streams in a landscape perspective: The importance of material processing on spatial patterns and temporal coherence. Freshwater Biology 43:477–497.

Kling, G. K., K. Hayhoe, L. B. Johnson, J. J. Magnuson, S. Polasky, S. K. Robinson, B. J. Shuter, M. M. Wander, D. J. Wuebbles, D. R. Zak, S. C. Lindroth, S. C. Moser, and M. L. Wilson. 2003. Confronting climate change in the Great Lakes Region: Impacts on our communities and ecosystems. Union of Concerned Scientists, Cambridge Massachusetts and the Ecological Society of America, Washington, D.C.

Knight, S. E. 1988. The ecophysical significance of carnivory in *Utricularia vulgaris*. Ph.D. Thesis. University of Wisconsin-Madison.

Knight, S. E. 1992. Costs of carnivory in the common bladderwort, *Utricularia macrorhiza*. Oecologia 89:349–355.

Knight, S. E., and T. M. Frost. 1991. Bladder control in *Utricularia macrorhiza*: Lake-specific variation in plant investment in carnivory. Ecology 72:728–734.

Koonce, J. F., T. B. Bagenal, R. F. Carline, K. E. F. Hokanson, and M. Nagiec. 1977. Factors influencing year-class strength of percids: A summary and a model of temperature effects. Journal of Fish Research Board of Canada 34:1900–1909.

Krabbenhoft, D. P., and K. E. Webster. 1995. Transient hydrogeological controls in the chemistry of a seepage lake. Water Resources Research 31:2295–2305.

Krabbenhoft, D. P., M. P. Anderson, and C. J. Bowser. 1990a. Estimating groundwater exchange with Sparkling Lake, Wisconsin, 2: Calibration of a three-dimensional, solute transport model to a stable isotope plume. Water Resources Research 26:2455–2462.

Krabbenhoft, D. P., M. P. Anderson, C. J. Bowser, and J. Valley. 1990b. Estimating groundwater exchange with Sparkling Lake, Wisconsin, 1: Use of the stable isotope mass-balance method. Water Resources Research 26:2445–2453.

Krabbenhoft, D. P., C. J. Bowser, C. Kendall, and J. R. Gat. 1994. Use of oxygen-18 and deuterium to assess the hydrology of ground-water/lake systems. Pages 67–90 in L. A. Baker, editor. Environmental chemistry of lakes and reservoirs. American Chemical Society, Washington, D.C.

Kratz, T. K., and T. M. Frost. 2000. The ecological organisation of lake districts: General introduction. Freshwater Biology 43:297–299.

Kratz, T. K., and V. L. Medland. 1989. Relationship of landscape position and groundwater input in northern Wisconsin kettle-hole peatlands. Pages 1141–1151 in R. R. Sharitz and J. W. Gibbons, editors. Freshwater wetlands and wildlife: Proceedings of a symposium held at Charleston, South Carolina, March 24–27, 1986. Office of Scientific and Technical Information, U.S. Dept. of Energy, Washington, D.C.

Kratz, T. K., T. M. Frost, and J. J. Magnuson. 1987a. Inferences from spatial and temporal variability in ecosystems: Long-term zooplankton data from lakes. American Naturalist 129:830–846.

Kratz, T. K., R. B. Cook, C. J. Bowser, and P. L. Brezonik. 1987b. Winter and spring pH depressions in northern Wisconsin lakes caused by increases in pCO_2. Canadian Journal of Fisheries and Aquatic Sciences 44:1082–1088.

Kratz, T. K., T. M. Frost, J. E. Elias, and R. B. Cook. 1991a. Reconstruction of a regional, 12,000-yr silica decline in lakes by means of fossil sponge spicules. Limnology and Oceanography 36:1244–1249.

Kratz, T. K., J. J. Magnuson, C. J. Bowser, and T. M. Frost. 1986. Rationale for data collection and interpretation in the Northern Lakes Long-Term Ecological Program. Pages 22–33 in B. G. Isom, editor. Rationale for sampling and interpretation of ecological data in the assessment of freshwater ecosystems: A symposium sponsored by ASTM Committee D-19 on Water, Philadelphia, Pennsylvania, 31 Oct.–1 Nov. 1983. American Society for Testing and Materials, Philadelphia, Pennsylvania.

Kratz, T. K., B. J. Benson, E. R. Blood, G. L. Cunningham, and R. A. Dahlgren. 1991b. The influence of landscape position on temporal variability in four North American ecosystems. American Naturalist 138:355–378.

Kratz, T. K., J. Schindler, D. Hope, J. L. Riera, and C. J. Bowser. 1997a. Average annual carbon dioxide concentrations in eight neighboring lakes in northern Wisconsin, USA. Verhandlungen Internationale Vereinigung für Limnologie 26:335–338.

Kratz, T. K., K. E. Webster, C. J. Bowser, J. J. Magnuson, and B. J. Benson. 1997b. The influence of landscape position on lakes in northern Wisconsin. Freshwater Biology 37:209–217.

Kratz, T. K., P. A. Soranno, S. B. Baines, B. J. Benson, J. J. Magnuson, T. M. Frost, and R. C. Lathrop. 1998. Inter-annual synchronous dynamics in north temperate lakes in Wisconsin, USA. Pages 273–287 in D. G. George, J. C. Jones, P. Puncochar, D. S. Reynolds, and D. W. Sutcliffe, editors. Management of lakes and reservoirs during global climate change. Kluwer Academic Publishers, The Netherlands.

Kratz, T. K., J. J. Magnuson, P. Bayley, B. J. Benson, C. W. Berish, C. S. Bledsoe, E. R. Blood, C. J. Bowser, S. R. Carpenter, G. L. Cunningham, R. A. Dahlgren, T. M. Frost, J. C. Halfpenny, J. D. Hansen, D. Heisey, R. Inouye, D. W. Kaufman, A. McKee, and J. Yarie. 1995. Temporal and spatial variability as neglected ecosystem properties: Lessons learned from 12 North American ecosystems. Pages 359–383 in D. Rapport and P. Calow, editors. Evaluating and monitoring the health of large-scale ecosystems. Springer-Verlag, New York.

Krohelski, J. T., and W. G. Batten. 1995. Simulation of state and hydrologic budget of Devils Lake, Sauk County, Wisconsin. U.S. Geological Survey, Middleton, Wisconsin.

Krohelski, J. T., D. T. Feinstein, and B. N. Lenz. 1999. Simulation of stage and hydrologic budget for Shell Lake, Washburn County, Wisconsin. Water Resources Investigations Report 02–4014 U.S. Geological Survey, Middleton, Wisconsin.

Krohelski, J. T., Y.-F. Lin, W. J. Rose, and R. J. Hunt. 2002. Simulation of Fish, Mud and Crystal Lakes and the shallow ground-water system, Dane County, Wisconsin. Water Resources Investigation Report 02–4014 U.S. Geological Survey, Middleton, Wisconsin.

Krueger, D. M. 2003. The control of exotic fish species through food web management: Implications for the recovery of native species. M.S. Thesis. University of Minnesota-Duluth.

Krug, W. R. 1999. Simulation of the effects of operating Lakes Mendota, Monona, and Waubesa, south-central Wisconsin, as multipurpose reservoirs to maintain dry-weather flow. Open-File Report 99-67 U.S. Geological Survey, Middleton, Wisconsin.

Kunkel, K. E., K. Andsager, and D. R. Easterling. 1999. Long-term trends in heavy precipitation events over North America. Journal of Climate 12:2513–2525.

Lassen, H. H. 1975. The diversity of freshwater snails in view of the equilibrium theory of island biogeography. Oecologia 19:1–8.

Lathrop, R. C. 1990. Response of Lake Mendota (Wisconsin, U.S.A.) to decreased phosphorus loadings and the effect on downstream lakes. Verhandlungen Internationale Vereinigung für Limnologie 24:457–463.

Lathrop, R. C. 1992a. Nutrient loadings, lake nutrients and water clarity. Pages 69–96 in J. F. Kitchell, editor. Food web management: A case study of Lake Mendota. Springer-Verlag, New York.

Lathrop, R. C. 1992b. Benthic macroinvertebrates. Pages 173–192 in J. F. Kitchell, editor. Food web management: A case study of Lake Mendota. Springer-Verlag, New York.

Lathrop, R. C. 1992c. Decline in zoobenthos densities in the profundal sediments of Lake Mendota (Wisconsin, U.S.A.). Hydrobiologia 235/236:353–361.

Lathrop, R. C. 1998. Water clarity responses to phosphorus and *Daphnia* in Lake Mendota. Ph.D. Thesis. University of Wisconsin-Madison.

Lathrop, R. C., and S. R. Carpenter. 1992a. Phytoplankton and their relationship to nutrients. Pages 97–126 in J. F. Kitchell, editor. Food web management: A case study of Lake Mendota. Springer-Verlag, New York.

Lathrop, R. C., and S. R. Carpenter. 1992b. Zooplankton and their relationship to phytoplankton. Pages 127–150 in J. F. Kitchell, editor. Food web management: A case study of Lake Mendota. Springer-Verlag, New York.

Lathrop, R. C., S. R. Carpenter, and D. M. Robertson. 1999. Summer water clarity responses to phosphorus, *Daphnia* grazing, and internal mixing in Lake Mendota. Limnology and Oceanography 44:137–146.

Lathrop, R. C., S. R. Carpenter, and D. M. Robertson. 2000. Interacting factors causing exceptional summer water clarity in Lakes Mendota and Monona. Verhandlungen Internationale Vereinigung für Limnologie 27:1776–1779.

Lathrop, R. C., S. R. Carpenter, and L. G. Rudstam. 1996. Water clarity in Lake Mendota since 1900: Responses to differing levels of nutrients and herbivory. Canadian Journal of Fisheries and Aquatic Sciences 53:2250–2261.

Lathrop, R. C., T. M. Lillesand, and B. S. Yandell. 1991. Testing the utility of simple multi-date Thematic Mapper calibration algorithms for monitoring turbid inland waters. International Journal of Remote Sensing 12:2045–2063.

Lathrop, R. C., S. B. Nehls, C. L. Brynildson, and K. R. Plass. 1992. The fishery of the Yahara Lakes. Technical Bulletin Number 181. Wisconsin Department of Natural Resources, Madison.

Lathrop, R. C., S. R. Carpenter, C. A. Stow, P. A. Soranno, and J. C. Panuska. 1998. Phosphorus loading reductions needed to control blue-green algal blooms in Lake Mendota. Canadian Journal of Fisheries and Aquatic Sciences 55:1169–1178.

Lathrop, R. C., B. M. Johnson, T. B. Johnson, M. T. Vogelsang, S. R. Carpenter, T. R. Hrabik, J. F. Kitchell, J. J. Magnuson, L. G. Rudstam, and R. S. Stewart. 2002. Stocking piscivores to improve fishing and water clarity: A synthesis of the Lake Mendota biomanipulation project. Freshwater Biology 47:2410–2424.

Lawrence, T. M., R. D. Fuller, and C. T. Driscoll. 1986. Spatial relationships of aluminum chemistry in the streams of the Hubbard Brook Experimental Forest, New Hampshire. Biogeochemistry 2:115–135.

Leathers, D. J., and A. W. Ellis. 1996. Synoptic mechanisms associated with snowfall increases to the lee of Lakes Erie and Ontario. International Journal of Climatology 16:1117–1135.

Lee, D. R. 1977. A device for measuring seepage flux in lakes and estuaries. Limnology and Oceanography 22:140–147.

Lee, R. G., D. R. Field, and W. R. Burch Jr. 1990. Community and forestry: Continuities in the sociology of natural resources. Westview Press, Boulder, Colorado.

Lehner, C. E. 1980. Report on the inventory and future use of historical water chemistry and plankton data of Birge, Juday, and associates. Center for Limnology, University of Wisconsin-Madison.

Lenters, J. D., M. T. Coe, and J. A. Foley. 2000. Surface water balance of the continental United States, 1963–1995: Regional evaluation of a terrestrial biosphere model and the NCEP/NCAR reanalysis. Journal of Geophysical Research 105:22, 393–322, 425.

Lenters, J. D., T. K. Kratz, and C. J. Bowser. 2005. Effects of climate variability on Lake evaporation: Results from a long-term energy budget study of Sparkling Lake, northern Wisconsin (U.S.A.). Journal of Hydrology.

Leopold, A. 1949. Thinking like a mountain. Pages 129–133 in A Sand County almanac and sketches here and there. Oxford University Press, New York.

Lewis, D. B. 2000. Niche-based and biogeographic constraints on benthic communities. Ph.D. Thesis. University of Wisconsin-Madison.

Lewis, D. B. 2001. Trade-offs between growth and survival: Responses of freshwater snails to predacious crayfish. Ecology 82:758–765.

Lewis, D. B., and J. J. Magnuson. 1999. Intraspecific gastropod shell strength variation among north temperate lakes. Canadian Journal of Fisheries and Aquatic Sciences 56:1687–1695.

Lewis, D. B., and J. J. Magnuson. 2000. Landscape spatial patterns in freshwater snail assemblages across Northern Highland catchments. Freshwater Biology 43:409–420.

Likens, G. E., F. H. Borman, R. S. Pierce, J. S. Eaton, and N. M. Johnson. 1977. Biogeochemistry of a forested ecosystem. Springer-Verlag, New York.

Likens, G. E., editor. 1985. An ecosystem approach to aquatic ecology: Mirror Lake and its environment. Springer-Verlag, New York.

Lillesand, T. M., and R. W. Kiefer. 2003. Remote sensing and image interpretation, 5th edition. John Wiley, New York.

Lindeman, R. L. 1942. The trophic-dynamic aspect of ecology. Ecology 23:399–418.

Livingstone, D. M. 2000. Large-scale climatic forcing detected in historical observations of lake ice break-up. Verhandlungen Internationale Vereinigung für Limnologie 27:2775–2783.

Lodge, D. M. 1991. Herbivory on freshwater macrophytes. Aquatic Botany 41:195–224.

Lodge, D. M., and A. M. Hill. 1994. Factors governing species composition, population size, and productivity of cool-water crayfishes. Nordic Journal of Freshwater Research 69:111–136.

Lodge, D. M., and J. G. Lorman. 1987. Reductions in submersed macrophyte biomass and species richness by the crayfish Orconectes rusticus. Canadian Journal of Fisheries and Aquatic Sciences 44:591–597.

Lodge, D. M., D. P. Krabbenhoft, and R. G. Striegl. 1989. A positive relationship between groundwater velocity and submersed macrophyte biomass in Sparkling Lake, Wisconsin. Limnology and Oceanography 34:235–239.

Lodge, D. M., T. K. Kratz, and G. M. Capelli. 1986. Long-term dynamics of three crayfish species in Trout Lake, Wisconsin. Canadian Journal of Fisheries and Aquatic Sciences 43:993–998.

Lodge, D. M., M. W. Kershner, J. E. Aloi, and A. P. Covich. 1994. Effects of an omniverous crayfish (Orconectes rusticus) on a freshwater food web. Ecology 75:1265–1281.

Lodge, D. M., K. M. Brown, S. P. Klosiewski, R. A. Stein, A. P. Covich, B. K. Leathers, and C. Brönmark. 1987. Distribution of freshwater snails: Spatial scale and the relative importance of physiochemical and biotic factors. American Malacological Bulletin 5:73–84.

Lodge, D. M., R. A. Stein, K. M. Brown, A. P. Covich, C. Bronmark, J. A. Garvey, and S. P. Klosiewski. 1998. Predicting impact of freshwater exotic species on native biodiversity: Challenges in spatial scaling. Australian Journal of Ecology 23:53–67.

Lofgren, B. M., F. H. Quinn, A. H. Clites, R. A. Assel, A. J. Eberhardt, and C. L. Luukkonen. 2002. Evaluation of potential impacts on Great Lakes water resources based on climate scenarios of two GCMs. Journal of Great Lakes Research 28:537–554.

Loftus, D. H., and P. F. Hulsman. 1986. Predation on larval lake whitefish (Coregonus *clupeaformis*) and lake herring (*Coregonus artedii*) by adult rainbow smelt (*Osmerus mordax*). Canadian Journal of Fisheries and Aquatic Sciences 43:812–818.

Lomolino, M. V. 1984. Mammalian island biogeography: Effects of area, isolation and vagility. Oecologia 61:376–382.

Lorman, J. G. 1980. Ecology of the crayfish *Orconectes rusticus* in northern Wisconsin. Ph.D. Thesis. University of Wisconsin-Madison.

Lorman, J. G., and J. J. Magnuson. 1978. The role of crayfishes in aquatic ecosystems. Fisheries 3:8–19.

Loucks, O. L., and W. E. Odum. 1978. Analysis of five North American lake ecosystems, I. A strategy for comparison. Verhandlungen Internationale Vereinigung für Limnologie 20:556–561.

Ludwig, H. R., and J. A. Leitch. 1996. Interbasin transfer of aquatic biota via anglers' bait buckets. Fisheries 21:14–18.

Luecke, C., M. J. Vanni, J. J. Magnuson, J. F. Kitchell, and P. T. Jacobson. 1990. Seasonal regulation of *Daphnia* populations by planktivorous fish: Implications for the spring clear-water phase. Limnology and Oceanography 35:1718–1733.

Lukaszewski, Y., S.E. Arnott, and T. M. Frost. 1999. Regional versus local processes in determining zooplankton community composition of Little Rock Lake, Wisconsin, U.S.A. Journal of Plankton Research 21:991–1003.

Luttenton, M. R., M. J. Horgan, and D. M. Lodge. 1998. Effects of three *Orconectes* crayfishes on epilithic microalgae: A laboratory experiment. Crustaceana 71:845–855.

Lyons, J. 1987. Prey choice among piscivorous juvenile walleyes. Canadian Journal of Fisheries and Aquatic Sciences 44:758–764.

Lyons, J. 1989. Changes in the abundance of small littoral-zone fishes in Lake Mendota, Wisconsin. Canadian Journal of Zoology 67:2910–2916.

Lyons, J., and J. J. Magnuson. 1987. Effects of walleye predation on the population dynamics of small littoral-zone fishes in a northern Wisconsin lake. Transactions of the American Fisheries Society 116:29–39.

MacArthur, R. H., and E. O. Wilson. 1963. An equilibrium theory of island biogeography. Evolution 17:373–387.

MacArthur, R. H., and E. O. Wilson. 1967. The theory of island biogeography. Princeton University Press, Princeton, New Jersey.

Machlis, G. E., J. E. Force, and W. R. Burch Jr. 1997. The human ecosystem, Part I: The human ecosystem as an organizing concept in ecosystem management. Society and Natural Resources 10:347–367.

MacIsaac, H. J., and J. J. Gilbert. 1990. Does exploitative or interference competition from *Daphnia* limit the abundance of *Keratella* in Loch Leven? A reassessment of May and Jones (1989). Journal of Plankton Research 12:1315–1322.

Magnuson, J. J. 1976. Managing with exotics—a game of chance. Transactions of the American Fisheries Society 105:1–9.

Magnuson, J. J. 1988. Two worlds for fish recruitment: Lakes and oceans. Early Life History Series Publications. Proceedings of the 11th Annual Larval Fish Conference. American Fisheries Society Symposium 5:1–6.

Magnuson, J. J. 1990. Long-term ecological research and the invisible present. BioScience 40:495–501.

Magnuson, J. J. 2002a. Signals from ice cover trends and variability. Fisheries in a changing climate. Pages 3–14 in vol. 32. American Fisheries Society Symposium, Bethesda, Maryland, American Fisheries Society.

Magnuson, J. J. 2002b. Three generations of limnology at the University of Wisconsin-Madison. Verhandlungen Internationale Vereinigung fur Limnologie 28:856–860.

Magnuson, J. J., and A. L. Beckel. 1985. Exotic species: A case of biological pollution. Wisconsin Academy Review 32:8–10.

Magnuson, J. J., and C. J. Bowser. 1990. A network for long-term ecological research in the United States. Freshwater Biology 23:137–143.

Magnuson, J. J., and T. K. Kratz. 2000. Lakes in the landscape: Approaches to regional limnology. Verhandlungen Internationale Vereinigung für Limnologie 27:74–87.

Magnuson, J. J., and R. C. Lathrop. 1992. Historical changes in the fish community. Pages 193–231 in J. F. Kitchell, editor. Food web management: A case study of Lake Mendota. Springer-Verlag, New York.

Magnuson, J. J., B. J. Benson, and T. K. Kratz. 1990. Temporal coherence in the limnology of a suite of lakes in Wisconsin, U.S.A. Freshwater Biology 23:145–149.

Magnuson, J. J., B. J. Benson, and T. K. Kratz. 2004. Patterns of coherent dynamics within and between lake districts at local to intercontinental scales. Boreal Environmental Research 9:359–369.

Magnuson, J. J., B. J. Benson, and A. S. McLain. 1994. Insights on species richness and turnover from long-term ecological research: Fishes in north temperate lakes. American Zoologist 34:437–451.

Magnuson, J. J., C. J. Bowser, and A. L. Beckel. 1983 (fall). The invisible present: Long-term ecological research on lakes. Pages 3–6. Letters and Science Magazine, College of Letters and Science, University of Wisconsin-Madison.

Magnuson, J. J., C. J. Bowser, and T. K. Kratz. 1984. Long-term ecological research (LTER) on north temperate lakes of the United States. Verhandlungen Internationale Vereinigung für Limnologie 22:533–535.

Magnuson, J. J., G. M. Capelli, J. G. Lorman, and R. A. Stein. 1975. Consideration of crayfish for macrophyte control. Pages 66–74 in P. L. Brezonik and J. L. Fox, editors. The proceedings of a symposium on water quality management through biological control. University of Florida, Gainesville.

Magnuson, J. J., J. T. Krohelski, K. E. Kunkel, and D. M. Robertson. 2003. Wisconsin's waters and climate: Historical changes and possible futures. Transactions of the Wisconsin Academy of Science, Arts and Letters 90:23–36.

Magnuson, J. J., C.A. Paszkowski, F. J. Rahel, and W. M. Tonn. 1989. Fish ecology in severe environments of small isolated lakes in northern Wisconsin. Pages 487–515 in Freshwater Wetlands and Wildlife. Conference 8603101, DOE Symposium Series No. 61, USDOE office of Scientific and Technical Information, Oak Ridge, Tennessee.

Magnuson, J. J., R. W. Wynne, B. J. Benson, and D. M. Robertson. 2000a. Lake and river ice as a powerful indicator of past and present climates. Verhandlungen Internationale Vereinigung für Limnologie 27:2749–2756.

Magnuson, J. J., T. K. Kratz, T. M. Frost, C. J. Bowser, B. J. Benson, and R. Nero. 1991. Expanding the temporal and spatial scales of ecological research and comparison of divergent ecosystems: Roles for the LTER in the United States. Pages 45–70 in P. G. Risser, editor. Long-term ecological research. John Wiley, Sussex, England.

Magnuson, J. J., W. M. Tonn, A. Banerjee, J. Toivonen, O. Sanchez, and M. Rask. 1998. Isolation versus extinction in the assembly of fishes in small northern lakes. Ecology 79:2941–2956.

Magnuson, J. J., T. F. Kratz, T. F. H. Allen, D. E. Armstrong, B. J. Benson, C. J. Bowser, D. W. Bolgrien, S. R. Carpenter, T. M. Frost, S. T. Gower, T. M. Lillesand, J. A. Pike, and M. G. Turner. 1997a. Regionalization of long-term ecological research (LTER) on north temperate lakes. Verhandlungen Internationale Vereinigung für Limnologie 26:522–528.

Magnuson, J. J., D. M. Robertson, B. J. Benson, R. H. Wynne, D. M. Livingstone, T. Arai,

R. A. Assel, R. G. Barry, V. Card, E. Kuusisto, N. G. Granin, T. D. Prowse, K. M. Stewart, and V. S. Vuglinski. 2000b. Historical trends in lake and river ice cover in the Northern Hemisphere. Science 289:1743–1746, *Errata 2001. Science 1291:1254.*

Magnuson, J. J., K. E. Webster, R. A. Assel, C. J. Bowser, P. J. Dillon, J. G. Eaton, H. E. Evans, E. J. Fee, R. I. Hall, L. R. Mortsch, D. W. Schindler, and F. H. Quinn. 1997b. Potential effects of climate changes on aquatic systems: Laurentian Great Lakes and Precambrian Shield region. Hydrological Processes 11:825–871.

Maguire, Jr., B. 1963. The passive dispersal of small aquatic organisms and their colonization of isolated bodies of water. Ecological Monographs 33:161–185.

Malley, D. F., P. S. S. Chang, D. L. Findlay, and G. A. Linsey. 1988. Extreme perturbation of the zooplankton community of a small Precambrian Shield lake by the addition of nutrients. Verhandlungen Internationale Vereinigung für Limnologie 23:2237–2247.

Marin, L. E., T. K. Kratz, and C. J. Bowser. 1990. Spatial and temporal patterns in the hydrogeochemistry of a poor fen in northern Wisconsin. Biogeochemistry 11:63–76.

Martinez, B. 2002. North Temperate Lakes LTER: Whole lake removal of exotic species. Network News 15:8.

Marzolf, G. R. 1982. Long-term ecological research in the United States: A network of research sites, 1982. Page i in J. C. Halfpenny, K. Ingraham, and J. Hardesty, editors. Long-Term Ecological Research (LTER) Network. Available in archives of the LTER at the Center for Limnology, University of Wisconsin-Madison.

Mayden, R. L., F. B. Cross, and O. T. Gorman. 1987. Distributional history of the rainbow smelt, *Osmerus mordax* (Salmoniformes: Osmeridae), in the Mississippi River basin. Copeia 1987:1051–1054.

McCartney, P. H., and M. B. Jones. 2002. Using XML-encoded metadata as a basis for advanced information systems for ecological research. Pages 379–384 in vol. VII: Informatics Systems Development II. Proceedings of the 6th World Multiconference on Systemics, Cybernetics and Informatics, Orlando, Florida.

McDonald, M. G., and A. W. Harbaugh. 1988. A modular three-dimensional finite-difference groundwater flow model. U.S. Geological Survey Techniques of Water-Resources Investigations Book 6, Ch. A1.

McKillop, W. B. 1985. Distribution of aquatic gastropods across the Ordovician dolomite-Precambrian granite contact in southeastern Manitoba, Canada. Canadian Journal of Zoology 63:278–288.

McLain, A. S. 1991. Conceptual and empirical analyses of biological invasion: Non-native fish invasion into north temperate lakes. Ph.D. Thesis. University of Wisconsin-Madison.

Meléndez-Colom, E. C., and K. S. Baker. 2002. Common information management framework: In practice. Pages 385–389 in vol. VII: Informatics Systems Development II. Proceedings of the 6th World Multiconference on Systemics, Cybernetics and Informatics, Orlando, Florida.

Merritt, M. L., and L. F. Konikow. 2000. Documentation of a computer program to simulate lake-aquifer interaction using the MODFLOW ground-water flow model and the MOC3D solute-transport model. U.S. Geological Survey Water Resources Investigations Report 00–4167.

Michaels, S. 1995. Regional analysis of lakes, groundwater and precipitation, northern Wisconsin: A stable isotope study. M.S. Thesis. University of Wisconsin-Madison.

Michener, W. K. 2002. Networking: Ecoinformatics training in Maputo, Mozambique. The Network Newsletter. vol. 15. LTER Network Office, Albuquerque, New Mexico. http://intrnet.lternet.edu/archives/documents/Newletters/NetworkNews/fall02/fall02_pg15.html.

Michener, W. K., and G. Bonito. 2003. The LTER community outreach: Ecoinformatics training for field stations. The Network Newsletter. vol. 16. LTER Network Office, Albuquerque, New Mexico. http://intrnet.lternet.edu/archives/documents/Newletters/NetworkNews/spring03/spring03_pg11.html.

Michener, W. K., J. W. Brunt, J. Helly, T. B. Kirchner, and S. G. Stafford. 1997. Non-geospatial metadata for the ecological sciences. Ecological Applications 7:614–624.

Michener, W. K., editor. 1986. Research data management in the ecological sciences. University of South Carolina Press, Columbia, South Carolina.

Michener, W. K., and J. W. Brunt, editors. 2000. Ecological data: Design, processing and management. Blackwell Science Limited, London, England.

Michener, W. K., J. W. Brunt, and S. G. Stafford, editors. 1994. Environmental information management and analysis: Ecosystem to global scales. Taylor and Francis, Bristol, Pennsylvania.

Michener, W. K., J. H. Porter, and S. G. Stafford, editors. 1998. Data and information management in the ecological sciences: A resource guide. LTER Network Office, University of New Mexico, Albuquerque, New Mexico.

Minns, C. K. 1989. Factors affecting fish species richness in Ontario lakes. Transactions of the American Fisheries Society 118:533–545.

Mittelbach, G. G., C. W. Osenberg, and P. C. Wainwright. 1992. Variation in resource abundance affects diet and feeding morphology in the pumpkinseed sunfish (*Lepomis gibbosus*). Oecologia 90:8–13.

Mortimer, C. H. 1941/42. The exchange of dissolved substances between mud and water in lakes. Journal of Ecology 29:280–329.

Muthen, B. O. 1994. Multilevel covariance structure analysis. Sociological Methods and Research 22:376–398.

Nakicenovic, N., and R. Swart. 2000. Special report on emission scenarios, 2000. IPCC Intergovernmental Panel on Climate Change. Cambridge University Press, Cambridge, England.

Nasar, J. L. 1987. Physical correlates of perceived quality in lakeshore development. Leisure Sciences 9:259–279.

National Research Council. 1993. Soil and water quality: An agenda for agriculture. Committee on long-range soil and water conservation, Board of Agriculture, National Research Council. National Academy Press, Washington, D.C.

National Science Foundation. 1977. Long-term ecological measurements. Report of a conference, Woods Hole, Massachusetts, March 16–18. National Technical Information Service, Springfield, Virginia.

Nduku, W. K., and A. D. Harrison. 1980. Cationic responses of organs and haemolymph of *Biomphalaria pfeifferi* (Krauss), *Biomphalaria glabrata* (Say) and *Helisoma trivolvis* (Say) (*Gastropoda: Planorbidae*) to cationic alterations of the medium. Hydrobiologia 68:119–138.

Neill, W. E., and A. Peacock. 1980. Breaking the bottleneck: Interactions of invertebrate predators and nutrients in oligotrophic lakes. Pages 715–724 in W. C. Kerfoot, editor. Evolution and ecology of zooplankton communities. American Society of Limnology and Oceanography Special Symposium 3. University Press of New England, Hanover, New Hampshire.

Nellbring, S. 1989. The ecology of smelts (Genus *Osmerus*): A literature review. Nordic Journal of Freshwater Research 65:116–145.

Newbold, J. D., J. W. Elwood, R. V. O'Neill, and W. Van Winkle. 1981. Measuring nutrient spiraling in streams. Canadian Journal of Fisheries and Aquatic Sciences 38:860–863.

Nichols, S. A., and R. C. Lathrop. 1994. Cultural impacts on macrophytes in the Madison lakes since the late 1800s. Aquatic Botany 47:225–247.

Nichols, S. A., R. C. Lathrop, and S. R. Carpenter. 1992. Long-term vegetation trends: A history. Pages 151–173 in J. F. Kitchell, editor. Food web management: A case study of Lake Mendota. Springer-Verlag, New York.

Novotny, V., and H. Olem. 1994. Water quality: Prevention, identification, and management of diffuse pollution. Van Nostrand Reinhold, New York.

Nowak, P., and F. Madison. 1998. Farmers and manure management: A critical analysis. Pages 1–32 in J. Hatfield and B. Stewart, editors. Animal waste utilization: Effective use of manure as a soil resource. Ann Arbor Press, Chelesa, Michigan.

Okwueze, E. E. 1983. Geophysical investigations of the bedrock and the groundwater-lake flow system in the Trout Lake region of Vilas County, northern Wisconsin. Ph.D. Thesis. University of Wisconsin-Madison.

Olsen, T. M., D. M. Lodge, G. M. Capelli, and R. J. Houlihan. 1991. Mechanisms of impact of introduced crayfish (*Orconectes rusticus*) on littoral congeners, snails, and macrophytes. Canadian Journal of Fisheries and Aquatic Sciences 48:1853–1861.

Osenberg, C. W., and G. G. Mittelbach. 1989. Effects of body size on the predator-prey interaction between pumpkinseed sunfish and gastropods. Ecological Monographs 59:405–432.

Padilla, D. K., M. A. Chokowski, and L. A. J. Buchan. 1996. Predicting the spread of zebra mussels (*Dreissena polymorpha*) to inland waters using boater movement patterns. Global Ecology and Biogeography Letters 5:353–359.

Parrott, T. M., editor. 1953. Shakespeare: 23 plays and sonnets. Henry V, Act I, Scene II, Line 209, Revised edition. Charles Scribner and Sons, New York.

Perry, W. L., D. M. Lodge, and J. L. Feder. 2002. Importance of hybridization between indigenous and nonindigenous freshwater species: An overlooked threat to North American biodiversity. Systematic Biology 51:255–275.

Perry, W. L., J. L. Feder, and D. M. Lodge. 2001a. Implications of hybridization between introduced and resident *Orconectes* crayfishes. Conservation Biology 15:1656–1666.

Perry, W. L., J. L. Feder, G. Dwyer, and D. M. Lodge. 2001b. Hybrid zone dynamics and species replacement between *Orconectes* crayfishes in a northern Wisconsin lake. Evolution 55:1656–1666.

Phelps, E. S. 1896. Chapters from a life. Houghton-Mifflin, Boston, Massachusetts.

Pickett, S. T. A., J. Kolasa, and C. G. Jones. 1994. Ecological understanding. Academic Press, San Diego, California.

Pierson, P. 1994. Dismantling the welfare state?: Reagan, Thatcher, and the politics of retrenchment. Cambridge University Press, New York.

Pint, C. D. 2002. Groundwater flow model of the Trout Lake basin: Calibration and lake capture zone analysis. M.S. Thesis. University of Wisconsin-Madison.

Pint, C. D., R. J. Hunt, and M. P. Anderson. 2003. Flow path delineation and ground water age, Allequash Basin, Wisconsin. Ground Water 41:895.

Pionke, H. B., W. J. Gburek, and A. N. Sharpley. 1996. Flow and nutrient export patterns for an agricultural hill-land watershed. Water Resources Research 32:1795–1804.

Poister, D. 1992. Nutrient sedimentation and recycling in three northern temperate lakes. M.S. Thesis. University of Wisconsin-Madison.

Poister, D. 1995. Effects of the community composition and vertical distribution of phytoplankton on pigment and phosphorus sedimentation in three Wisconsin lakes. Ph.D. Thesis. University of Wisconsin-Madison.

Poister, D., D. E. Armstrong, and J. P. Hurley. 1994. A 6-year record of nutrient element sedimentation and recycling in three north temperate lakes. Canadian Journal of Fisheries and Aquatic Sciences 51:2457–2466.

Pollard, A. I. 2002. Patterns of benthic invertebrate distribution in connected lake and stream ecosystems. Ph.D. Thesis. University of Wisconsin-Madison.

Porte, J. 1982. Emerson in his journals/selected and edited. Belknap Press of Harvard University, Cambridge, Massachusetts.

Porter, K. G. 1977. The plant-animal interface in aquatic ecosystems. American Scientist 65:159–170.

Porter, J. H. 2002. EML: Augmenting research tools and capabilities. LTER Databits, Information management newsletter of the Long-Term Ecological Research Network. Spring 2002. Available at Intranet.lternet.edu/archives/documents/Newsletters/DataBits/02spring/.

Porter, J. H., B. P. Hayden, and D. L. Richardson. 1996. Data and information management at the Virginia Coast Reserve Long-Term Ecological Research Site. Pages 731–736 in Eco-Informa '96: Global Networks for Environmental Information. Environmental Research Institute of Michigan, Ann Arbor.

Post, E. 2003. Large-scale climate synchronizes the timing of flowering by multiple species. Ecology 84:277–281.

Preissing, J., D. W. Marcouiller, G. P. Green, S. C. Deller, and N. R. Sumathi. 1996. Recreational homeowners and regional development: A comparison of two northern Wisconsin counties. Staff paper 96.4. Center for Community Economic Development, University of Wisconsin-Extension.

Proctor, V. W., and C. R. Malone. 1965. Further evidence of the passive dispersal of small aquatic organisms via the intestinal tract of birds. Ecology 46:728–729.

Quinlan, R. A., M. Paterson, R. I. Hall, P.J. Dillon, A. N. Wilkinson, B. F. Cumming, M. S. V. Douglas, and J. P. Smol. 2003. A landscape approach to examining spatial patterns of limnological variables and long-term environmental change in a southern Canadian lake district. Freshwater Biology 48:1676–1697.

Rahel, F. J. 1984. Factors structuring fish assemblages along a bog lake successional gradient. Ecology 65:1276–1298.

Rahel, F. J. 1986. Biogeographic influences on fish species composition of northern Wisconsin lakes with applications for lake acidification studies. Canadian Journal of Fisheries and Aquatic Sciences 43:124–134.

Rahel, F. J., and J. J. Magnuson. 1983. Low pH and the absence of fish species in naturally acidic Wisconsin lakes: Inferences for cultural acidification. Canadian Journal of Fisheries and Aquatic Sciences 40:3–9.

Ramcharan, C. W., D. K. Padilla, and S. I. Dodson. 1992. Models to predict potential occurrence and density of the zebra mussel, *Dreissena polymorpha*. Canadian Journal of Fisheries and Aquatic Sciences 49:2611–2620.

Rasmussen, P. W., D. M. Heisey, E. V. Nordheim, and T. M. Frost. 1993. Time-series intervention analysis: Unreplicated large-scale experiments. Pages 138–158 in S. M. Scheiner and J. Gurevitch, editors. Design and analysis of ecological experiments. Chapman and Hall, New York.

Rawson, D. S. 1939. Some physical and chemical factors in the metabolism of lakes. Pages 9–26 in E. R. Moulton, editor. Problems in lake biology. AAAS, Washington, D.C.

Real, L. A., and J. H. Brown. 1991. Foundations of ecology: Classic papers with commentaries. University of Chicago Press, Chicago, Illinois.

Redfield, A. C. 1958. The biological control of chemical factors in the environment. American Scientist 46:205–221.

Reed-Andersen, T., S. R. Carpenter, and R. C. Lathrop. 2000a. Phosphorus flow in a watershed-lake ecosystem. Ecosystems 3:561–573.

Reed-Andersen, T., S. R. Carpenter, D. K. Padilla, and R. C. Lathrop. 2000b. Predicted impact of zebra mussel invasion on water clarity in Lake Mendota. Canadian Journal of Fisheries and Aquatic Sciences 57:1617–1626.

Reed-Andersen, T., E. M. Bennett, B. S. Jorgensen, G. Lauster, D. B. Lewis, D. Nowacek, J. L. Riera, B. L. Sanderson, and R. Stedman. 2000c. The distribution of recreational boating across lakes: Do landscape variables affect recreational use? Freshwater Biology 43:439–448.

Reisser, W. 1992. Algae and symbioses: Plant, animals, fungi, viruses, interaction explored. Biopress Limited, Bristol, England.

Reynolds, C. S. 1997. Vegetation processes in the pelagic: A model for ecosystem theory. In Kinne, O., editor. Excellence in ecology. Book 9. International Ecology Institute, Olendorf/Luhe.

Ricker, W. E. 1975. Computation and interpretation of biological statistics of fish populations. Bulletin 191, Fisheries Research Board of Canada, Department of the Environment, Ottawa.

Riera, J. L., J. E. Schindler, and T. K. Kratz. 1999. Seasonal dynamics of carbon dioxide and methane in two clear-water lakes and two bog lakes in northern Wisconsin, USA. Canadian Journal of Fisheries and Aquatic Sciences 56:265–274.

Riera, J. L., J. J. Magnuson, T. K. Kratz, and K. E. Webster. 2000. A geomorphic template for the analysis of lake districts applied to the Northern Highland Lake District, Wisconsin, USA. Freshwater Biology 43:301–318.

Riera, J. L., J. J. Magnuson, J. R. Vande Castle, and M. D. MacKenzie. 1998. Analysis of large-scale spatial heterogeneity in vegetation indices among North American landscapes. Ecosystems 1:268–282.

Risser, P. G., and J. Lubchenco. 1993. Ten-year review of the National Science Foundation Long Term Ecological (LTER) Program. Report to NSF of the ten-year review committee. Available at LTER Network Web site, intranet.lternet.edu/archives/documents/reports/Advisory-reports/ten-year-review-of-LTER.html.

Robertson, D. M. 1989. The use of lake water temperature and ice cover as climatic indicators. Ph.D. Thesis. University of Wisconsin-Madison.

Robertson, D. M., and R. A. Ragotzkie. 1990a. Changes in the thermal structure of moderate to large sized lakes in response to changes in air temperature. Aquatic Science 52:360–380.

Robertson, D. M., and R. A. Ragotzkie. 1990b. Thermal structure of a multibasin lake: Influence of morphometry, interbasin exchange, and groundwater. Canadian Journal of Fisheries and Aquatic Sciences 47:1206–1212.

Robertson, D. M., R. A. Ragotzkie, and J. J. Magnuson. 1992. Lake ice records used to detect historical and future climatic changes. Climatic Change 21:407–427.

Robertson, D. M., R. W. Wynne, and Y. B. Chang. 2000. Influences of El Niño on lake and river ice cover in the Northern Hemisphere from 1900 to 1995. Verhandlungen Internationale Vereinigung für Limnologie 27:2784–2788.

Roden, E. E., and J. W. Edmonds. 1997. Phosphate mobilization in iron-rich anaerobic sediments: Microbial Fe(III) oxide reduction versus iron sulfate formation. Archives of Hydrobiology 139:347–378.

Rodgers, C. 1998. Map produced by Vilas County, Office of Planning and Zoning, Eagle River, Wisconsin.

Rooke, J. B., and G. L. Mackie. 1984. Mollusca of six low-alkalinity lakes in Ontario. Canadian Journal of Fisheries and Aquatic Sciences 41:777–782.

Rose, W. J. 1993. Hydrology of Little Rock Lake in Vilas County, north-central Wisconsin. 93-4139 U.S. Geological Survey Water Resources Investigations Report, Madison, Wisconsin.

Roth, B. M. 2001. The role of competition, predation, and their interaction in invasion dynamics: Predator accelerated replacement. M.S. Thesis. University of Wisconsin-Madison.

Rudstam, L. G., and B. M. Johnson. 1992. Development, evaluation and transfer of new technology. Pages 507–523 in J. F. Kitchell, editor. Food web management: A case study of Lake Mendota. Springer-Verlag, New York.

Rudstam, L. G., and J. J. Magnuson. 1985. Predicting the vertical distribution of fish populations: Analysis of cisco, *Coregonus artedii*, and yellow perch, *Perca flavescens*. Canadian Journal of Fisheries and Aquatic Sciences 44:811–821.

Rudstam, L. G., C. S. Clay, and J. J. Magnuson. 1987. Density and size estimates of cisco (*Coregonus artedii*) using analysis of echo peak PDF from a single-transducer sonar. Canadian Journal of Fisheries and Aquatic Sciences 53:1409–1417.

Rudstam, L. G., R. C. Lathrop, and S. R. Carpenter. 1993. The rise and fall of a dominant planktivore: Direct and indirect effects on zooplankton. Ecology 74:303–319.

Rusak, J. A., C. S. Brock, and P. R. Levitt. Unpublished manuscript. Local, landscape, and regional regulation of zooplankton communities of the northern Great Plains.

Rusak, J. A., N. D. Yan, K. M. Somers, and D. J. McQueen. 1999. The temporal coherence of zooplankton population abundances in neighboring north-temperate lakes. American Naturalist 153:46–58.

Ryden, K. C. 1993. Mapping the invisible landscape: Folklore, writing, and the sense of place. University of Iowa Press, Iowa City.

Ryder, R. A. 1982. The morphoedaphic index—use, abuse, and functional concepts. Transactions of the American Fisheries Society 111:154–164.

Sagova, M., and M. S. Adams. 1993a. Aggregation of numbers, size and taxa of benthic animals at four levels of spatial scale. Archive für Hydrobiologie 128:329–352.

Sagova, M., and M. S. Adams. 1993b. Relationship between plant roots and benthic animals in three sediment types of a dimictic mesotrophic lake. Archiv für Hydrobiologie 128:423–436.

Sampson, C. J. 1999. Aquatic chemistry of Little Rock Lake, Wisconsin, during acidification and recovery. Ph.D. Thesis. University of Minnesota-Minneapolis.

Sampson, C. J., and P. L. Brezonik. 2003a. Ion budgets and sediment-water interactions during the experimental acidification and recovery of Little Rock Lake, Wisconsin. Environmental Science and Technology 37:5625–5635.

Sampson, C. J., and P. L. Brezonik. 2003b. Responses of nutrients to experimental acidification in Little Rock Lake, U.S.A. Water, Air, and Soil Pollution 142:1–19.

Sampson, C. J., P. L. Brezonik, T. M. Frost, K. E. Webster, and T. D. Simonson. 1995. Experimental acidification of Little Rock Lake, Wisconsin: The first four years of chemical and biological recovery. Water, Air, and Soil Pollution 85:1713–1719.

Sanderson, B. L. 1998. Factors regulating water clarity in northern Wisconsin lakes. Ph.D. Thesis. University of Wisconsin-Madison.

Sanderson, B. L., and T. M. Frost. 1996. Regulation of dinoflagellate populations: Relative importance of grazing, resource limitation and recruitment from sediments. Canadian Journal of Fisheries and Aquatic Sciences 53:1409–1417.

Sanderson, B. L., T. R. Hrabik, J. J. Magnuson, and D. M. Post. 1999. Cyclic dynamics of a yellow perch (*Perca flavescens*) population in an oligotrophic lake: Evidence for the role of intraspecific interactions. Canadian Journal of Fisheries and Aquatic Sciences 56:1534–1542.

Schafran, G. C., and C. T. Driscoll. 1987. Spatial and temporal variations in aluminum chemistry of a dilute, acidic lake. Biogeochemistry 3:105–109.

Schindler, D. E., S. R. Carpenter, J. J. Cole, J. F. Kitchell, and M. L. Pace. 1997. Influence of food web structure on carbon exchange between lakes and the atmosphere. Science 277:248–251.

Schindler, D. W. 1974. Eutrophication and recovery in experimental lakes: Implications for lake management. Science 184:897–899.

Schindler, D. W. 1977. Evolution of phosphorus limitation in lakes. Science 196:260–262.

Schindler, D. W. 1987. Detecting ecosystem responses to anthropogenic stresses. Canadian Journal of Fisheries and Aquatic Sciences 44:6–25.

Schindler, D. W. 1988. Effects of acid rain on freshwater ecosystems. Science 239:149–157.

Schindler, D. W. 1990. Experimental perturbations of whole lakes as tests of hypotheses concerning ecosystem structure and function. Oikos 57:25–41.

Schindler, D. W. 1998. Replication versus realism: The need for ecosystem-scale experiments. Ecosystems 1:323–334.

Schindler, D. W., P. J. Curtis, B. R. Parker, and M. P. Stainton. 1996a. Consequences of climate warming and lake acidification for UV-B penetration in North American boreal lakes. Nature 379:705–708.

Schindler, D. W., S. E. Bayley, B. R. Parker, K. G. Beaty, D. R. Cruikshank, E. J. Fee, E. U. Schindler, and M. P. Stanton. 1996b. The effects of climate warming on the properties of boreal lakes and streams in the Experimental Lakes Area, northwestern Ontario. Limnology and Oceanography 41:1004–1017.

Schindler, D. W., K. H. Mills, D. F. Malley, D. L. Findlay, J. A. Shearer, I. J. Davies, M. A. Turner, G. A. Linsey, and D. R. Cruikshank. 1985. Long-term ecosystem stress: The effects of years of experimental acidification on a small lake. Science 228:1395–1401.

Schindler, D. W., T. M. Frost, K. H. Mills, P. S. S. Chang, I. J. Davies, L. Findlay, D. F. Malley, J. A. Shearer, M. A. Turner, P. J. Garrison, C. J. Watras, K. E. Webster, J. M. Gunn, P. L. Brezonik, and W. A. Swenson. 1991. Comparisons between experimentally and atmospherically acidified lakes during stress and recovery. Proceedings of the Royal Society of Edinborough 97B:193–226.

Schindler, J. E., and D. P. Krabbenhoft. 1998. The hyporheic zone as a source of dissolved organic carbon and carbon gases to a temperate forested stream. Biogeochemistry 43:157–174.

Schmidt-Nielsen, K. 1998. The camel's nose: Memoirs of a curious scientist. Island Press/Shearwater Books, Washington, D.C.

Schnaiberg, J., J. Riera, M. G. Turner, and P. R. Voss. 2002. Explaining human settlement patterns in a recreational lake district: Vilas County, Wisconsin, U.S.A. Environmental Management 30:24–34.

Schneider, D. W. 1990. Direct assessment of the independent effects of exploitative and interference competition between *Daphnia* and rotifers. Limnology and Oceanography 35:916–922.

Schneider, D. W., and T. M. Frost. 1996. Habitat duration and community structure in temporary ponds. Journal of the North American Benthological Society 15:64–86.

Schrameyer, R. B. 1997. Wisconsin's statewide lake organizations—a history. Lake Tides 22:2–3.

Scurlock, J. M., R. J. Olson, R. A. McCord, and W. K. Michener. 2002a. Data banks: Archiving ecological data and information. Pages 248–259 in vol. 2. Encyclopedia of global environmental change, John Wiley, New York.

Scurlock, J. M., P. Kanciruk, R. A. McCord, R. J. Olson, and W. K. Michener. 2002b. Metadata. Pages 409–411 in vol. 2. Encyclopedia of global environmental change. John Wiley, New York.

Senge, P. M. 1990. The fifth discipline, the art and practice of the learning organization. Doubleday/Currency, New York.

Shafer, M. M., and D. E. Armstrong. 1994. Mass fluxes and recycling of phosphorus in Lake Michigan: Role of major particle phases in regulating the annual cycle. Pages 285–322 in L. A. Baker, editor. Advances in Chemistry Series 237. Environmental chemistry of lakes and reservoirs. American Chemical Society, Washington, DC.

Shapiro, J., V. Lamarra, and M. Lynch. 1975. Biomanipulation: An ecosystem approach to lake restoration. Pages 85–96 in P. L. Brezonik and J. L. Fox, editors. Proceedings of a symposium on water quality management through biological control. University of Florida, Gainesville, Florida.

Sharpley, A. N. 1995. Identifying sites vulnerable to phosphorus loss in agricultural runoff. Journal of Environmental Quality 24:947–951.

Shirane, H. 1998. Traces of dreams; landscape, cultural memory and poetry of Bashō. Stanford University Press, Stanford, California.

Sierszen, M. E., and T. M. Frost. 1992. Selectivity in suspension feeders: Food quality and the cost of being selective. Archiv für Hydrobiologie 123:257–273.

Sierszen, M. E., and T. M. Frost. 1993. Response of predatory zooplankton populations to the experimental acidification of Little Rock Lake, Wisconsin. Journal of Plankton Research 15:553–562.

Smith, D. J., B. J. Benson, and D. F. Balsiger. 2002. Designing web database applications for ecological research. Pages 408–413 in vol. VII: Informatics Systems Development II. Proceedings of the 6th World Multiconference on Systemics, Cybernetics and Informatics, Orlando, Florida.

Sonzogni, W. C., and G. F. Lee. 1974. Diversion of wastewaters from Madison lakes. Journal of Environmental Engineering 100:153–170.

Soranno, P. A., S. R. Carpenter, and R. C. Lathrop. 1997. Internal phosphorus loading in Lake Mendota: Response to external loads and weather. Canadian Journal of Fisheries and Aquatic Sciences 54:1883–1893.

Soranno, P. A., S. L. Hubler, S. R. Carpenter, and R. C. Lathrop. 1996. Phosphorus loads to surface waters: A simple model to account for spatial pattern of land use. Ecological Applications 6:865–878.

Soranno, P. A., K. E. Webster, J. L. Riera, T. K. Kratz, J. S. Baron, P. A. Bukaveckas, G. W. Kling, D. S. White, N. Caine, R. C. Lathrop, and P. R. Leavitt. 1999. Spatial variation among lakes within landscapes: Ecological organization along lake chains. Ecosystems 2:395–410.

Sousounis, P. J., and J. M. Bisanz. 2000. Great Lakes overview. Preparing for a changing climate, the potential consequences of climate variability and change. Great Lakes. U.S. Global Change Research Program, USEPA, Washington, D.C.

Sousounis, P. J., and E. K. Grover. 2002. Potential future weather patterns over the Great Lakes region. Journal of Great Lakes Research 28:496–520.

Spycher, G., J. B. Cushing, D. L. Henshaw, S. G. Stafford, and N. Nadkarm. 1996. Solving problems for validation , federation, and migration of ecological databases. Pages 695–700 in Eco-Informa '96: Global Networks for Environmental Information. Environmental Research Institute of Michigan, Ann Arbor.

Stafford, S. G., J. W. Brunt, and B. J. Benson. 1996. Training environmental information managers of the future. Pages 111–116 in Eco-Informa '96: Global Networks for Environmental Information. Environmental Research Institute of Michigan, Ann Arbor.

Staggs, M. D. 1992. Benefits on a larger scale. Pages 525–537 in J. F. Kitchell, editor. Food web management: A case study of Lake Mendota. Springer-Verlag, New York.

State Board of Forestry. 1910. Minutes of the meeting of the State Board of Forestry. Wisconsin Historical Society Archives, Madison.

Stauffer, R. E., and G. F. Lee. 1973. The role of thermocline migration in regulating algal blooms. Pages 73–81 in E. J. Middlebrooks, editor. Modeling the eutrophication process. Ann Arbor Science, Ann Arbor, Michigan.

Stedman, R. C. 2000. "Up north": A social psychology of place. Ph.D. Thesis. University of Wisconsin-Madison.

Stedman, R. C. 2002. Toward a social psychology of place: Predicting behaviors from place-based beliefs, attitude, and identity. Environment and Behavior 34:561–582.

Stedman, R. C., and R. B. Hammer. 2006. Environmental perception in a rapidly growing, amenity–rich region: The effects of lakeshore development on perceived water quality in Vilas County, Wisconsin. Society and Natural Resources.

Stefan, H. G., M. Hondzo, J. G. Eaton, and J. H. McCormick. 1995. Predicted effects of global climate on fishes in Minnesota lakes. Pages 1124–1135 in R. J. Beamish, editor. Climate change and northern fish populations. Canadian Special Publication in Fisheries and Aquatic Sciences 121, National Research Council of Canada, Ottawa.

Stein, R. A. 1977. Selective predation, optimal foraging, and the predator-prey interaction between fish and crayfish. Ecology 58:1237–1253.

Stein, R. A., and J. J. Magnuson. 1976. Behavioral response of crayfish to a fish predator. Ecology 57:751–761.

Stenseth, N. C., K. S. Chan, H. Tong, R. Boonstra, S. Boutin, C. Krebs, E. Post, M. O'Donaghue, N. G. Yoccozz, M. C. Forchhammer, and J. W. Hurrell. 1999. Common dynamic structure of Canada lynx populations within three climatic regions. Science 285:1071–1073.

Stow, C. A., S. R. Carpenter, K. E. Webster, and T. M. Frost. 1998. Long-term environmental monitoring: Some perspectives from lakes. Ecological Applications 8:269–276.

Strahler, A. N. 1964. Quantitative geomorphology of drainage basins and channel networks. Pages 4/39–34/76 in V. T. Chow, editor. Handbook of applied hydrology. McGraw-Hill, New York.

Straile, D. 2002. North Atlantic Oscillation synchronizes food-web interactions in central European lakes. Proceedings of the Royal Society of London B 269:391–395.

Straile, D., D. M. Livingstone, G. A. Weyhenmeyer, D. G. George. 2003. The response of freshwater ecosystems to climate variability associated with the North Atlantic Oscillation. Pages 263–279 in J. W. Hurrell, Y. Kushnir, G. Ottersen, M. Visbeck, editors. The North Atlantic Oscillation, Climatic significance and environmental impact. Geophysical Monograph 134. American Geophysical Union, Washington, D.C.

Stubbs, M., and B. J. Benson. 1996. Query access to relational databases via the World Wide Web. Pages 105–109 in Eco-Informa '96: Global Networks for Environmental Information. Environmental Research Institute of Michigan, Ann Arbor.

Stumborg, B. E., K. A. Baerenklau, and R. C. Bishop. 2001. Nonpoint source pollution and present values: A contingent valuation of Lake Mendota. Review of Agricultural Economics 23:120–132.

Swanson, F. J., and R. E. Sparks. 1990. Long-term ecological research and the invisible place. BioScience 40:502–508.

Swenson, W. A. 2002. Demographic changes in a largemouth bass population following closure of the fishery. American Fisheries Society Symposium 31:627–637.

Swenson, W. A., J. H. McCormick, T. D. Simonson, K. M. Jensen, and J. G. Eaton. 1989. Experimental acidification of Little Rock Lake (Wisconsin): Fish research approach and early response. Archives of Environmental Contamination and Toxicology 18:167–184.

Syme, G. J., B. E. Nancarrow, and B. S. Jorgensen. 2002. The limits of environmental responsibility: A stormwater case study. Environment and Behavior 34:836–848.

Tansley, A. G. 1935. The use and abuse of vegetational concepts and terms. Ecology 16:284–307.

Tarrow, S. G. 1994. Power in movement: Social movements, collective action and politics. Cambridge University Press, New York.

Thayer, C. B. 1983. The battle of Wisconsin Heights, an eye-witness account of the Black Hawk war of 1832. Banta, Menasha, Wisconsin.

Thoreau, H. D. 1854. Walden, or, Life in the Woods. Ticknor and Fields, Boston.

Thorp, J. H., and M. D. Delong. 1994. The riverine productivity model—an heuristic view of carbon sources and organic processing in large river ecosystems. Oikos 70:305–308.

Thwaites, F. T. 1929. Glacial geology of part of Vilas County, Wisconsin. Transactions of the Wisconsin Academy of Sciences, Arts and Letters 24:109–125.

Tilman, D. 1996. Biodiversity: Population versus ecosystem stability. Ecology 77:350–363.

Tonn, W. M., and J. J. Magnuson. 1982. Patterns of species composition and richness of fish assemblages in northern Wisconsin lakes. Ecology 63:1149–1166.

Tonn, W. M., J. J. Magnuson, and A. M. Forbes. 1983. Community analysis in fishery management: An application with northern Wisconsin lakes. Transactions of the American Fisheries Society 112:368–377.

Tonn, W. M., J. J. Magnuson, M. Rask, and J. Toivonen. 1990. Intercontinental comparison of small-lake fish assemblages: The balance between local and regional processes. American Naturalist 136:345–375.

Townley, L. R., and M. G. Trefry. 2000. Surface water-groundwater interaction near shallow circular lakes. Water Resources Research 36:935–949.

Trebitz, A. S., S. A. Nichols, S. R. Carpenter, and R. C. Lathrop. 1993. Patterns of vegetation change in Lake Wingra following a *Myriophyllum spicatum* decline. Aquatic Botany 46:325–340.

Trenberth, K. E. 2001. Stronger evidence of human influences on climate: The 2001 IPCC assessment. Environment 43:8–19.

Tuan, Y.-F. 1977. Space and place: The perspective of experience. University of Minnesota Press, Minneapolis.

Turner, M. G., S. L. Collins, A. L. Lugo, J. J. Magnuson, T. S. Rupp, and F. J. Swanson. 2003. Disturbance dynamics and ecological response: The contribution of long-term ecological research. BioScience 53:46–56.

Urban, N., P. L. Brezonik, L. A. Baker, and L. A. Sherman. 1994. Rates of sulfate reduction and diffusion in sediments of Little Rock Lake, Wisconsin. Limnology and Oceanography 39:797–815.

Urban, N., C. J. Sampson, P. L. Brezonik, and L. A. Baker. 2001. Sulfur cycling in the water column of Little Rock Lake, Wisconsin. Biogeochemistry 52:41–77.

Van Oosten, J. 1947. Mortality of smelt, *Osmerus mordax* (Mitchill), Lakes Huron and Michigan during the fall and winter of 1942–1943. Transactions of the American Fisheries Society 68:152–162.

Vande Castle, J. R., J. J. Magnuson, M. D. MacKenzie, and J. L. Riera. 1995. Regional ecosystem comparison using a standardized NDVI approach. Pages 797–804 in Ninth Annual Symposium on Geographic Information Systems, Vancouver, British Columbia, Canada.

Vander Zanden, M. J., J. M. Casselman, and J. B. Rasmussen. 1999. Stable isotope evidence for the food web consequences of species invasions in lakes. Nature 401:464–467.

Vanderbilt, K. 2001. Information management outreach in the East Asia-Pacific Region (EAPR-ILTER). Network Newsletter. vol. 14. LTER Network Office, Albuquerque, New Mexico. Available at http://intrnet.lternet.edu/archives/documents/Newletters/NetworkNews/fall01/fall01_pg11.html.

Vanni, M. J., J. Temte, Y. Allen, R. Dodds, P. J. Howard, P. R. Leavitt, and C. Luecke. 1992. Herbivory, nutrients and phytoplankton dynamics in Lake Mendota, 1987–1989. Pages 243–273 in J. F. Kitchell, editor. Food web management: A case study of Lake Mendota. Springer-Verlag, New York.

Vannote, R. L., G. W. Minshall, K. W. Cummins, J. R. Sedell, and C. E. Cushing. 1980. The river continuum concept. Canadian Journal of Fisheries and Aquatic Sciences 37:130–137.

Veen, C., C. Federer, D. Buso, and T. Siccama. 1994. Structure and function of the Hubbard Brook data management syestem. Bulletin of the Ecological Society of America 75:45–48.

Vilas County News-Review. 1934. Vilas County zoning map. Page 3. Office of the County Clerk, Eagle River, Wisconsin.

Vollenweider, R. A. 1968. Scientific fundamentals of eutrophication of lakes and flowing waters, with particular reference to nitrogen and phosphorus as factors in eutrophication. Technical Report DAS/CSJ/68.27 Organization for Economic Cooperation and Development, Paris, France.

Voss, P. V., and G. V. Fuguitt. 1979. Turnaround migration in the Upper Great Lakes Region. Applied Population Laboratory, University of Wisconsin-Extension, Madison, Wisconsin.

Wakefield, J. A. 1834. History of the war between the United States and the Sac and Fox Nations of Indians. Calvin Goudy, Jacksonville, Illinois.

Walters, C. J. 1986. Adaptive management of renewable resources. Macmillan, New York.

Walters, C. 1997. Challenges in adaptive management of riparian and coastal ecosystems. Conservation Ecology 1(2):1. Available at www.consecol.org/vol1/iss2/art1/.

Ward, J. V., and J. A. Stanford. 1983. The serial discontinuity concept of lotic systems. Pages 29–42 in T. D. Fontaine and S. M. Bartell, editors. Dynamics of lotic ecosystems. Ann Arbor Science Publishers, Ann Arbor, Michigan.

Wasser, C. 1996. Dynamic data transfer via the World Wide Web: Increasing your visitors' understanding of ecological data. Pages 737–742 in vol. 11. Proceedings of Eco–Informa '96: Global Networks for Environmental Information, Lake Buena Vista, Florida. Research Institute of Michigan. Ann Arbor.

Watras, C. J., and N. S. Bloom. 1992. Mercury and methyl mercury in individual zooplankton: Implications for bioaccumulation. Limnology and Oceanography 37:1313–1318.

Watras, C. J., and N. S. Bloom. 1994. The vertical distribution of mercury species in Wisconsin lakes: Accumulation in plankton layers. Pages 137–152 in C. J. Watras and J. W. Huckabee, editors. Mercury pollution: Integration and synthesis. Lewis Publishers. Chelsea, Michigan.

Watras, C. J., and T. M. Frost. 1989. Little Rock Lake (Wisconsin): Perspectives on an experimental ecosystem approach to seepage lake acidification. Archives of Environmental Contamination and Toxicology 18:157–165.

Watras, C. J., K. A. Morrison, R. J. M. Hudson, T. M. Frost, and T. K. Kratz. 2000. Decreasing mercury in northern Wisconsin: Temporal patterns in bulk precipitation and a precipitation-dominated lake. Environmental Science and Technology 34:4051–4057.

Watras, C. J., N. S. Bloom, J. M. Hudson, S. Gherini, R. Munson, S. A. Class, K. A. Morrison, J. Hurley, J. G. Wiener, W. F. Fitzgerald, R. Mason, G. Vandal, D. Powell, R. Rada, L. Rislov, M. Winfrey, J. Elder, D. Krabbenhoft, A. W. Andren, C. Babiarz, D. B.

Porcella, and J. W. Huckabee. 1994. Sources and fates of mercury and methylmercury in Wisconsin lakes. Pages 153–177 in C. J. Watras and J. W. Huckabee, editors. Mercury pollution: Integration and synthesis. Lewis Publishers, Chelsia, Michigan.

Webster, K. E., and P. L. Brezonik. 1995. Climate confounds detection of chemical trends related to acid deposition in upper Midwest lakes in the U.S.A. Water, Air, and Soil Pollution 85:1575–1580.

Webster, K. E., A. D. Newell, L. A. Baker, and P. L. Brezonik. 1990. Climatically induced rapid acidification of a softwater seepage lake. Nature 347:374–376.

Webster, K. E., T. M. Frost, C. J. Watras, W. A. Swenson, M. J. Gonzalez, and P. J. Garrison. 1992. Complex biological responses to the experimental acidification of Little Rock Lake, Wisconsin, U.S.A. Environmental Pollution 78:73–78.

Webster, K. E., T. K. Kratz, C. J. Bowser, J. J. Magnuson, and W. J. Rose. 1996. The influence of landscape position on lake chemical responses to drought in northern Wisconsin. Limnology and Oceanography 41:977–984.

Webster, K. E., P. A. Soranno, S. B. Baines, T. K. Kratz, C. J. Bowser, P. J. Dillon, P. Campbell, E. J. Fee, and R. E. Hecky. 2000. Structuring features of lake districts: Landscape controls on lake chemical responses to drought. Freshwater Biology 43:499–515.

Wegener, M. 2001. Long-term land use/cover change patterns in the Madison Lakes Area and their impacts on runoff volume to Lake Mendota. M.S. Thesis. University of Wisconsin-Madison.

Weightman, J., and D. Weightman, translators, 1969. The Raw and the Cooked. Harper and Row, New York.

Weir, M., A. S. Orloff, and T. Skocpol, editors. 1988. The politics of social policy in the United States. Princeton University Press, Princeton, New Jersey.

Wentz, D. A., and W. J. Rose. 1989. Interrelationships among hydrologic-budget components of a northern Wisconsin seepage lake and implications for acid-deposition modeling. Archives of Environmental Contamination and Toxicology 18:147–155.

Wentz, D. A., W. J. Rose, and K. E. Webster. 1995. Long-term hydrologic and biogeochemical responses of a soft water seepage lake in north central Wisconsin. Water Resources Research 31:199–212.

Wetzel, R. G. 1983. Limnology, 2d Edition. Saunders, Philadelphia.

Wiener, J. G., W. F. Fitzgerald, C. J. Watras, and R. Rada. 1990. Partitioning and bioavailability of mercury in an experimentally acidified Wisconsin lake. Environmental Toxicology and Chemistry 9:909–918.

Wilbur, K. M. 1976. Recent studies of invertebrate mineralization. Pages 79–108 in N. Watabe and K. M. Wilbur, editors. The mechanisms of mineralization in the invertebrates and plants. University of South Carolina Press, Columbia.

Wiley, M. J., L. L. Osborne, and R. W. Larimore. 1990. Longitudinal structure of an agricultural prairie river system and its relationship to current stream ecosystem theory. Canadian Journal of Fisheries and Aquatic Sciences 47:373–384.

Williams, D. R., M. E. Patterson, J. W. Roggenbuck, and A. E. Watson. 1992. Beyond the commodity metaphor: Examining emotional and symbolic attachment to place. Leisure Sciences 14:29–46.

Williamson, C. E. 1991. Copepoda. Pages 787–822 in J. H. Thorp and A. P. Covich, editors. Ecology and classification of North American freshwater invertebrates. Academic Press, New York.

Williamson, C. E., R. S. Stemberger, D. P. Morris, T. M. Frost, and S. G. Paulsen. 1996. Ultraviolet radiation in North American lakes: Attenuation estimates from DOC measurements and implications for plankton communities. Limnology and Oceanography 41:1024–1034.

Willis, T. V. 2003. Dynamics of fish communities and populations in five northern Wisconsin lakes from 1981–2001. Ph.D. Thesis. University of Wisconsin-Madison.

Wilson, K. A. 2002. Impacts of the invasive rusty crayfish (*Orconectes rusticus*) in northern Wisconsin lakes. Ph.D. Thesis. University of Wisconsin-Madison.

Wilson, K. A., J. J. Magnuson, D. M. Lodge, A. H. Hill, T. K. Kratz, W. L. Perry, and T. V. Willis. 2005. A long–term rusty crayfish (*Orconectes rusticus*) invasion: Dispersal patterns and community change in a north temperate lake. Canadian Journal of Fisheries and Aquatic Sciences 61:2255–2266.

Wilson, M. A., and S. R. Carpenter. 1999. Economic valuation of freshwater ecosystem services in the United States, 1977–1997. Ecological Applications 9:772–783.

Winter, T. C. 2000. The vulnerability of wetlands to climate change: A hydrologic landscape perspective. Journal of the American Water Resources Association 36:305–311.

Winter, T. C. 2001. The concept of hydrologic landscapes. Journal of the American Water Resources Association 37:335–349.

Winter, T. C., J. W. Harvey, O. L. Franke, and W. M. Alley. 1998. Ground water and surface water: A single resource. U.S. Geological Survey Circular 1139, Denver, Colorado.

Winter, T. C. 1981. Uncertainties in estimating the water balance of lakes. Water Resources Bulletin 17:82–115.

Wisconsin Administrative Code. Enacted 1983. Chapter NR 19.27(4). Revisor of Statutes Bureau, Madison, Wisconsin.

Wisconsin Department of Natural Resources. 1995a. What is a lake district? Publication WR-402–95. Available at www.uwsp.edu/cnr/uwexlakes.

Wisconsin Department of Natural Resources. 1995b. What is a lake association? Publication WR-403–95. Available at www.uwsp.edu/cnr/uwexlakes.

Wisconsin Department of Natural Resources. 1995c. What is a qualified lake association? Publication WR-404–95. Available at www.uwsp.edu/cnr/uwexlakes.

Wisconsin Department of Natural Resources. 1996. Northern Wisconsin's lakes and shorelines; examining a resource under pressure. Available at www.uwsp.edu/cnr/uwexlakes.

Wisconsin Department of Natural Resources. 2004. Wisconsin lakes partnership. Publication WR-405–04. Available at www.uwsp.edu/cnr/uwexlakes.

Wright, G. 1990. The origins of American industrial success. American Economic Review 80:651–668.

Wynne, R. H. 1995. Satellite monitoring of lake ice breakup on the Laurentian shield as a robust climate indicator. Ph.D. Thesis. University of Wisconsin-Madison.

Wynne, R. H. 2000. Statistical modeling of lake ice phenology: Issues and implications. Verhandlungen Internationale Vereinigung für Limnologie 27:2820–2825.

Wynne, R. W., J. J. Magnuson, M. K. Clayton, T. M. Lillesand, and D. C. Rodman. 1996. Determinants of temporal coherence in the satellite-derived 1987–1994 ice breakup dates of lakes on the Laurentian Shield. Limnology and Oceanography 41:832–838.

Yan, N. D., C. J. LaFrance, and G. G. Hitchin. 1982. Planktonic fluctuations in a fertilized, acidic lake: The role of invertebrate predators. Pages 137–154 in R. E. Johnson, editor. Acid rain/fisheries. American Fisheries Society, Bethesda, Maryland.

Yan, N. D., R. W. Nero, and D. C. Lasenby. 1985. Are Chaoborus more abundant in acidified than nonacidified lakes in central Canada? Holarctic Ecology 8:93–99.

Yan, N. D., W. Keller, N. M. Scully, D. R. S. Lean, and P. J. Dillon. 1996. Increased UV-B penetration in a lake owing to drought-induced acidification. Nature 381:141–143.

Index

Note: page numbers followed by *f* and *t* indicate figures and tables.